INFRARED ASTRONOMY

INTERNATIONAL ASTRONOMICAL UNION
UNION ASTRONOMIQUE INTERNATIONALE

SYMPOSIUM No. 96

HELD IN KONA, HAWAII, JUNE 23–27, 1980

INFRARED ASTRONOMY

EDITED BY

C. G. WYNN-WILLIAMS

and

D. P. CRUIKSHANK

Institute for Astronomy, University of Hawaii

With the Assistance of
Deborah Weiner

D. REIDEL PUBLISHING COMPANY

DORDRECHT : HOLLAND / BOSTON : U.S.A. / LONDON : ENGLAND

Library of Congress Cataloging in Publication Data

Main entry under title:

Infrared astronomy.

(Symposium - International Astronomical Union ; no. 96)
'Co-sponsored by IAU Commissions, 16, 28, and 34.'
1. Infra-red astronomy–Congresses. I. Wynn-Williams, C. G.
II. Cruikshank, Dale P. III. Weiner, Deborah. IV. International
Astronomical Union. Commission 16. V. International Astronomical Union.
Commission 28. VI. International Astronomical Union. Commission 34.
VII. Series: International Astronomical Union. Symposium ; no. 96.
QB470.A1I52 523.01'5012 80-27353
ISBN 90-277-1227-1
ISBN 90-277-1228-X (pbk.)

Published on behalf of
the International Astronomical Union
by
D. Reidel Publishing Company, P. O. Box 17, 3300 AA Dordrecht, Holland

Sold and distributed in the U.S.A. and Canada
by Kluwer Boston Inc.,
190 Old Derby Street, Hingham, MA 02043, U.S.A.

In all other countries, sold and distributed
by Kluwer Academic Publishers Group,
P. O. Box 322, 3300 AH Dordrecht, Holland

D. Reidel Publishing Company is a member of the Kluwer Group.

Printed in The Netherlands

DEDICATION

This symposium is dedicated to the memory of Harold L. Johnson, an early pioneer in the field of infrared astronomy, who died on April 2, 1980, in Mexico City at the age of 58.

Dr. Johnson was known as the founder of the UBV photometric system which is universal in astronomy today. He later expanded this system into the near infrared region with the addition of the R, I, J, K and L photometric colors which extend out to a wavelength of 4 μm. He was awarded the Helen B. Warner Prize by the American Astronomical Society in 1956 for his work in photometry and extinction measurements of standard stars. Harold Johnson's career was marked by his ability as an innovator. He had a major influence in introducing modern electronic techniques into the field of astronomy. Later in his career he made important contributions to the development of infrared Fourier transform spectroscopy as a reliable and productive astronomical technique.

Harold Johnson in many ways made the connection between the new field of infrared astronomy and traditional optical astronomy. As a respected practitioner of optical astronomy he was able to extend optical techniques into the infrared. This, as well as his instrumental innovations, made infrared techniques an essential part of modern astronomical observations.

Harold Johnson will be remembered by many friends as a person who was always willing to help, especially those just starting their careers. He would always give his honest opinion, a trait which offended some but was sincerely valued by many more. Astronomy has truly lost a person of integrity and a giant in the field. As a measure of this loss, the International Astronomical Union's Symposium 96 on Infrared Astronomy is dedicated to his memory.

TABLE OF CONTENTS

PREFACE

IAU Symposium 96 on "Infrared Astronomy" was held at the Kona Lagoon Hotel, Hawaii, during 23-27 June 1980. The meeting, which was attended by 203 astronomers from 16 countries, was timed to coincide with the emergence of Mauna Kea Observatory as a major infrared facility, with four large telescopes available to the international community for infrared observations. The meeting was generously supported organizationally and financially by the Office of Space Sciences of NASA, by the Science Research Council of the United Kingdom, by the Canada-France-Hawaii Telescope Corporation and by the Institute for Astronomy of the University of Hawaii. It was co-sponsored by IAU Commissions 16, 28 and 34.

The Scientific Organizing Committee consisted of D. P. Cruikshank and C. G. Wynn-Williams (co-chairmen), M. Combes, G. G. Fazio, U. Fink, R. D. Gehrz, A. R. Hyland, R. A. McLaren, A. F. M. Moorwood, G. Neugebauer, G. H. Rieke, J. Ring, and R. van Duinen. The Local Organizing Committee consisted of B. Campbell, A. J. Longmore, and C. M. Telesco.

The topics covered in the conference were the infrared aspects of planetary astronomy, of interstellar clouds and star-forming regions, of Galactic structure, and of galaxies. Papers on normal and evolved stars and on instrumentation were excluded from the verbal sessions of the conference, although a few posters were accepted on these topics.

During the course of the conference, 21 review papers and 143 contributed papers were presented. Of the contributed papers, 78 were presented verbally, and 65 in the form of posters. Two special sessions were held during the conference: M. W. Werner organized a discussion on infrared astronomy from space, and M. Harwit chaired a session on the future of airborne astronomy. This volume contains all of the review papers, plus a record of the discussions which followed the reviews. A list of the contributed papers is included at the end of the volume. The decision to exclude the latter was made primarily to keep the size and the purchase price of this volume within reasonable limits.

C. G. Wynn-Williams and D. P. Cruikshank (eds.), Infrared Astronomy, xv–xvi.

The volume was prepared from camera-ready typescripts prepared, in most cases, by the authors. In Honolulu we checked the text and all readily-accessible references for errors, and made the appropriate corrections in the typescripts. The discussions, which have in some cases been reworded and compressed for clarity, were drafted from a combination of the questionnaire forms completed by those who made comments and a tape recording of the proceedings. The draft discussions were then checked by the speakers. We would like to express our appreciation to Mary Missbach and JoAnn Yuen for their help in completing the typing of the volume.

In the final review paper in this volume Malcolm Longair expresses the opinion that IAU Symposium 96 will surely be the last major conference entitled, "Infrared Astronomy." If this proves to be correct, we commend to our readers a book that will perhaps be the last one entitled, "*Infrared Astronomy*."

C. G. Wynn-Williams
Dale P. Cruikshank

August 1980
Honolulu, Hawaii.

THE COMPOSITION OF PLANETARY ATMOSPHERES

Th. Encrenaz and M. Combes
Observatoire de Meudon
F-92190 Meudon, France

During the past 10 years, great progress has been made in our knowledge of the planetary atmospheres, mainly due to the recent development of infrared techniques and space astronomy.

Three steps can be considered in the research development of planetary atmospheres from infrared spectroscopy: (1) detection of molecules; (2) determination of thermal profiles; (3) determination of abundance ratios. The third point, which requires the achievement of both (1) and (2), is the most important from an astrophysical point of view, since abundance ratios are a basic tool for cosmogonical and cosmological studies.

Infrared spectroscopy has led to the discovery of minor molecules and isotopes on Mars and Venus. Nevertheless the most accurate abundance ratios have been derived from *in situ* measurements (mass spectrometry, chromatography).

IR observations have proven to be efficient for the detection of thin atmospheres on faint objects (Triton, Pluto, Io). Also, they were especially successful for the determination of the chemical atmospheric composition of the giant planets and for the related determination of abundance ratios in these planets. Jupiter is the most extensively studied, and IR spectroscopy has led to the discovery of a large number of molecules and to the determination of major abundance ratios: H_2/He, D/H, $^{12}C/^{13}C$, $^{15}N/^{14}N$, N/H, P/N, C/H. The authors' conclusion is that these ratios are close to the cosmic values while other authors conclude to a significant enrichment of most of the elements with respect to H. This controversy is reviewed and analyzed. Concerning the other outer planets, the H_2/He ratio, in spite of large uncertainties, seems to be solar, too, and the C/H ratio, following most of the studies, is significantly enriched on Uranus and Neptune.

C. G. Wynn-Williams and D. P. Cruikshank (eds.), Infrared Astronomy, 1–33.

TABLE I

Observed Molecules in the Atmosphere of Jupiter

Molecule	Spectral Range	Reference
He	584 Å	Judge and Carlson (1974)
HD	7460 Å	Trauger et al (1973)
H_2	0.8 μm	Kiess et al. (1960)
	2.5 μm	Martin et al. (1976)
	1.25 μm	De Bergh et al. (1977)
CH_4	0.8 μm	Wildt (1932)
	1 - 2 μm	Kuiper (1952)
	1.1 μm	Walker and Hayes (1967), Maillard et al. (1973)
	8 μm	Gillett et al. (1969)
$^{13}CH_4$	1.1 μm	Fox et al. (1972), De Bergh et al. (1976)
CH_3D	5 μm	Beer et al. (1972)
NH_3	Visible	Wildt (1932)
	1 - 2 μm	Kuiper (1952)
	10 μm	Aitken and Jones (1972)
	50 - 200 μm	Furniss et al. (1978), Erickson et al. (1978), Baluteau et al.(1978,1980)
$^{15}NH_3$	10 μm	Encrenaz et al. (1978)
H_2O	5 μm	Larson et al. (1975)
CO	5 μm	Beer (1975)
GeH_4	5 μm	Fink et al. (1978)
PH_3	2 μm	Larson and Fink (1977)
	5 μm	Larson et al. (1977)
	10 μm	Ridgway (1974)
C_2H_2	13 μm	Ridgway (1974), Combes et al. (1974), Aumann and Orton (1976)
C_2H_6	12 μm	Ridgway (1974), Combes et al (1974), Tokunaga et al. (1976)

I. INTRODUCTION

Two main reasons can explain the huge progress achieved during the past decade in our knowledge of the composition of planetary atmospheres. First, in situ measurements from space vehicles (Mariner, Pioneer, Venera,....) have provided basic data on the atmospheric composition of terrestrial planets. Second, the recent and spectacular development of infrared astronomy, from the ground and in space, has drastically improved our knowledge of the atmospheric composition of such fainter objects as the Giant Planets and their satellites.

In the case of Jupiter, infrared spectroscopy is responsible for the discovery of most of the minor molecules (see Table I). While H_2, CH_4, and NH_3 were the only detected constituents known in 1970, a dozen new molecules have been identified during the past decade. Most of these molecules have been detected in the near IR range ($1\mu m<\lambda<3\mu m$) or in the far infrared range ($\lambda>5\mu m$). Infrared astronomy is also responsible for the discovery of an internal source of energy on Jupiter, Saturn, and Neptune (Armstrong et al. 1972; Gautier and Courtin 1979). In the case of Titan our present understanding of its atmosphere is mainly built upon infrared spectroscopic observations.

The success of infrared astronomy for the study of planetary atmospheres is easy to understand if we keep in mind that the planets of the solar system are typical infrared objects. There are two kinds of infrared radiation coming from the planets; a solar photon in the visible and near IR range entering the planetary atmosphere can be either scattered by the molecules and/or the particles and sent back to the Earth at the same wavelength, or absorbed by one of the atmospheric constituents and converted into thermal heat. Knowing the size of the planet, its mean visible albedo and its distance to the Sun, it is easy to calculate the effective temperature expected for the planet (Table II) and thus the spectrum of its infrared radiation (Figure 1). In the case of the giant planets, the value of the albedo is close to 0.5 which means that approximately one half of the solar incoming flux is scattered back in the visible and near IR range, and the other half is thermal radiation, approximately corresponding to the backbody curve of its effective temperature. As seen on Table II and Figure 1, the wavelength of maximum thermal emission ranges from 13 μm in the case of the Earth, to 110 μm in the case of Neptune.

Apart from the favorable shape of their spectra, the planets have another advantage: they are extended objects. Their diameters can reach one arcmin for Venus, 44 arcsec for Jupiter. For these two reasons Venus, Mars, and Jupiter are, apart from the Sun and the Moon, the brightest infrared sources in the sky, and are consequently observed in all the astronomical infrared experiments, at least as calibration sources.

Concerning the infrared experiments devoted to planetary observations, various types of instruments are used. In the terrestrial

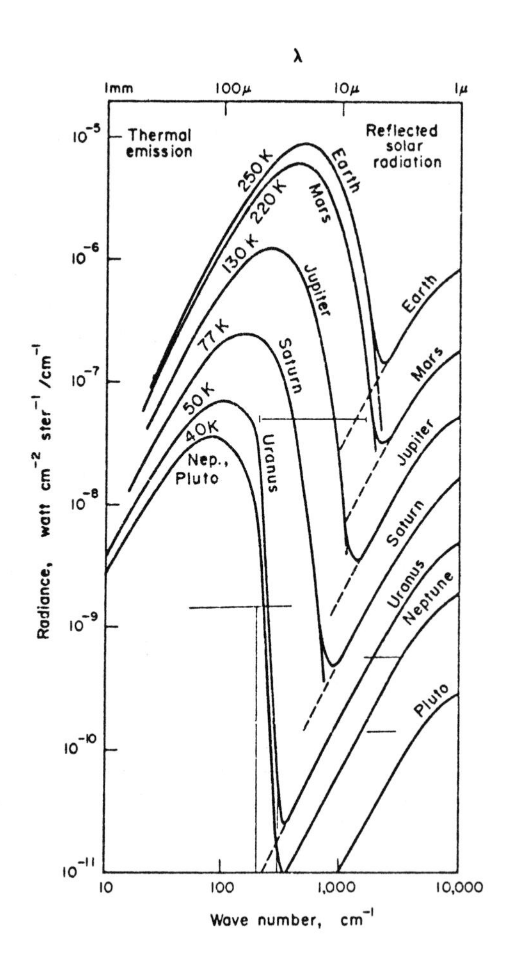

Figure 1. Thermal emission and reflected solar radiation of the planets (from Encrenaz 1979).

TABLE II

Thermal Emission of the Giant Planets

(NB: The difference in Columns I and III indicates the existence of an internal source of energy on Jupiter, Saturn, and Neptune.)

Planet	T_e(expected) (°K)	λ_{max}(expected) (μ)	T_e(observed) (°K)
Jupiter	105	50	125
Saturn	77	66	90 - 95
Uranus	55	80	57
Neptune	45	110	∿ 55

atmospheric windows (5μm, 10μm, 20μm), the use of ground-based large telescopes, associated with high sensitivity detectors, gives--in the case of the brightest planets--a signal-to-noise ratio high enough to allow high spectral resolution. With Fourier Transform Spectrometers, the resolving power can reach 10^5 in the near IR range (Maillard et al. 1973; Lecacheux et al. 1976) and 10^4 at 5 μm (Larson 1980) and 10 μm (Tokunaga et al. 1979b). An even higher resolving power (10^6) is obtained with the heterodyne technique at 10 μm, which has been developed during the past 5 years (Abbas et al. 1976) and applied to planetary observations (Mumma et al. 1979). The development of airborne observations has opened the available spectral range, especially toward longer wavelengths (λ>50μm). However, the resolving power is most often limited to 100 or less, due to the smaller telescope, the shorter integration time and the decrease in flux relative to shorter wavelengths. Nevertheless these far infrared planetary observations have proven to be very useful in particular for the determination of the H_2/He ratio on the giant planets (Gautier et al. 1977; Courtin et al. 1978, 1979). In contrast, most of the detection of minor molecules has been obtained in the near infrared range or in the 5 μm and 10 μm atmospheric window with high resolution spectroscopy.

In this paper, we first discuss the astrophysical information which can be derived from infrared planetary spectra (Part II). In Part III, we present the results obtained in the identification of atmospheric constituents. Part IV briefly reports the contribution of IR spectroscopy in the determination of the thermal profile T(P). In Part V we present and analyze the astrophysical implications: determination of abundance ratio, physical processes involved in the vertical profiles of non-uniformly mixed constituents. Perspectives and conclusions are discussed in Part VI.

II. THE ASTROPHYSICAL OBJECTIVES

The planetary atmospheres constitute a basic tool for our understanding of the origin and the evolution of our Solar System. The terrestrial planets, because of their small size and relative proximity to the Sun, have been unable to retain their primitive atmospheres; a comparative study of the elemental and isotopic relative abundances derived from the composition of their atmospheres and/or their surfaces is the only way to understand the early stages of planetary formation in the vicinity of the Sun. In contrast, the giant planets, which are cold and large objects, must have retained in their atmospheres the composition of the primordial solar nebula. The study of their atmospheric composition is of major interest for understanding the composition and physical conditions of the primordial nebula at the time of the solar system formation.

The first astrophysical result which can be derived from the reduction of infrared planetary spectra is the detection of minor molecules. This determination requires only the knowledge of the frequencies of the lines of multiplets. The infrared range is especially well adapted for this work, because most of the molecules exhibit strong vibration-rotation bands (mostly fundamentals) between 1 and 20 μm. In contrast with the visible range, the structure of these bands is relatively simple and they have been extensively studied in the laboratory. They are sufficiently separated from one another and they present a large number of individual lines, which allows unambiguous determinations.

The mechanism of line formation in a planetary atmosphere depends upon the kind of radiation which is considered. In the near infrared range, the solar radiation is either absorbed or scattered by a molecule or a particle. Thus, in this spectral range, molecular lines in a planetary spectrum usually appear in absorption, apart from the case of nonthermal effects. In contrast, in the far infrared range where thermal emission takes place the measured flux refers in a first approximation to the atmospheric level where the optical depth is equal to one; the probed atmospheric regions most often correspond to pressures larger than 1-10 bars, where L.T.E. takes place. The far IR measured flux is thus a strong function of the thermal atmospheric profile, and the molecular lines or bands can appear either in emission or in absorption, as a function of the temperature lapse rate in the formation region.

It clearly appears that, even when the absolute intensities of the molecular bands are known in the laboratory, it is usually not possible to derive the abundance of an atmospheric constituent from a planetary spectrum. In the far infrared range, this determination requires the knowledge of the thermal profile; in the near-infrared range, it requires the T(P) profile and also all the scattering parameters (scattering probabilities, scattering phase functions) which are usually unknown. However, in the latter case, a rough estimate of the absorber abundance can be derived with the hypothesis of the "reflecting layer model" which assumes no scattering above a purely reflecting layer (cloud or surface). This assumption is reasonable in the case of a tenuous atmosphere, as Mars; it has been used also for a long time in the case of the giant planets, in the absence of any other information; the validity of the method in this case will be discussed in more detail below.

The second step in the interpretation of planetary spectra is thus the determination of the thermal profile (e.g., Figure 2), which is needed in all cases to derive abundance measurements and, later on, abundance ratios. Various methods can be used for the determination of the temperature-pressure (T(P)) relationship. In the case of Mars, Venus, and Jupiter, thermal profiles have been derived from radio-occultation experiments aboard space vehicles (Mariner, Pioneer: Kliore et al. 1972; Kliore and Woiceshyn 1976), and also, on Mars and Venus, from *in situ* measurements. In the case of the giant planets, most of our information about the T(P) profile comes from their infrared spectrum, as is discussed in detail by Dr. G. Orton in the present

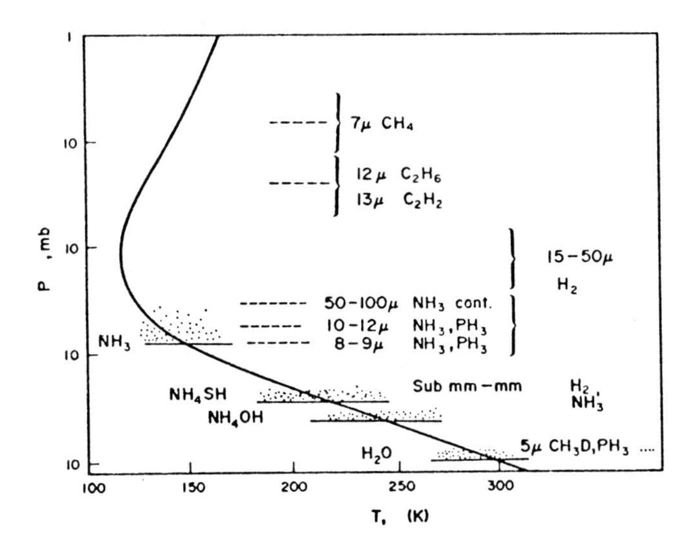

Figure 2. Jovian atmospheric layers probed in the far IR range (from Encrenaz 1979).

volume. On Jupiter, Hanel et al. (1979a,b) have obtained a set of thermal profiles on various points of the Jovian disk, from interferometric measurements at 4 cm^{-1} resolution between 200 and 2000 cm^{-1} (5–50 µm). For the other giant planets all the information comes from ground-based observations (Gautier et al. 1977a; Courtin et al. 1978, 1979).

Once the first two steps are achieved, it becomes possible to derive from infrared planetary spectra information which is of much higher astrophysical interest. Indeed, once the thermal profile is known, abundance ratios can be derived and, in the case of non-uniformly mixed constituents, vertical distribution profiles can be obtained. First, these determinations allow us to study the physical and chemical processes involved in the planetary atmospheres (photochemistry, condensation, circulation, chemical reactions,...), and to learn the nature and the evolution of the cloud structure, and then to start understanding the climate of these planets. But these planetary studies also have a more general implication. The elemental and isotopic ratios derived in the giant planets actually reflect, in many cases, the composition and physical conditions of the primordial nebula at the time of the solar system formation. Comparison with data obtained in the interstellar medium can give information about the general problem of star formation. On the other hand, comparative studies of the basic ratios on the four giant planets may show evidence for an inhomogeneous distribution of the elements as a function of their distance from the Sun. The study of some stable isotopes ($^{12}C/^{13}C$, $^{14}N/^{15}N$) can give information about the chemical evolution of the Galaxy; finally, precise determinations of D/H, H/He, $^{3}He/^{4}He$ in the solar system (and in particular in the giant planets) are basic tools for discriminating among different cosmological models.

III. DETECTION OF MOLECULES

In spite of the major role of in situ measurements for the determination of the atmospheric composition on Mars or Venus, it is worthwhile to recall what has been the contribution of infrared spectroscopy in the identification of molecular species. Tables III and IV give a list of molecules detected on Venus and Mars by infrared spectroscopy.

TABLE III

Observed Molecules in the Atmosphere of Venus

Molecule	Spectral Range	Resolution	Reference
CO_2	0.9 - 1.8 μm	5×10^{-3} μm	Kuiper (1952)
	750 - 1250 cm^{-1}	1.25 cm^{-1}	Hanel et al. (1968)
CO_2,H_2SO_4	450 - 1250 cm^{-1}	0.67 cm^{-1}	Samuelson et al. (1975)
	" "	0.25 cm^{-1}	Kunde et al. (1977)
	500 - 800 cm^{-1}	3.2 cm^{-1}	Orton and Aumann (1977)
CO_2	900 - 1100 cm^{-1}	10^{-3} cm^{-1}	Mumma et al. (1980)
HCl, HF, CO	5000 - 6000 cm^{-1}	0.1 cm^{-1}	Connes et al. (1967) Connes et al. (1968)
H_2O	35 - 55 μm		Taylor et al. (1980)
CO_2,H_2SO_4	1 - 4 μm	0.04 μm	Pollack et al. (1974)

Carbon dioxide was first identified spectroscopically in the near infrared range (Kuiper 1952) on Mars and Venus where it is the most abundant element. Other determinations were made later on at 8700 Å on Mars (Kaplan et al. 1964; Spinrad et al. 1966) with a first measurement of the CO_2 abundance. In the case of Venus, the amount of carbon dioxide was obtained from in situ measurements (Venera 4, 5, and 6: Vinogradov et al. 1968; Avduevsky et al. 1970).

Water vapor was first spectroscopically identified on Mars at 8700 Å (Kaplan et al. 1964); these observations already indicated a variation of H_2O with the seasons on Mars. The amount of H_2O on Mars was later monitored by the IRIS experiment during the Mariner 9 mission (Conrath et al. 1973). On Venus, water vapor was detected from the earth in the red range (Schorn et al. 1969), but did not appear in the near IR (Connes et al. 1967) nor in the 8-22 μm range (Kunde et al. 1977). Further information on H_2O was provided by the infrared radiometer experiment (VORTEX) of Pioneer Venus which has identified H_2O in its far infrared channel (35-55 μm) (Taylor et al. 1980; Beer 1980; McCleese 1980).

TABLE IV

Observed Molecules in the Martian Atmosphere

Molecule	Resolution (cm^{-1})	Spectral Range (cm^{-1})	Abundance	Reference
CO_2	0.2	9530	68±26 m-atm	Belton and Hunten (1966)
"		700 - 1200		Sinton and Strong (1960)
"	20	"	65±30 m-atm	Verdet et al.(1972)
"		4000 - 12500		Kuiper (1952)
"	2.4	200 - 2000		Conrath et al.(1973)
CO_2, $C^{16}O^{18}O$, $C^{16}O^{18}O$, $^{13}CO_2$	"	"		Maguire (1977)
$^{13}CO_2$	10^{-3}	1000		Peterson et al. (1974)
CO_2	"	"		Mumma et al.(1975); Betz (1975)
H_2O	2.4	200 - 2000		Conrath et al.(1973)
CO	0.1	4000 - 6500	5.6 cm-am	Kaplan et al.(1969)
O_2 (O_3)	0.02	7850 - 7900		Noxon et al. (1976) Traub et al. (1977)

Infrared spectroscopy has proven to be especially useful in the near infrared (1 - 2.5 μm) with the use of the high resolution Fourier Transform spectrometer built by P. and J. Connes (Connes and Connes 1966): with this instrument, HCl, HF, CO, H_2SO_4 have been identified on Venus, CO and O_2 on Mars. Another powerful instrument has been the IRIS Michelson interferometer working between 5 μm and 50 μm, flown on the Mariner 9 mission: H_2O, CO_2 and its isotopes have been identified on Mars (Maguire 1977). A third kind of result which has to be mentioned is the high resolution (10^6) profiles of CO_2 obtained at 10 μm with a new heterodyne spectrometer (Mumma et al. 1975; Betz 1975; Mumma et al. 1980).

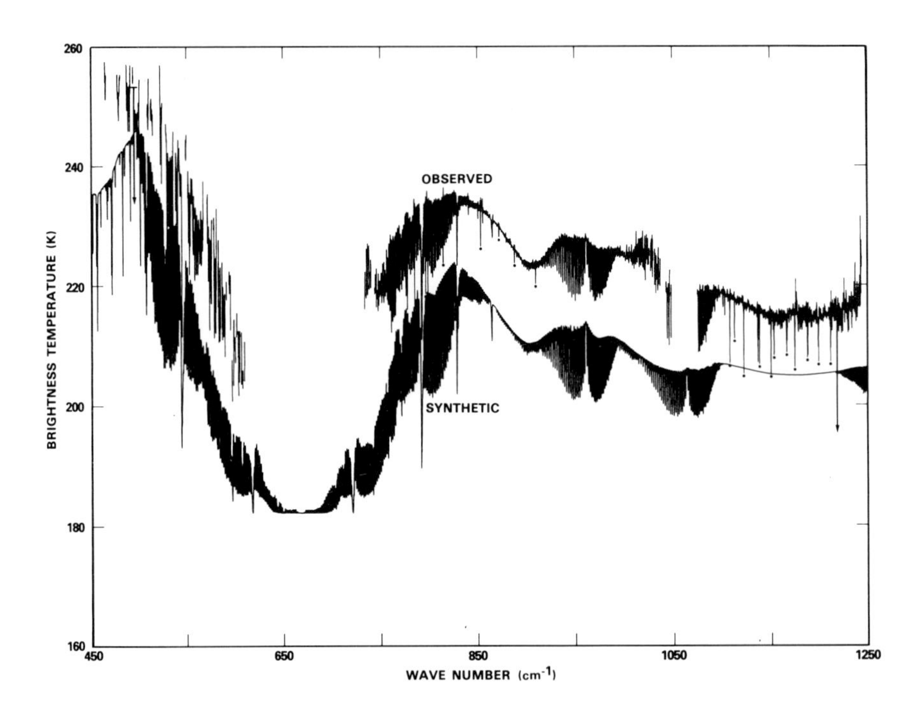

Figure 3. Synthetic and observed spectra of Venus, by Kunde et al. (1977). The broad absorption feature around 900 cm^{-1} is attributed to H_2SO_4 haze.

The infrared range is also most suitable for searching for particles and droplets. Spectroscopic experiments can identify the nature of these particles and droplets by the observation of their broad infrared absorptions, while polarimetry measurements can give information upon their size. A haze of H_2SO_4 has been identified on Venus between 870 and 930 cm^{-1} (Figure 3) (Kunde et al. 1977). On Mars, the silicate absorption at 10 μm is present in a ground-based spectrum recorded during the dust storm (Verdet et al. 1972) and on the Mariner 9 IRIS spectra (Figure 4) (Hanel et al. 1972). The IRIS experiment also identified around 800 cm^{-1} broad absorption features due to H_2O ice clouds on some points of the Martian disk. In the case of giant planets, all the molecular identifications are the results of infrared spectroscopic experiments, apart from He and HD on Jupiter

(Table I). Most of these observations have been made with Fourier Transform spectrometers with a resolving power better than 10^3 (Figure 5). On Jupiter H_2, $^{13}CH_4$, CH_3D, C_2H_2, C_2H_6, CO, PH_3, $^{15}NH_3$ have been detected from the ground while H_2O and GeH_4 were observed from the Kuiper Airborne Observatory at 5 μm. NH_3 ice was also identified on Jupiter at 8 - 9 μm (Orton 1975b; Encrenaz et al. 1980). This illustrates the major role played by ground-based astronomy in the terrestrial atmospheric windows, especially 5 μm, 7 - 13 μm (Figure 6), and the near IR range. In the case of Saturn (Table V) the interpretation of its spectrum at 10 μm has been controversial: the features appearing between 930 and 990 cm^{-1} were attributed either to PH_3 in absorption (Bregman et al. 1975) or to C_2H_4 in emission (Encrenaz et al. 1975). The recent spectrum recorded by Tokunaga et al. (1980) demonstrates that Bregman et al.'s interpretation was right. The absence of C_2H_4 on Saturn can be related to the absence of C_2H_4 on Jupiter (Encrenaz et al. 1978). In contrast, the presence of C_2H_4 has been suggested on Titan from the observation of an emission feature at 10.5 μm (Gillett 1975).

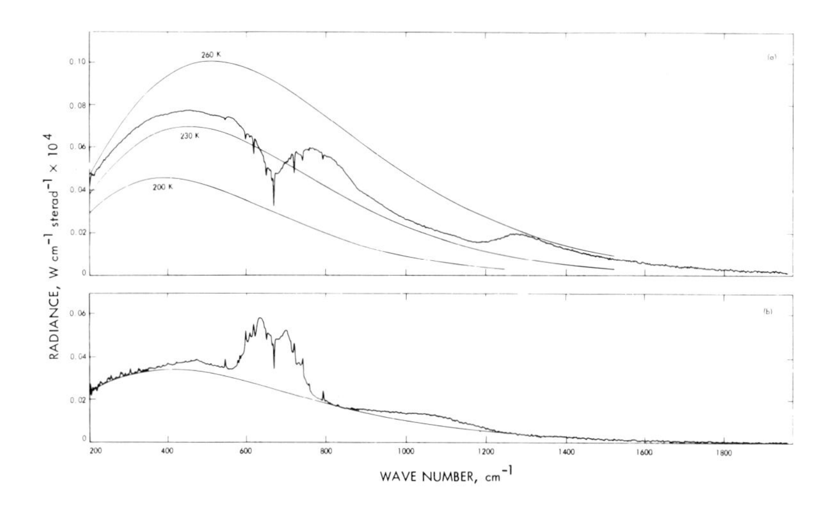

Figure 4. Thermal spectra of Mars (A: non polar; B: polar), with the Mariner 9 IRIS experiment. The figure is taken from Hanel et al. (1972).

Concerning Uranus and Neptune (Table VI), the detection of molecules comes from the visible and near infrared range when strong CH bands are present. In addition, an emission feature has been detected at 8 μm and 12 μm on Neptune, corresponding to CH_4 and C_2H_6 respectively, but not on Uranus (Gillett and Rieke 1977; Macy and Sinton 1977). As will be discussed below, this difference in the spectra of Uranus and Neptune in the 8 - 12 μm range is explained by different thermal profiles in the upper atmospheres of Uranus and Neptune.

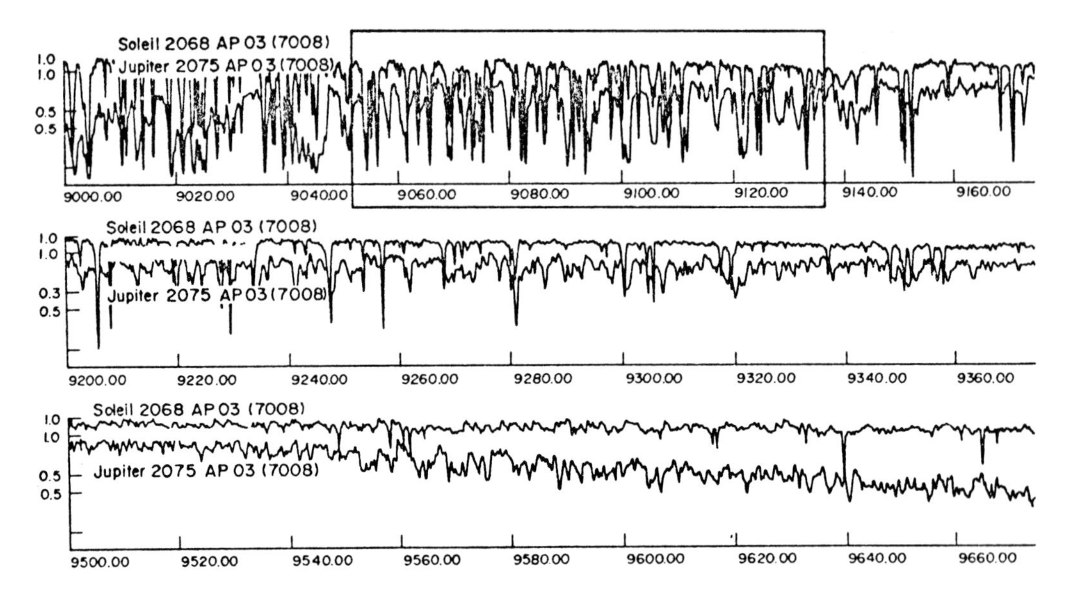

Figure 5. The spectrum of Jupiter (heavy line) and the solar comparison spectrum (light line) in the region of the $3\nu_3$ CH_4 band (9000-9700 cm^{-1}. This figure is taken from Maillard et al. (1973).

TABLE V

Observed Molecules in the Atmosphere of Saturn

Molecule	Spectral Range	Reference
H_2	0.8 μm	Kiess et al. (1960)
	1.25 μm	DeBergh et al. (1977)
HD	6064 Å	Macy and Smith (1978)
CH_4	0.8 μm, 1-2 μm	Wildt (1932); Kuiper (1952)
	1.1 μm	DeBergh et al. (1973)
$^{13}CH_4$	1.1 μm	Combes et al. (1975)
CH_3D	5 μm	Fink and Larson (1977)
NH_3	6450 Å	Encrenaz et al. (1974)
PH_3	3 μm, 5 μm	Fink and Larson (1977)
	10 μm	Bregman et al. (1975); Tokunaga et al. (1980)
C_2H_6	12 μm	Tokunaga et al. (1975)

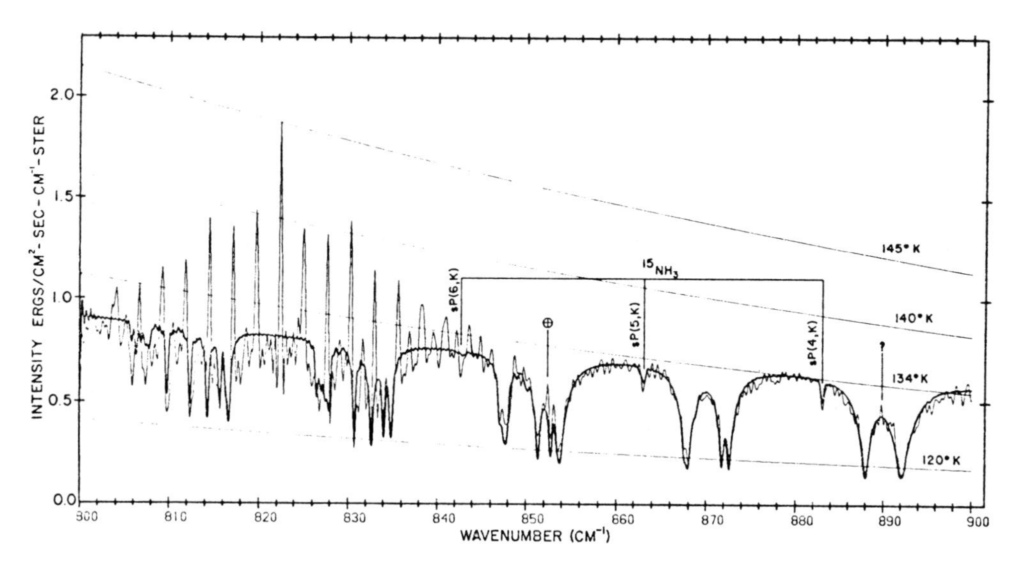

Figure 6. The spectrum of Jupiter in the 11-12.5 μm range, from Tokunaga et al. (1979b).

TABLE VI

Observed Molecules on Uranus, Neptune, and Titan

Planet/Satellite	Molecule	Spectral	Reference
Uranus	H_2	0.6 - 0.8 μm	Giver and Spinrad 1966
	CH_4	Visible	Kuiper (1952)
	HD	Visible	Macy and Smith (1978)
Neptune	H_2	0.6 - 0.8 μm	Trafton (1972)
	CH_4	Visible	Kuiper (1952
	"	8 μm	Macy and Sinton (1977)
	C_2H_6	12 μm	Gillett and Rieke (1977)
Titan	H_2	6400 Å	Trafton (1974)
	C_2H_6	Visible	Kuiper (1944)
	"	1.1 μm	Trafton (1975)
	"	8 μm	Gillett et al. (1973)
	C_2H_2, C_2H_6, CH_3D, C_2H_4	10 - 13 μm	Gillett (1975)

There are three satellites of major planets where an atmosphere has been detected at the present time. The most fascinating of them is probably Titan which fills at least two conditions favorable for the development of the early stages of life: a surface and a reducing atmosphere. The atmosphere of Titan has been known since 1944 (Kuiper 1944). Methane was detected in the visible, then the $3\nu_3$ band of CH_4 at 1.4 μm was observed by Trafton (1975). The identification of H_2, first suggested by Trafton (1972), is still tentative (Münch et al. 1977). The other molecular determinations have been done in the 7 - 13 μm region by Gillett (1975): emission features have been tentatively identified as C_2H_2, C_2H_6, C_2H_4, and CH_3D. The far infrared spectrum of Titan (16 - 30 μm) recorded by McCarthy et al. (1980) has been interpreted by the authors in terms of a thin dust haze high in the atmosphere. The peculiar interest presented by Io and its volcanoes has been spectacularly demonstrated by the imaging experiment of the Voyager mission. At the same time, a detection of SO_2 has been made at 1350 cm^{-1} with the IRIS interferometer (Pearl et al. 1979). The apparent discrepancy with the upper limit of SO_2 derived from IUE in the ultraviolet range (Butterworth et al. 1980) can be resolved if the SO_2 atmosphere is confined to the vicinity of the volcanoes, or if the SO_2 abundance on Io is variable with time. It is interesting to mention that the volcano eruptions on Io seem to be correlated with 5 μm outbursts, which could be due to vapor eruptions of liquid sulfur (Sinton 1979).

The first indication of methane atmospheres on Triton, satellite of Neptune, and Pluto, was given by Cruikshank and Silvaggio (1979, 1980) who reported the observation of broad absorption features at 2.2 μm and 1.7 μm respectively. The Pluto absorption was attributed to a combination of gaseous and solid methane, in agreement with the previous CH_4ice identification of Cruikshank et al. (1976). Trafton, using model calculations, reported later that the existence of gaseous CH_4 on Pluto requires the presence of a heavier and more abundant gas (Trafton 1979b). The Triton absorption was attributed to a tenuous CH_4 atmosphere. In both cases, more measurements are required for determining the nature and the density of these tenuous atmospheres.

IV. DETERMINATION OF THE THERMAL PROFILE

In this part, we give a few examples of the contribution of infrared spectroscopy in the determination of the thermal planetary structure, especially in the case where this information is absolutely needed for our understanding of the atmospheric composition.

The first method consists in the determination of the T(P) profile using the thermal infrared spectrum of the planet in a spectral range where the main atmospheric constituent is absorbing. The method consists of choosing a set of frequencies corresponding to well separated atmospheric levels where the radiation comes from (in a first approximation the levels of optical depth equal to 1). From the measurement of the thermal flux at these wavelengths, it is possible to generate an iterative method which converges toward the true profile T(P). This method has been used in the case of Venus, in the CO_2 absorption region at 15 μm, by the VORTEX experiment on the Pioneer Orbiter mission (Taylor et al. 1980). In the case of the giant planets, the method has been used in a slightly different way: two absorbers are responsible for most of the far IR spectrum between 15 and 50 μm: hydrogen and helium, in a relative abundance ratio which must be determined also. Since the absorption coefficient is a function of this H_2/He ratio, it is possible to generate a double-iterative method, which converges toward both the true T(P) profile and the true H_2/He ratio. This method is presented and discussed in detail by Orton (1980) in the present symposium. It was first applied to Jupiter (Orton 1975a,1977; Wallace and Smith 1977; Gautier et al. 1977a) then to the other giant planets (Gautier et al. 1977b; Courtin et al. 1978, 1979) from ground-based observations. The same method was used on Jupiter by the IRIS experiment during the Voyager mission (Hanel et al. 1979a,b). This determination has the advantage of requiring no high spectral resolution: for the giant planets, a resolving power of 10 to 50 is sufficient. The measurement of the H_2/He mixing ratio derived from this method will be discussed in more detail below. Figure 2 shows a typical thermal profile in the case of Jupiter. It can be shown that the four profiles are characterized by a convective zone where the measured lapse rate seems very close to the adiabatic value, a minimum value at a pressure around 0.03 bar, and apart from Uranus, a strong increase of temperature as a function of altitude in the lower stratosphere.

In another way, spectroscopic observations in the thermal range (λ>5 μm) can provide constraints on the temperature profile. The reason is, as explained above, the close relationship between the measured brightness temperature (T_B) and the true temperature of a given atmospheric level. As an example, the measured T_B cannot be smaller than the minimum value of the temperature profile: this remark has been used to derive an upper limit of the Jovian minimum temperature from the observation of the NH_3 absorption line centers at 10 μm (Combes et al. 1976). Another example is given by the emission bands of CH_4 at 7.7 μm and, at 12 - 13 μm, C_2H_2 and C_2H_6. The first observation of the high flux measured at 8 μm on Jupiter by Gillett et al. (1969) was the first element in favor of a temperature inversion on Jupiter. Later on, the observation of C_2H_2 and C_2H_6 gave another experimental support to this idea (Ridgway 1974; Combes et al. 1974; Orton and Aumann 1977). Information upon T(P) from the lower Jovian stratosphere has also been derived from the far infrared Jovian spectrum in the NH_3 rotational spectrum (Vapillon et al. 1977; Goorvitch et al. 1979); however, there is another unknown parameter which is the upper NH_3 density distribution. In the case of Saturn, infrared observations at 20 μm along the central meridian were used to derive T(P) distributions at the equator and at the poles (Tokunaga et al. 1978).

TABLE VII

Abundance Ratios on Venus

Molecule/Ratio	Measured Value	Atmospheric Level	Reference
$^{12}C/^{13}C$	Terrestrial		Connes et al. (1968) (Ground-based Sp.)
Upper Atmosphere			
CO_2	6×10^{9} cm^{-3}	135 km	Von Zahn et al. (1979) Mass spectr.
He	5×10^{6} cm^{-3}		
CO_2	1.1×10^{9} cm^{-3}	> 155 km	"
CO	2.4×10^{8} cm^{-3}		
N_2	2.1×10^{8} cm^{-3}		
O	6.6×10^{8} cm^{-3}		
He	2×10^{6} cm^{-3}		
$^{12}C/^{13}C$, $^{16}O/^{17}O$, $^{16}O/^{18}O$	Terrestrial		Niemann et al. (1979) Mass spectr. (probe)
Lower Atmosphere			
$^{36}Ar/^{12}C$	10^{-4}		Hoffman et al. (1979) Mass spectr. (probe)
S_2, S_8, COS, H_2S			
CO_2	96.4±1.03%	24 km	Oyama et al. (1979) Gas chromat. (probe)
N_2	3.41±0.02%		
H_2O	< 0.06%		
O_2	69.3±1.3 ppm		
Ar	18.6±2.4		
Ne	4.3 (+5.5,-3.9)		
SO_2	186 (+349,-156)		

V. ATMOSPHERIC COMPOSITION AND ABUNDANCE RATIOS

A. Venus and Mars

Results on abundance ratios measured in the atmospheres of Venus and Mars are listed in Tables VII and VIII. In the case of Venus, apart from a $^{12}C/^{13}C$ determination from infrared ground-based spectroscopy, the information comes entirely from space missions. Abundances were measured by mass spectrometers and gas chromatographs aboard the probes of the Pioneer Venus mission and the Venera 11 and 12 missions. These *in situ* measurements have shown that the Venus atmosphere is governed by sulfur chemistry. In the upper oxidizing atmosphere (40 - 90 km), sulfur is present in H_2SO_4 and, at lower levels, SO_2 and O_2 are detected (Sill 1979). In the lower reducing atmosphere near the surface S_2, S_8, COS, and H_2S have been detected by the Pioneer mass spectrometer (Hoffman et al. 1979). Sulfur is expected to condense in droplets near 32 km. The CO molecule has been detected in the upper atmosphere above 150 km (Niemann et al. 1979), in agreement with millimetric ground-based measurements at 2.6 μm (Schloerb et al. 1979; Muhleman et al. 1979; Wilson and Klein 1979). In the case of Mars, the isotopic ratios previously estimated by ground-based experiments and by Mariner 9 IRIS data have been remeasured with high accuracy with the Viking mass spectrometer experiment (Nier et al. 1976; Biemann et al. 1976). In particular, a ^{15}N enhancement by a factor of 75% has been found (Table VIII). This enrichment is attributed to selective escape from the Martian upper atmosphere, implying that Mars must have lost an appreciable amount of N_2 to space during its history (McElroy et al. 1976).

TABLE VIII

Abundance Ratios on Mars

Ratio	Method	Measured Value	Reference
$^{12}C/^{13}C$	Ground-based Sp.	Terrestrial	Kaplan et al. (1969)
"	"	"	Young (1971)
"	Mariner 9 (IRIS)	89±13	Maguire (1977)
"	Viking Mass Sp.	85±9	Biemann et al.(1976)
	"	87±3	Nier et al. (1976)
$^{16}O/^{17}O$ $^{16}O/^{18}O$	Mariner 9 (IRIS)	Terrestrial	Maguire (1977)
$^{16}O/^{18}O$	Viking Mass Sp	526±50	Biemann et al(1976)
$^{14}N/^{15}N$	"	175±25 (75% ^{15}N enrich.)	"
$^{36}Ar/^{38}Ar$	"	4 - 7	"

B. The Giant Planets and Their Satellites

In what follows, two cases can be considered. If the constituents are homogeneously mixed with hydrogen, their vertical distribution is defined by a single parameter, which is the ratio of the corresponding element to hydrogen. In the case of nonhomogeneously mixed constituents, we have to determine their vertical distributions independently at different atmospheric levels, in order to understand the physical processes involved.

TABLE IX. ABUNDANCE RATIOS IN THE GIANT PLANETS

Ratio	Spectral Range	Jupiter	Saturn	Uranus	Other*
$\frac{H_2}{H_2+He}$	thermal radiation	0.897±0.030 (1)†	>0.5 (2)	0.9±0.1 (3)	(⊙) 0.89 (4) (PN) 0.871±0.02(5)
$\frac{C}{H}$	scattering model (visible+NIR)	$2-3\times10^{-3}$ (6)		$3-10\times10^{-3}$(7)	(⊙) $4.7^{+1.2}_{-1.0}\times10^{-4}$(9)
	scattering model (1–2 μm)	8×10^{-4} (8)	1.15×10^{-3}(8)	2×10^{-3}(8)	(Neptune) 9×10^{-4}(8)
	scattering model (1.1 μm)	$1.5\pm0.7\times10^{-3}$(10)	$2\pm1\times10^{-3}$(10)		
	visible+NIR	$6.2\pm2.1\times10^{-4}$(11)		$\sim10^{-2}$(12)	(Neptune) $\sim10^{-2}$(12)
	thermal radiation	7×10^{-4}(13) $<10^{-3}$(14) $7.0\pm2.2\times10^{-4}$(15)			
$\frac{D}{H}$	HD/H_2 (visible)	$5.1\pm0.7\times10^{-5}$(16) $<2.3\times10^{-5}$(11)	$5.5\pm2.9\times10^{-5}$(17)	$3.0\pm1.2\times10^{-5}$(17) $1-2\times10^{-5}$(19)	(PN) 2.5×10^{-5}(18) (LIM) $1-2\times10^{-5}$(20)
$\frac{CH_3D}{H_2}$	5 μm	5×10^{-7}(21) 2.5×10^{-7}(22)			
	10 μm	$\leq2\times10^{-7}$(14) 5×10^{-7}(15)			
$^{12}C/^{13}C$	1.1 μm	89^{+12}_{-10} (23)	89^{+25}_{-18} (23)		(⊙) 89±5 (24)
$^{15}N/^{14}N$	10 μm	0.0037±0.0015 (25) 0.003 – 0.006 (26)			(⊕) 0.0037 (4)

*Other: (⊙)=Sun; (PN)=Primordial Nebula; (LIM)=Local Interstellar Medium; (⊕)=Earth.

†REFERENCES:
(1) Gautier et al. 1980
(2) Gautier et al. 1977b
(3) Courtin et al. 1978
(4) Cameron 1974
(5) Lequeux et al. 1979
(6) Wallace and Hunten 1978
(7) Benner and Fink 1980
(8) Fink and Larson 1979
(9) Lambert 1978
(10) Buriez and De Bergh 1980
(11) Combes and Encrenaz 1979
(12) Lutz et al. 1976
(13) Orton 1977
(14) Encrenaz et al. 1980
(15) Kunde et al. 1980
(16) Trauger et al. 1977
(17) Macy and Smith 1978
(18) Geiss and Reeves 1972
(19) Encrenaz and Combes 1978
(20) Laurent 1978
(21) Beer and Taylor 1978
(22) Kunde et al. 1979
(23) Combes et al. 1977
(24) Hall et al. 1972
(25) Encrenaz et al. 1978
(26) Tokunaga et al. 1979b

a) The case of homogeneously mixed components

The abundance ratios which have been determined from homogeneously mixed constituents are H_2/He, C/H, D/H, $^{12}C/^{13}C$ and $^{14}N/^{15}N$.

H_2/He. The H_2/He ratio has been derived on Jupiter by the inversion of the Jovian far infrared spectrum, as explained above (Part IV). The first ground-based determinations resulted in a solar value of H_2/He (Orton 1975a,1977; Wallace and Smith 1977; Gautier et al. 1977a): He/H_2 + He) = 0.89 ± 0.06. After the Voyager IRIS measurement, the error bar was divided by two (Rouan et al. 1980; Gautier et al. 1980). As shown on Table IX, the Jovian He/H_2 value is equal to the solar value, and--what is more significant--slightly smaller than the primordial value estimated by Lequeux et al. (1979) (Gautier et al. 1980). In the case of the other giant planets, the same method could apply in theory; however, in the case of Saturn and Neptune, the uncertainty in the present data is too large so that no information is obtained on H_2/He; only T(P) is derived (Gautier et al. 1977b; Courtin et al. 1979). In contrast, H_2/He was obtained on Uranus, in good agreement with Jupiter's value: $H_2/(H_2 + He) = 0.9 \pm 0.1$ (Courtin et al. 1978).

C/H. Since the CH_4 spectrum extends from the visible to 8 μm, it is possible to obtain estimates of C/H from both the near IR and the far IR range, at least on Jupiter where both sets of data exist.

The first C/H estimates on Jupiter were made some 10 years ago in the visible and near infrared range, where data were first available. As mentioned above, the simple "reflecting layer model" (RLM) was used, in the absence of any information about the Jovian atmospheric scattering. However, it is easy to demonstrate that the RLM approximation is not valid for the dense atmospheres of the giant planets: the ratios derived from different lines or bands of CH_4 vary by a factor of 3. Another attempt to solve the problem is the use of a scattering model; however, too many parameters, still unknown, are involved in this model, so that the solution is not unique: as summarized by Wallace and Hunten (1978) the derived C/H ratio ranges from the solar value to 3 or 4 times the solar value; in particular the recent study by Buriez and de Bergh (1980a), from the $3\nu_3 CH_4$ band at 1.1 μm, concludes that carbon is enriched by a factor 2 to 4. A different approach is used by Fink and Larson (1979) who prefer to use the CH_4 weak bands at longer wavelengths (1.2 and 1.6 μm) because they may be used for the four giant planets and Titan, and also because the effect of scattering is minimized. They derive a C/H ratio enriched by a factor 2 relative to the solar value. A third method is defined by Combes and Encrenaz (1979), which consists of eliminating, as much as possible, the effects of scattering by defining appropriate conditions in the selection of the lines or bands used for the abundance ratio. Applying their method to Jupiter, they conclude that carbon is enriched by a factor 1.3.

Simultaneously, C/H estimates of Jupiter have been derived from its 7 - 8 μm spectrum. The first tentative estimate was made by Orton

(1977) corresponding to an enrichment of 1.5. This result is confirmed by the upper limit derived from the absence of the CH_4 absorp-feature at 8 - 9 μm, corresponding to a carbon enrichment smaller than 2 on Jupiter (Encrenaz et al. 1980). Finally, the carbon enrichment by a factor 1.5 on Jupiter is also confirmed by the Voyager IRIS experiment (Rouan et al. 1980; Kunde et al. 1980).

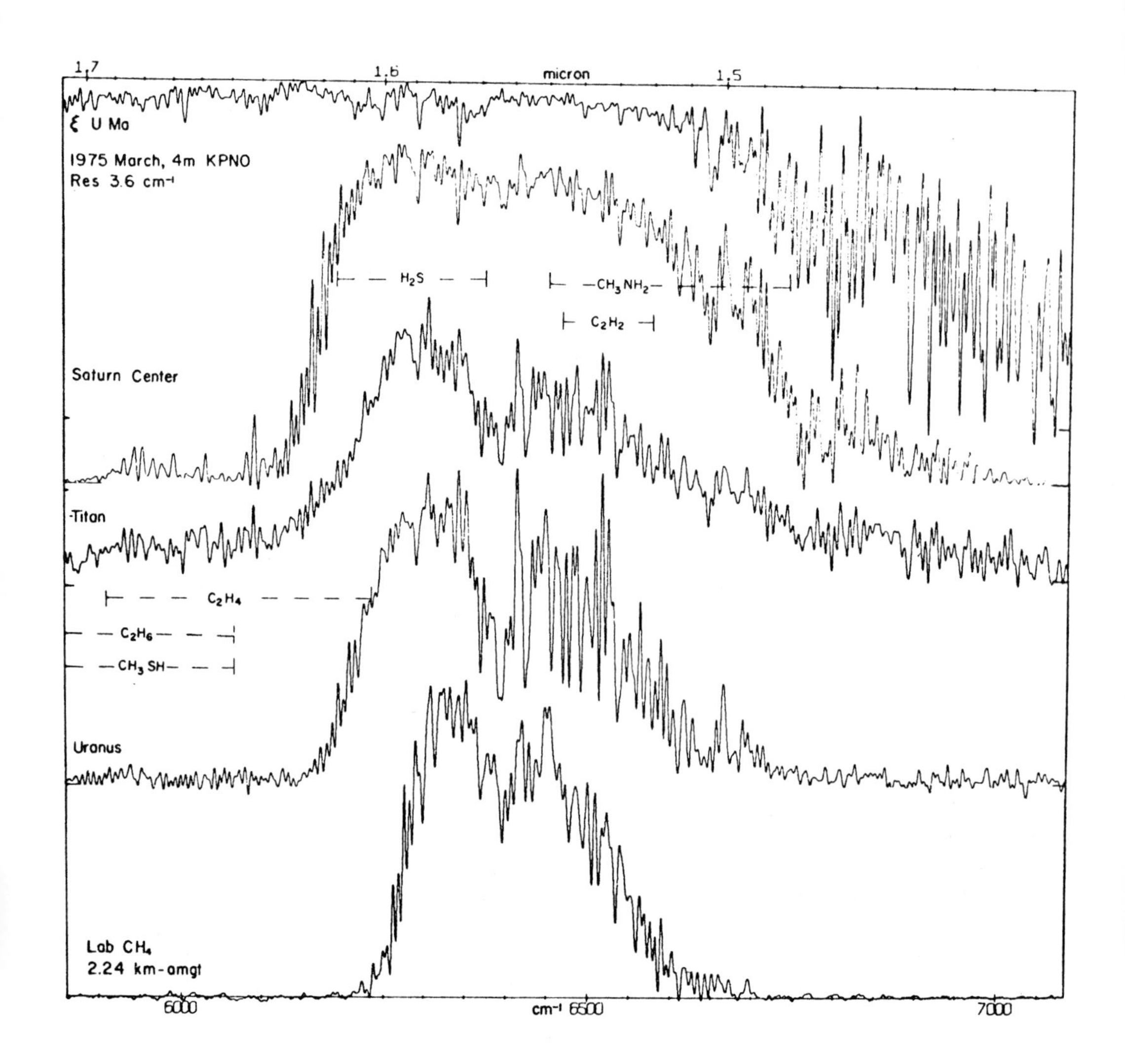

Figure 7. Spectra of Saturn, Uranus and Titan in the 6400 cm^{-1} methane window. The figure is taken from Fink and Larson (1979).

In conclusion, in spite of the apparent contradiction among the various determinations of the Jovian C/H ratio, some general considerations can be drawn. As shown in Table IX, there are basically two classes of results: (1) the Jovian C/H values derived from visible and near IR data, based on scattering model calculations, ranging from 2 to 5 times the solar value; (2) the values corresponding to a Jovian carbon enrichment less than a factor 2, which are based upon two methods: the use of the thermal spectrum between 7 and 9 μm, and the use of visible or near IR range data without scattering models, where the lines and bands have been chosen in order to eliminate as much as possible scattering effects. We have no explanation at the present time to explain this discrepancy. However, we strongly favor the second set of C/H values--implying a small carbon enrichment on Jupiter--for the following reasons. First, this low enrichment is derived from two completely independent methods. Second, the thermal infrared range is, in our opinion, more reliable because the physics involved is much simpler, since the effect of scattering is much less important than in the visible and near infrared ranges. Third, the variety of C/H results obtained from scattering models illustrates the non-uniqueness of the solution, because too many unknown parameters are involved in the calculations.

In the case of the other giant planets, the only data come from the visible and near infrared ranges. Here again a discrepancy exists between the values derived by Fink and Larson (1979) from the 1 - 2 μm range (see Figure 7), implying a constant carbon enrichment on the four giant planets, and other estimates implying higher carbon enrichments on Saturn (Buriez and de Bergh 1980b) or Uranus and Neptune (Lutz et al. 1976). With the improvement in quality of far infrared spectra, it should be possible to derive in the near future C/H estimates based upon thermal radiation, at least for Saturn and Neptune.

Finally, the case of Titan is especially puzzling. While methane has been detected in large amounts, the H_2 identification is too tentative to allow a reliable C/H ratio to be derived. Another difficulty comes from the very high CH_4abundances derived on Titan from the near infrared bands, relative to the CH_4amounts obtained from the visible CH_4 bands. In any case the H_2 abundance seems to be small compared to methane but another inert gas, such as N_2, could be present in large abundance so that the uncertainty about the surface pressure ranges from 11 mb (Fink and Larson 1979) to 1 bar or more, according to various models (Trafton 1979a; Rages et al. 1979; Giver et al. 1979; Hunten 1978).

In conclusion, it should be mentioned that the solar C/H ratio itself is very uncertain: the derived value has been varying by a factor of 2 during the past 5 years (Cameron 1974; Mount and Linsky 1975; Pagel 1977; Lambert 1978). In the preceding discussion we have adopted the most recent value (4.7×10^{-4}), in agreement with Pagel (1977) and Lambert (1978).

$^{12}C/^{13}C$. The $^{13}CH_4$ molecule has been discovered in the $3\nu_3$ band at 1.1 μm on both Jupiter and Saturn (Fox et al. 1972; de Bergh et al. 1976; Combes et al. 1975). The first $^{12}C/^{13}C$ values were derived with the RLM assumption and had no internal consistency: the choice of different lines led to different results: for Jupiter, 110 ± 35 (Fox et al. 1972) and 70^{+30}_{-15} (de Bergh et al. 1976); for Saturn, 35^{+40}_{-15} (Lecacheux et al. 1976).

Combes et al. (1977) applied to $^{12}C/^{13}C$ the method described later in detail by Combes and Encrenaz (1979) for eliminating the effect of scattering by a proper selection of the lines used in the ratio determination. This method led to new $^{12}C/^{13}C$ determinations on Jupiter and Saturn, 89^{+12}_{-10} and 89^{+25}_{-18}, respectively. As in the case of Mars and Venus, these ratios are in good agreement with the solar value. There is no determination of $^{12}C/^{13}C$ on Uranus and Neptune at the present time.

$^{14}N/^{15}N$. In this case, the only determination comes from the thermal radiation at 10 μm. The presence of $^{15}NH_3$ on Jupiter was first suggested by Lacy (1977, private communication) and definitely identified by Encrenaz et al. (1978). The derived $^{14}N/^{15}N$ Jovian ratio was found in agreement with the terrestrial value, but with a large uncertainty, mainly due to the noise of the spectrum. Tokunaga et al. (1979b) recorded later a high quality spectrum of the same spectral range and derived a ^{15}N enrichment on Jupiter with an error bar which still includes the terrestrial value. The error bar is still large because the authors show that it is not possible to find a $^{15}N/^{14}N$ ratio which fits the observations for all the $^{15}NH_3$ multiplets. More $^{15}NH_3$ laboratory measurements are needed to solve this problem. There is no $^{15}N/^{14}N$ determination at the present time on the other outer planets.

D/H. The D/H ratio on the giant planets is probably the most interesting number to determine, in view of its astrophysical implications. As in the case of C/H, its measurement is still strongly controversial.

Basically two methods can be used for determining the D/H ratio. The HD molecule can be used in the visible, associated with H_2 or CH_4 measurements in the same spectral range. The CH_3D molecule can also be used in the thermal range; the CH_3D/H_2 ratio is thus derived from thermal models.

From the HD observation on Jupiter at 7460 Å, Trauger et al. (1973, 1977) derived for the Jovian D/H, with the RLM approximation, a value of $5.1 \pm 0.7 \times 10^{5}$, which corresponds to a deuterium enrichment around 2 relative to the expected value of the primordial nebula (estimated from other measurements in the local interstellar medium). This value was later reestimated by Combes et al. (1978) and Combes and Encrenaz (1979) in order to avoid the RLM approximation and minimize the effects of scattering. Instead of determining the Jovian D/H directly, they choose to measure D/C, because the visible HD and CH_4 lines were more appropriate to give a reliable ratio. By using their determination of C/H (see above) they derived D/H $\leq 2.3 \times 10^{-5}$, which implies no deuterium enrichment. This method has the advantage of avoiding the use of the H_2 quadrupole lines, which are very difficult to measure in the laboratory. In the case of Saturn and Uranus, D/H values have been obtained by Macy and Smith (1978), in the RLM approximation, in the same range as Trauger et al.'s value for Jupiter. The Combes-Encrenaz method could not be used on these planets, in the absence of appropriate data.

In the thermal emission range, estimates of the Jovian CH_3D/H_2 ratio were obtained at 5 μm and 10 μm. From 5 μm observations, Beer and Taylor (1978) derived a value of 5×10^{-7} in the deep Jovian atmosphere. Since the CH_3D/H_2 is proportional to the product (D/H)x(C/H), their result implied a significant enrichment in deuterium and/or carbon. However, from Voyager IRIS data at 5 μm, the best fit corresponded to $CH_3D/H_2 = 2.5 \times 10^{-7}$ (Kunde et al. 1979). At 8 - 9 μm, from the absence of CH_3D features in the Jovian spectrum, Encrenaz et al. (1980) found an upper limit of 2×10^{-7} for CH_3D/H_2 which implies, with their C/H value, an upper limit of 4×10^{-5} for D/H. More recently Kunde et al. (1980) have derived a CH_3D/H_2 ratio of 5×10^{-7} in the same spectral range from the Voyager IRIS data. Discrepancies between the results obtained at 5 μm and 10 μm may be explained by the differences of atmospheric levels: the 5 μm measurements refer to a much deeper level than the 10 μm observations do, and the conditions of line formation may not be identical if the cloud structure is different. We cannot explain the differences in the results which have been done in the same range. It can be mentioned, however, that at 10 μm the determination is difficult because the $CH_3D\nu_2$ band is located in the wings of NH_3 and PH_3 absorption bands, which may vary over the disk and with time. In both cases, the CH_3D band itself is not observed, and the derived CH_3D/H_2 is chosen by finding the best fit for the continuum at medium resolution. We must also notice that when we use the CH_3D/H_2 ratio another uncertainty in the D/H determination comes from the fractionation factor for deuterium exchange between methane and hydrogen, which is strongly a function of the temperature. The value used for Jupiter has been estimated by Beer and Taylor (1978). In conclusion the determination of the D/H ratio--on Jupiter and on the other planets--is not solved, and more measurements are needed, both in the visible and in the far infrared ranges.

b) The case of nonuniformly mixed components

Various physical processes may be at the origin of a departure from the hydrostatic law in the vertical distribution of an atmospheric constituent. Condensation can deplete a density distribution according to the corresponding saturation law: it occurs in the case of NH_3 on all the giant planets, CH_4 and C_2H_2 probably on Uranus and Neptune. Photodissociation by solar UV radiation is active in the upper atmospheres and strongly depletes the CH_4 and NH_3 upper profiles. Vertical circulation and chemical reactions may also alter the distribution of a constituent, as may be the case of PH_3.

$\underline{NH_3}$. NH_3 follows its saturation law below the level of minimum temperature on the four giant planets. Above this level ammonia would follow a hydrostatic law in the absence of photodissociation. In the case of Jupiter, it is now demonstrated that NH_3 is strongly depleted in the upper atmosphere. This information comes from Jovian spectra recorded in the rotational NH_3 band (40 - 110 μm) and in the 10 μm NH_3 band, where no thermal emission appears in the center of the NH_3 emission multiplets (Figure 8) (Goorvitch 1978; Goorvitch et al. 1979; Baluteau et al. 1978,1979; Gautier et al. 1979; Marten et al. 1980). The NH_3 photodissociation has also been demonstrated from UV observations with IUE (Combes et al. 1980). It has to be mentioned that emission lines of NH_3 were observed at Jupiter's pole by heterodyne technique at 10 μm (Kostiuk et al. 1977) with a line width smaller than the Doppler width. However, this NH_3 emission which is of non-thermal origin cannot be simply correlated to the NH_3 distribution of the upper atmosphere.

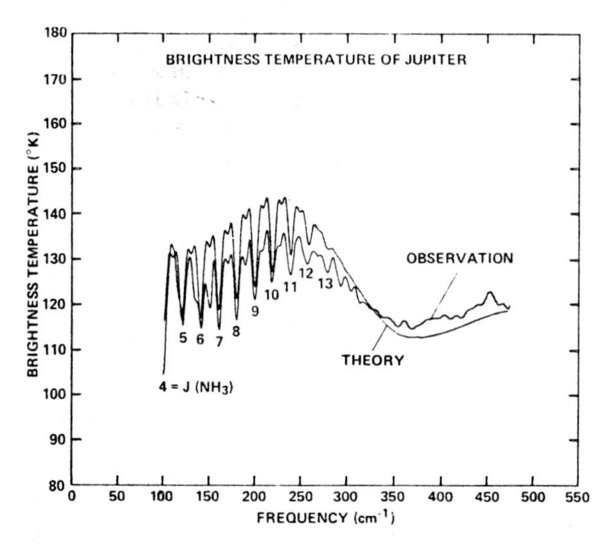

Figure 8. The far IR spectrum of Jupiter in the NH_3 rotational band (Erickson et al. 1978; Goorvitch et al. 1979).

Below the temperature minimum of Jupiter, the NH_3 distribution is probed in the NH_3 ν_2 band at 9 μm and 10 μm. Combes et al. (1976) and Encrenaz et al. (1978,1980) found a good agreement with the NH_3 saturation curve, while Tokunaga et al. (1979b), with an improved Jovian spectrum at 10 - 13 μm, find a better fit if the NH_3 density in the troposphere is 0.5 times the saturated vapor pressure density. The same conclusion is reached by Marten et al. (1980).

Below the NH_3 cloud level around 145 K, different results of the NH_3/H_2 ratio are derived. From visible data and using a scattering model, Sato and Hansen (1979) conclude that nitrogen is enriched by a factor around 1.5 - 2. From near infrared data, Encrenaz and Combes (1977) and Combes and Encrenaz (1979) derive an N/H value depleted by a factor 2 and suggest that nitrogen may be trapped at lower levels in some icy components (NH_4SH, NH_4OH). A similar conclusion is reached by Marten et al. (1980) who observe a difference in the NH_3/H_2 ratios below and above the region 250-330 K, and interpret this difference in terms of a cloud of nitrogen compounds at this level.

In the case of Saturn, very little is known about its NH_3 distribution. Ammonia was observed in the visible range (Encrenaz et al. 1974) but not in the near infrared range where the NH_3 bands are stronger (Owen et al. 1977; Larson et al. 1979): this illustrates again the effects of scattering in Saturn's atmosphere. Marten et al. (1980) derived, below the loud level, a NH_3/H_2 ratio which includes the solar value. On Uranus and Neptune, Gulkis et al. (1977) and Olsen and Gulkis (1978) found an NH_3 depletion in the deep atmosphere from the millimetric spectra.

$\underline{PH_3}$. The presence of phosphine in the atmospheres of the gaint planets is of major interest for the study of dynamics on these planets. According to thermochemical equilibrium calculations (Lewis 1969; Barshay and Lewis 1978) phosphine should not be observable on the outer planets, because PH_3 is expected to react with H_2O below a temperature of 2000 K. Its discovery on Jupiter and Saturn at 2 μm, 5 μm, and 10 μm (Ridgway 1974; Larson et al. 1977; Ridgway et al. 1976; Tokunaga et al. 1980) was entirely unexpected. It was suggested by Prinn and Lewis (1975) that PH_3 was probably carried from deep atmospheric levels where it is stable, up to the atmospheric regions where it has been observed: for Jupiter, the 200-230 K level at 5 μm and the 130 K-145 K level at 9 - 10 μm; the time needed for the transportation would have to be short enough for PH_3 to be observed at the top of the current before it has completed reacted with H_2O. The same mechanism could be also responsible for the observation of GeH_4 and CO. If this explanation is right, we would eventually expect to measure a P/H ratio smaller than the solar value. In the case of Jupiter, various estimates have been derived: at 10 μm, Ridgway et al. (1976) found a P/H close to the solar value, Tokunaga et al. (1979b) estimated a P/H value depleted by a factor 4, in agreement with Encrenaz et al. (1980), Fink and Larson (1977) at 2 μm, and Beer and Taylor (1979) at 5 μm. In the case of Saturn, the first observation of Bregman was consistent with a solar P/H value, but the recent result of Tokunaga et al. (1980) implies a phosphorus enrichment by a factor 3 at least. Similarly, Larson et al. (1980) derive a PH enhancement by a factor of 2 from 2 - 5 μm observations. As suggested by the authors, rapid convection is probably not sufficient to account for this large amount of phosphine. Larson et al. (1980) indicate some mechanisms which could explain the observed PH_3 amount: the reaction of PH_3 with H_2O may be slower than previously thought (Sill 1979); moreover, according to

Strobel (1977) photodissociation of PH_3 is expected to be inhibited by the absence of gaseous NH_3, as is the case on Saturn. This last argument could possibly explain the quantitative difference between the PH_3 abundances on Jupiter and Saturn.

C_2H_2, C_2H_6. Another surprising result was the discovery of C_2H_2 and C_2H_6 in the infrared spectrum of Jupiter (Ridgway 1974). This result gave the first observational evidence for CH_4 photodissociation by the solar UV radiation in the upper Jovian atmosphere. C_2H_6 was present in all the following observations (Combes et al. 1974; Tokunaga et al. 1976; Encrenaz et al. 1976; Aumann and Orton 1976; Encrenaz et al. 1978) but C_2H_2 seemed to be absent on some observations (Tokunaga et al. 1976; Encrenaz et al. 1976) which may suggest a possible variation of C_2H_2 with time. This variation could be explained by condensation of C_2H_2 in the region of the temperature minimum.

Density profiles of C_2H_2 and C_2H_6 have been obtained by the Voyager IRIS experiment (Hanel et al. 1979b), on different points of the Jovian disk from the equator to the South Pole; the C_2H_2 amount appears to be significantly depleted at the South Pole, which could again be due to an atmospheric cooling. Moreover, Hanel et al. (1979b) report a significant change in the C_2H_2/C_2H_6 ratio between the two encounters; this result may be a confirmation of the possible C_2H_2 variation suggested from ground-based observations.

On Saturn, only C_2H_6 was detected (Tokunaga et al. 1975); the absence of acetylene is probably due to condensation. In contrast, C_2H_2 has been tentatively identified on Titan, with C_2H_6 and C_2H_4 (Gillett 1975); this seems to imply a higher temperature on Titan at the level of formation.

The presence of C_2H_6 on Neptune was especially useful because it provided evidence for a warm upper stratosphere in contrast with Uranus. The difference of thermal profiles on these two planets has to be correlated with the absence of an internal heat source on Uranus, which itself remains unexplained.

VI. PERSPECTIVES AND CONCLUSIONS

In conclusion, we can try to summarize what is known and what has still to be studied in the composition of planetary atmospheres. The atmospheric composition and structure of Mars and Venus have been studied extensively by *in situ* experiments. In the case of Jupiter we now have a good knowledge of the mean atmospheric parameters (temperature, pressure) and we start with the results of Voyager 1 and 2 to have some spatial resolution of the thermal and cloud structure. However a large uncertainty still exists on some basic elemental and isotopic ratios, as C/H and D/H. This uncertainty is even larger on the other outer planets. However, in the case of Jupiter, in view of the results summarized in Table IX, we believe that the enrichment in helium, deuterium, and carbon is moderate and not sufficient at the present time to definitely imply an inhomogeneous interior of Jupiter.

What progress can we expect in our knowledge of the giant planets within the next decade? In the case of Jupiter the main event will be the Galileo mission. Apart from the mass spectrometer experiment aboard the probe, information on the atmospheric composition will come, in particular, from the Near Infrared Mapping Spectrometer which will record 0.6 - 5.2 μm scans with moderate spectral resolution (∿ 100-200) but high spatial resolution. In the case of Saturn--and possibly Uranus--the Voyager mission will provide, as in the case of Jupiter, the T(P) distribution and the H_2/He ratio. Day and night measurements of Titan with the IRIS experiment will give valuable information about the atmospheric cooling of Titan, and thus about its density. The same kind of information will possibly be obtained from the reduction of observations of Titan during the 1980 eclipses.

In the case of Uranus and Neptune, which will probably not be objectives of space missions in the near future, the contribution of Voyager and ground-based infrared spectroscopy should be of major interest within the next decade. From aircraft measurements, a better H_2/He value will be derived while estimates of C/H and D/H will probably be obtained from the ground in the near infrared (as already started by Fink and Larson (1979) and at 5 μm and 10 μm. The upper atmospheres of Titan, Uranus, and Neptune will be studied in more detail with better observations of the CH_4 and C_2H_6 emission lines around 10 μm. All this progress will help us to solve two important questions: (1) the evidence for a difference in the atmospheric abundance ratios from Jupiter to Neptune, and (2) the nature of Titan's atmosphere. In conclusion we can expect that infrared astronomy will still provide us new and exciting results on planetary atmospheres within the next few years.

REFERENCES

Abbas, M. M., Mumma, M. J., Kostiuk, T., and Buhl, D.: 1976, Applied Optics 15, 427.
Aitken, D., and Jones, B.: 1972, Nature 240, 230.
Armstrong, K. R., Harper, D. A., and Low, F. J.: 1972, Astrophys. J. 178, L89.
Aumann, H. H., and Orton, G. S.: 1976, Science 194, 107.
Aumann, H. H., and Orton, G. S.: 1979, Icarus 38, 251.
Avduevsky, V. S., Marov, M. Y., and Rozhdestvensky, M. K.: 1970, J. Atmospheric Sci. 27, 561.
Baluteau, J. P., Marten, A., Bussoletti, E., Anderegg, M., Moorwood, A. F. M., Beckman, J. E., and Coron, N.: 1978, Astron. Astrophys. 64, 61.
Baluteau, J. P., Marten, A., Moorwood, A. F. M., Anderegg, M., Biraud, Y., Coron, N., and Gautier, D.: 1980, Astron. Astrophys. 81, 152.

Barshay, S., and Lewis, J. S.: 1978, Icarus 33, 593.
Beer, R.: 1975, Astrophys. J. Letters 200, L167.
Beer, R.: 1980, Communication presented at the IAU Symposium 96, Hawaii.
Beer, R., Farmer, C. B., Norton, R. H., Martonchik, J. V., and Barnes, T. G.: 1972, Science 175, 1360.
Beer, R., and Taylor, F. W.: 1978, Astrophys. J. 219, 763.
Beer, R., and Taylor, F. W.: 1979, Icarus 40, 189.
Belton, M. J. S., and Hunten, D. M.: 1966, Astrophys. J. 146, 307.
Benner, D. C., and Fink, U.: 1980, Icarus (to be published).
Betz, A.: 1975, Space Sci. Rev. 17, 659.
Biemann, K., Owen, T., Rushneck, D. R., LaFleur, A. L., and Howarth, D. W.,: 1976, Science 194, 76.
Bregman, J., Lester, D. F., and Rank, D. M.: 1975, Astrophys. J. Letters 202, L55.
Buriez, J. C., and de Bergh, C.: 1980a, Astron. Astrophys. 83, 149.
Buriez, J. C., and de Bergh, C.: 1980b, Astron. Astrophys. (to be published).
Butterworth, P. S., Caldwell, J., Moore, V., Owen, T., Rivolo, A. R., and Lane, A. L.: 1980, Nature 285, 308.
Cameron, A. G. W.: 1974, Space Sci. Rev. 15, 121
Combes, M., Encrenaz, Th., Vapillon, L., Zeau, Y.,and Lesqueren, C.: 1974, Astron. Astrophys. 34, 33.
Combes, M., de Bergh, C., Lecacheux, J., and Maillard, J. P.: 1975, Astron. Astrophys. 40, 81.
Combes, M., Encrenaz, Th., Berezne, J., Vapillon, L., and Zeau, Y.: 1976, Astron. Astrophys. 50, 287.
Combes, M., Maillard, J. P., and de Bergh, C.: 1977, Astron. Astrophys. 61, 531.
Combes, M., Encrenaz, Th., and Owen, T.: 1978, Astrophys. J. 221, 378.
Combes, M., and Encrenaz, Th.: 1979, Icarus 39, 1.
Combes, M., Courtin, R., Caldwell, J., Encrenaz, Th., and Fricke, K. H.: 1980, Proceedings of the COSPAR meeting (to be published).
Connes, J., and Connes, P.: 1966, J. Opt. Soc. Am. 56, 896.
Connes, P., Connes, J., Benedict, W. S., and Kaplan, L. D.: 1967, Astrophys. J. 147, 1230.
Connes, P., Connes, J., Kaplan, L. D., and Benedict, W. S.: 1968, Astrophys. J. 152, 731.
Conrath, B., Curran, R., Hanel, R., Kunde, V., Maguire, W., Pearl, J., Pirraglia, J., and Welker, J.: 1973, J. Geophys. Res. 78, 4267.
Courtin, R., Gautier, D., and Lacombe, A.: 1978, Astron. Astrophys. 63, 97.
Courtin, R., Gautier, D., and Lacombe, A.: 1979, Icarus 37, 236.
Cruikshank, D. P., Pilcher, C. B., and Morrison, D.: 1976, Science 194, 835.
Cruikshank, D. P., and Silvaggio, P.: 1979, Astrophys. J. 233, 1016.
Cruikshank, D. P., and Silvaggio, P.: 1980, Icarus 41, 96.
Curran, R., Conrath, B., Hanel, R., Kunde, V., and Pearl, J.: 1973, Science 182, 381.
De Bergh, C., Vion, M., Combes, M., Lecacheux, J., and Maillard, J. P.: 1973, Astron. Astrophys. 28, 457.

De Bergh, C., Lecacheux, J., and Maillard, J. P.: 1974, Astron. Astrophys. 56, 227.
De Bergh, C., Maillard, J. P., Lecacheux, J., and Combes, M.: 1976, Icarus, 29, 307.
De Bergh, C., Lecacheux, J., and Maillard, J. P.: 1977, Astron. Astrophys. 56, 227.
Encrenaz, T.: 1979, Infrared Phys. 19, 353.
Encrenaz, Th., Owen, T., and Woodman, J. H.: 1974, Astron. Astrophys. 37, 49.
Encrenaz, Th., Combes, M., Zeau, Y., Vapillon, L., and Berezne, J.: 1975, Astron. Astrophys. 42, 355.
Encrenaz, Th., and Combes, M.: 1977, Bull. Am. Astron. Soc. 9, 477.
Encrenaz, T., and Combes, M.: 1978, Bull. Am. Astron. Soc. 10, 576.
Encrenaz, Th., Gautier, D., Michel, G., Zeau, Y., Lecacheux, J., Vapillon, L., and Combes, M.: 1976, Icarus 29, 311.
Encrenaz, Th., Combes, M., and Zeau, Y.: 1978, Astron. Astrophys. 70, 29.
Encrenaz, Th., Combes, M., and Zeau, Y.: 1980, Astron. Astrophys. 84, 148.
Erickson, E. F., Goorvitch, D., Simpson, J. P., and Strecker, D. W.: 1978, Icarus 35, 61.
Fink, U., and Larson, H. P.: 1977, Bull. Am. Astron. Soc. 9, 535 (abstract).
Fink, U., Larson, H. P., and Treffers, R. R.: 1978, Icarus 34, 344.
Fink, U., and Larson, H. P.: 1979, Astrophys. J. 233, 1021.
Fox, K., Owen, T., Mantz, A. W., and Rao, K. N.: 1972, Astrophys. J. Letters 176, L81.
Furniss, I., Jennings, R. E., and King, K. J.: 1978, Icarus 35, 74.
Gautier, D., Lacombe, A., and Revah, I.: 1977a, J. Atmospheric Sci. 34, 1130.
Gautier, D., Lacombe, A., and Revah, I.: 1977b, Astron. Astrophys. 61, 149.
Gautier, D., Marten, A., Baluteau, J. P., and Lacombe, A.: 1979, Icarus 37, 214.
Gautier, D., and Courtin, R.: 1979, Icarus 39, 28.
Gautier, D., Conrath, B. J., Hanel, R. A., Kunde, V. G., Chedin, A., and Scott, N.: 1980, J. Geophys. Res. (in press).
Geiss, J., and Reeves, H.: 1972, Astron. Astrophys. 18, 126.
Gillett, F. C.: 1975, Astrophys. J. Letters, 201, L41.
Gillett, F. C., Forrest, W. J., and Merrill, K. M.: 1973, Astrophys. J. Letters, 184, L93.
Gillett, F. C., and Rieke, G. H.: 1977, Astrophys. J. Letters 218, L141.
Gillett, F. C., Low, F. J., and Stein, W. A.: 1969, Astrophys. J. 157, 925.
Giver, L. P., and Spinrad, H.: 1966, Icarus 5, 586.
Giver, L. P., Trafton, L. M., Podolak, M., and Rages, K.: 1979, Bull. Am. Astron. Soc. 11, 564.
Goorvitch, D.: 1978, Icarus 36, 127.
Goorvitch, D., Erickson, E. F., Simpson, J. P., and Tokunaga, A.: 1979, Icarus 40, 75.
Grandjean, J., and Goody, R. M.: 1955, Astrophys. J. 121, 548.

Gulkis, S., Jansen, M., and Olsen, E. T.: 1977, Bull. Am. Astron. Soc. 9, 472.
Hall, D. N. B., Noyes, R. W., and Ayres, T. R.: 1972, Astrophys. J. 171, 615.
Hanel, R., Forman, M., Stambach, G., and Meilleur, T.: 1968, J. Atmospheric Sci. 25, 586.
Hanel, R., Conrath, B., Hovis, W., Kunde, V., Lowman, P., Pearl, J., Prabhakara, C., Schlachman, B., and Levin, G.: 1972, Science 175, 305.
Hanel, R., Conrath, B., Flasar, M., Kunde, V., Lowman, P., Maguire, W., Pearl, J., Pirraglia, J., Samuelson, R., Gautier, D., Gierasch, P., Kumar, S., and Ponnamperuma, C.:1979a, Science 204, 972.
Hanel, R., Conrath, B., Flasar, M., Herath, L., Kunde, V., Lowman, P., Maguire, W., Pearl, J., Pirraglia, J., Samuelson, R., Gautier, D., Gierasch, P., Horn, L., Kumar, S., and Ponnamperuma, C.: 1979b, Science 206, 952.
Hoffman, J. H., Hodges, R. R., McElroy, M. B., Donahue, T. M., and Kolpin, M.: 1979, Science 203, 800.
Hunten, D. M.: 1978, NASA, JPL Saturn System Workshop.
Judge, D. L., and Carlson, R. S.: 1974, Science 183, 317.
Kaplan, L. D., Münch, G., and Spinrad, H.: 1964, Astrophys. J.
Kaplan, L. D., Connes, J., and Connes, P.: 1969, Astrophys. J. Letters 157, L187.
Kiess, C. C., Corliss, C. H., and Kiess, H. K.: 1960, Astrophys. J. 132, 221.
Kliore, A. J., Cain, D. L., Fjeldbo, G., Seidel, B. L., and Rasool, I.: 1972, Science 183, 323.
Kliore, A. J., and Woiceshyn, P. M.: 1976, in Jupiter, T. Gehrels, ed., University of Arizona Press, p. 216.
Kostiuk, T., Mumma, M. J., Hillman, J. J., Bulh, D., Brown, L. W., Faris, J. L., and Spears, D. L.: 1977, Infrared Physics 17, 431.
Kuiper, G. P.: 1944, Astrophys. J. 100, 378.
Kuiper, G. P.: 1952, The Atmospheres of the Earth and Planets, G. Kuiper, ed., University of Chicago Press.
Kunde, V. G., Hanel, R. A., and Herath, L. W.: 1977, Icarus 32, 210.
Kunde, V. G., Hanel, R. A., Conrath, B. J., and Maguire, W. C.: 1979, Bull. Am. Astron. Soc. 11, 587.
Kunde, V. G., Gautier, D., Maguire, W. C., Hanel, R. A., Marten, A., Baluteau, J. P., Rouan, D., Chedin, A., Scott, N., and Husson, N.: 1980, J. Geophys. Res. (in press).
Lambert, D.: 1978, Monthly Notices Roy. Astron. Soc. 182, 249.
Larson, H. P.: 1980, Ann. Rev. Astron. Astrophys., Vol. 18.
Larson, H. P., and Fink, U.: 1977, Bull. Am. Astron. Soc. 9, 515.
Larson, H. P., Fink, U., Smith, H. A., and Davis, D. C.: 1980, Astrophys. J. (in press).
Larson, H. P., Fink, U., Treffers, R. R., and Gautier, T. N.: 1975, Astrophys. J. Letters, 197, L137.
Larson, H. P., Treffers, R. R., and Fink, U.: 1977, Astrophys. J. 211, 972.
Laurent, C.: 1978, These de Doctorat d'Etat, Universite de Paris VII.

Lecacheux, J., de Bergh, C., Combes, M., and Maillard, J. P.: 1976, Astron. Astrophys. 53, 29.
Lequeux, J., Peimbert, M., Rayo, J. F., Serrano, A., and Torres-Peimbert, S.,: 1979, Astron. Astrophys. 80, 155.
Lewis, J. S.: 1969, Icarus 10, 365.
Lutz, B. L., Owen, T., and Cess, R. D.: 1976, Astrophys. J. 203, 541.
Macy, W., and Sinton, W. M.: 1977, Astrophys. J. Letters 218, L79.
Macy, W., and Smith, W. H.: 1978, Astrophys. J. Letters 222, L73.
McCarthy, J. F., Pollack, J. B., Houck, J. R., and Forrest, W. J.: 1980, Astrophys. J. (in press).
McCleese, .: 1980, Communication presented at the IAU Symposium No. 96, Hawaii.
McElroy, M. B., Yung, Y. L., and Nier, A. O.: 1976, Science 194, 70.
Maguire, W. C.: 1977, Icarus 32, 85.
Maillard, J. P., Combes, M., Encrenaz, Th., and Lecacheux, J.: 1973, Astron. Astrophys. 25, 219.
Marten, A., Courtin, R., Gautier, D., and Lacombe, A.: 1980, Icarus (in press).
Martin, T. Z., Cruikshank, D. P., Pilcher, C. B., and Sinton, W. M.: 1976, Icarus 27, 391.
Mount, G. H., and Linsky, J. L.: 1975, Astrophys. J. Letters 202, L51.
Muhleman, D. O., Clancy, T., Knapp, G. R., and Phillips, T. G.: 1979, Bull. Am. Astron. Soc. 11, 540.
Mumma, M. J., Kostiuk, T., Cohen, S., Buhl, D., and Von Thuna, P. C.: 1975, Space Sci. Rev. 17, 661.
Mumma, M. J., Kostiuk, T., Buhl, D., Chin, G., Abbas, M., and Zipoy, D.: 1979, Bull. Am. Astron. Soc. 11, 541.
Mumma, M. J., Buhl, D., Chin, G., Deming, D., Espenak, F., and Kostiuk, T.,: 1980, Communication presented at the IAU Symposium No. 96, Hawaii.
Münch, G., Trauger, J. T., and Roesler, F. L.: 1977, Astrophys. J. 216, 963.
Niemann, H. B., Hartle, R. E., Kasprzak, W. T., Spencer, N. W., Hunten, D. M., and Carignan, G. R.: 1979, Science 203, 770.
Nier, A. O., McElroy, M. B., and Yung, Y. L.: 1976, Science 194, 68.
Noxon, J. F., Traub, W. A., Carleton, N. P., and Connes, P.: 1976, Astrophys. J. 207, 1025.
Olsen, E. T., and Gulkis, S.: 1978, Bull. Am. Astron. Soc. 10, 577.
Orton, G. S.: 1975a, Icarus 26, 125.
Orton, G. S.: 1975b, Icarus 26, 142.
Orton, G. S.: 1977, Icarus 32, 41.
Orton, G. S.: 1980 (this volume).
Orton, G. S., and Aumann, H. H.: 1977, Icarus 32, 431.
Owen, T., McKellar, A. R. W., Encrenaz, Th., Lecacheux, J., de Bergh, C., and Maillard, J. P.: 1977, Astron. Astrophys. 54, 291.
Oyama, V. I., Carle, G. C., Woeller, F., and Pollack, J. B.: 1979, Science 205, 52.
Pagel, B. J. E.: 1977, 2nd Symposium on the Origin and Distribution of the Elements (IAGC), Paris.
Pearl, J., Hanel, R., Kunde, V. G., Maguire, W., Fox, K., Gupta, S., Ponnamperuma, C., and Raulin, F.: 1979, Nature 280, 755.

Peterson, D. W., Johnson, M. A., and Betz, A.: 1974, Nature 250, 128.
Pollack, J. B., Erickson, E. F., Witteborn, F. C., Chackerian, C., Jr., Summers, A. L., Van Camp, W., Baldwin, B. J., Augason, G. C., and Caroff, L. J.: 1974, Icarus 23, 8.
Prinn, R., and Lewis, J. S.: 1975, Science 190, 274.
Rages, K., Pollack, J. B., and Giver, L. P.: 1979, Bull. Am. Astron. Soc. 11, 563.
Ridgway, S. T.: 1974, Astrophys. J. Letters 187, L41.
Ridgway, S. T., Wallace, L. and Smith, G. R.: 1976, Astrophys. J. 207, 1002.
Rouan, D., Gautier, D., Baluteau, J. P., Marten, A., Chedin, A., Scott, N., Husson, N., Conrath, B. J., Hanel, R. A. Kunde, V. G., and Maguire, W. C.: 1980, Communication presented at the IAU Symposium No. 96, Hawaii.
Samuelson, R. E., Hanel, R. A., Herath, L. W., Kunde, V. G., and Maguire, W. C.: 1975, Icarus 25, 49.
Sato, M., and Hansen, J. E.: 1979, J. Atmospheric Sci. 36, 1133.
Schloerb, F. P., Robinson, S. E., and Irvine, W. M.: 1979, Bull. Am. Astron. Soc. 11, 540.
Schorn, R. A., Farmer, C. B., and Little, S. J.: 1969, Icarus 11, 283.
Sill, G.: 1979, Private communication, quoted by Larson et al. (1980).
Sinton, W. M.: 1979, Bull. Am. Astron. Soc. 11, 598.
Sinton, W. M., and Strong, J.: 1960, Astrophys. J. 131, 459.
Spinrad, H., Schorn, R. A., Moore, R., Giver, L. P., and Smith, H. J.: 1966, Astrophys. J. 146, 331.
Strobel, D. F.: 1977, Astrophys. J. Letters 214, L97.
Taylor, F. W., Beer, R., Chahine, M. T., Diner, D. J., Elson, L. S., Haskins, R. D., McCleese, D. J., Martonchik, J. V., Reichley, P. E., Bradley, S. P., Delderfield, J., Schofiel, J. T., Farmer, C. B., Froidevaux, L., Leung, J., Coffey, M. T., and Gille, J. C.: 1980, J. Geophys. Res. (in press).
Tokunaga, A., Knacke, R. F., and Owen, T.: 1975, Astrophys. J. Letters 197, L77.
Tokunaga, A., Knacke, R. F., and Owen, T.: 1976, Astrophys. J. 209, 294.
Tokunaga, A., Caldwell, J., Gillett, F. C., and Nolt, I. G.: 1978, Icarus 36, 216.
Tokunaga, A., Caldwell, J., Gillett, F. C., and Nolt, I. G.: 1979a, Icarus 39, 46.
Tokunaga, A., Knacke, R. F., Ridgway, S. T., and Wallace, L.: 1979b, Astrophys. J. 232, 603.
Tokunaga, A., Dinerstein, H. L., Lester, D. F., and Rank, D. M.: 1980, Icarus (in press).
Trafton, L. M.: 1972, Astrophys. J. 175, 285.
Trafton, L. M.: 1975, Astrophys. J. 195, 805.
Trafton, L. M.: 1979a, Bull. Am. Astron. Soc. 11, 563.
Trafton, L. M.: 1979b, Bull. Am. Astron. Soc. 11, 570.
Trafton, L. M.: 1974, in Exploration of the Planetary System, Woszczyk and Iwaniczewka, eds., IAU
Traub, W. A., Carleton, N. P., and Connes, P.: 1977, Bull. Am. Astron. Soc. 9, 513.

Trauger, J. T., Roesler, F. L., Carleton, N. P., and Traub, W. A.: 1973, Astrophys. J. Letters 184, L137.
Trauger, J. T., Roesler, F. L., and Mickelson, M. E.: 1977, Bull. Am. Astron. Soc. 9, 516.
Vapillon, L., Encrenaz, Th., Gautier, D., and Combes, M.: 1977, Astron. Astrophys. 58, 113.
Verdet, J. P., Zeau, Y., Gay, J., Encrenaz, Th., and Sevre, F.: 1972, Astron. Astrophys. 19, 159.
Vinogradov, A. P., Surkov, U. A., and Florensky, C. P.: 1968, J. Atmospheric Sci. 25, 535.
Von Zahn, U., Krankowsky, D., Mauersberger, K., Nier, A. O., and Hunten, D. M.: 1979, Science 203, 768.
Walker, M. F., and Hayes, S.: 1967, Publ. Astron. Soc. Pacific 79, 464.
Wallace, L., and Smith, G. R.: 1977, Astrophys. J. 212, 252.
Wallace, L., and Hunten, D. M.: 1978, Rev. Geophys. Space Phys. 16, 289.
Wilson, W. J., and Klein, M. J.: 1979, Bull. Am. Astron. Soc. 11, 540.
Wildt, R.: 1932, Veroff Univsternw. Gottingen 22, 171.
Young, L. D. G.: 1971, J. Quant. Spectr. Rad. Tran. 11, 1075.

DISCUSSION FOLLOWING PAPER DELIVERED BY T. ENCRENAZ

ORTON: A correlation of the Pioneer 11 Infrared Radiometer results at Saturn with the Radio Science occultation experiment to determine the structure of the neutral atmosphere has given a mixing ratio for H_2 of 0.90 ± 0.03 (assuming the rest to be He). This determination used a technique similar to that used by Gautier et al. (1980) to determine the same parameters from Voyager data.

JOSEPH: Is it possible to describe, simply, how thermal radiation is used to derive the C/H ratio?

ENCRENAZ: We use the CH_3D band and thermal profile models. For a given temperature profile, we can estimate at which level the CH_3D lines are formed. The observed band is then fit to synthetic spectra calculated for given CH_3D to H_2 ratios.

R. CAYREL: How important are considerations of non-LTE in the interpretation of observational data for planetary atmospheres?

ENCRENAZ: Fortunately, the atmospheric region we are probing has pressures above 10 or 100 mbar, and we are not bothered by problems of non-LTE. In contrast, in the UV region non-LTE must be accounted for.

ATMOSPHERIC STRUCTURE OF THE OUTER PLANETS FROM THERMAL EMISSION DATA

Glenn S. Orton
Jet Propulsion Laboratory
California Institute of Technology
Pasadena, California 91103

ABSTRACT

Determination of atmospheric temperature structure is of paramount importance to the understanding of planetary atmospheric structure. The most powerful methods for determining atmospheric structure exploit the opacities provided by the collision induced H_2 dipole and the ν_4 fundamental of CH_4. In addition to earth-based observations, useful measurements of thermal emission from Jupiter and Saturn have been or soon will be made by several spacecraft, with results cross-checked with independent radio occultation results. For Uranus and Neptune, only a limited set of whole-disk earth-based data exists. All the outer planets show evidence for stratospheric temperature inversions; temperature minima range from about 105 K for Jupiter and 87 K for Saturn, to roughly 55 K for Uranus and Neptune. In addition to better data, remaining problems may be resolved by better quantitative understanding of gas and aerosol absorption and scattering properties, chemical composition, and non-LTE source functions. Ultimately, temperature structure results must be supplemented by quantitative energy equilibrium models which will allow some meaning to be given to the relationships between such characteristics as temperature, clouds, incident solar and planetary radiation, and chemical composition.

1. INTRODUCTION

Direct information about the structure of planetary atmospheres has been obtained through *in situ* probes for the earth, Mars and Venus. While the Galileo mission will investigate the Jovian atmosphere by a direct probe in the 1980's, current information on the structure of atmospheres in the outer solar system comes only from remote measurements. For this reason, the review presented here will be concerned with the large outer planets only, ignoring satellites with atmospheres for convenience alone.

Atmospheric structure, in the most general sense, includes consideration of vertical and horizontal variations of temperature, gaseous constituents, aerosols, and winds. For the outer planets, aerosol and wind structure are poorly understood. Chemical composition is reviewed by Encrenaz

C. G. Wynn-Williams and D. P. Cruikshank (eds.), Infrared Astronomy, 35–56.

and Combes in this volume. This review will concentrate on temperature structure. Information about temperature structure is very important for several reasons: (1) definition of basic structure by discrimination between regions dominated by radiative vs. convective energy transport, (2) interpretation of thermal emission to determine properties of gases and aerosols, (3) accurate interpretation and modelling of temperature-dependent physical processes, (4) determination of the strength and direction of thermally driven winds, and (5) definition of boundary conditions for models of thermal history and of the planetary interior.

2. TECHNIQUES

Temperature structure information from *in situ* probes is costly and the information gathered is extremely localized. On the other hand, vertical resolution can surpass what is available from remote infrared measurements. Furthermore, simultaneous remote and direct measurements offer a means for calibrating information derived from remote measurements.

Other remote techniques rely on occultation of a radiation source by the planet to sound atmospheric properties. Occultation of spacecraft radio signals by Jupiter (Lindal et al., 1980) and Saturn (Kliore et al., 1980) have been used to determine temperature structure, as has the occultation of α Leonis by Jupiter observed in the ultraviolet (Atreya et al., 1980). Occultation techniques offer the advantage of better vertical resolution than available from infrared sounding. However, there are several disadvantages, besides the relative infrequency of the appropriate geometric conditions. Foremost among these are the poor horizontal resolution and the need to assume homogeneity of the atmospheric structure along all points of the ray path. Another disadvantage for the radio occultation technique, the requirement that composition be known, has been used, in concert with "simultaneous" infrared coverage, to determine the bulk compositions of Jupiter (Gautier et al., 1980) and Saturn (Kliore et al., 1980; Orton and Ingersoll, 1980).

Energy (i.e. radiative-convective) equilibrium models could also be considered a form of temperature sounding. Indeed, results from a variety of other techniques must be analyzed in the context of such models in order to facilitate quantitative physical interpretation. The disadvantages of such techniques lie in their heavy dependence on *a priori* assumptions about the deposition of solar energy, as well as their lack of direct constraint to the observations. This review will use the work of Appleby and Hogan (1980) for comparison with direct sounding results, as their models take advantage of the most recent available data base. Earlier classic work by Trafton (1967), as well as the work of Hogan et al. (1969), Wallace et al. (1974), Wallace (1975), Cess and Khetan (1973), Trafton and Stone (1974), Cess and Chen (1975), Tokunaga and Cess (1977), and Danielson (1977) should not go without mention, however.

Temperature sounding, the direct inversion of thermal emission measurements to determine temperature structure (also known as temperature recovery or temperature retrieval), is used quite successfully in terrestrial applications with data supplied by earth-orbiting satellites. The technique requires a vertical interval to be defined from which the radiation contributing to the available thermal emission data has originated. For a clear atmosphere of infinite depth, one is required to invert a set of measured intensities, $I(\mu,\nu)$ in

$$I(\mu,\nu) = \int_{x_{bottom}}^{x_{top}} B[\nu,T(x)] \frac{\partial\tau(x,\nu)}{\partial x} dx \quad (1)$$

to recover the temperature structure $T(x)$. In equation 1, ν is frequency, μ is emission angle cosine, B is the Planck brightness function, x is any independent vertical parameter (e.g. altitude or logarithm of pressure), and $\tau(x)$ is the transmission between x and x_{top}; x_{top} is sufficient to approximate conditions of clear space and x_{bottom} those of infinite depth. Measurements and inversions also make use of intensities with a filtered integration over some spectral interval $\Delta\nu$; in such a case the parameters in the integrand must be integrated over $\Delta\nu$ explicitly, although for $\Delta\nu$ sufficiently small, an average value of B may be used with explicit computation of τ averaged over $\Delta\nu$.

The kernel of equation 1, $\partial\tau/\partial x$, is known as the weighting function and it must be known *a priori*, implying that the abundance of some opaque chemical constituent must be known--preferably a uniformly mixed gas. The recovery is influenced by assumptions about boundary conditions (Orton, 1977), such as the temperature structure above and below the vertical interval where temperatures are being determined. Vertical resolution is limited by data noise and width of the weighting function (Conrath, 1972). For the outer planets, a resolution element is never better than half an atmospheric scale height. A substantial literature exists on numerical techniques for inverting equation 1 (e.g., Deepak, 1977). This author has used the techniques described by Chahine (1972, 1975) because of their simplicity and stability.

3. APPLICATION

For the outer planets, the most important source of atmospheric opacity in the far infrared is that provided by the collision induced dipole of H_2, due to its great abundance as an atmospheric constituent and its dominance of almost all the opacity for wavelengths longer than 14 µm ($\nu \leq 700$ cm^{-1}), from which most of the thermal flux from the outer planets originates. As a collision induced phenomenon, its absorption is proportional to the square of the pressure and its weighting functions are therefore quite narrow. Figure 1 demonstrates the pressure corresponding to the location of unit optical depth for the H_2 dipole as a function of frequency for the outer planets. The broad translational band at shortest frequencies and two broad rotational lines, S(0) and S(1), respectively centered at 370 and 600 cm^{-1}, can easily be seen. A further advantage of this opacity source is that the broad spectral features allow useful data to be sampled over relatively broad spectral bandpasses without a debilitating loss of vertical resolution. For the 100 - 600 cm^{-1} region the vertical coverage extends from approximately 700 to 50 mb. At lower frequencies, coverage can sometimes extend down to higher pressure regions before another opacity source such as NH_3 influences the atmospheric opacity; this, in fact, appears to be the case for Uranus and Neptune.

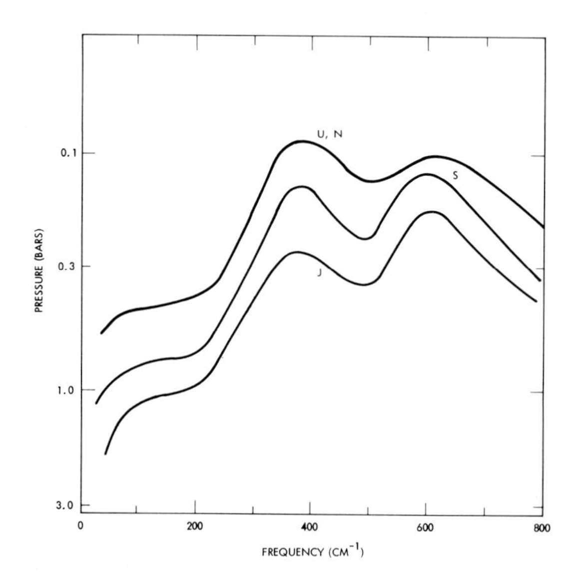

Figure 1. Pressure at which unit optical depth is reached in the atmospheres of the outer planets due to the opacity of the collision induced H_2 dipole, for a mixing ratio of 90%. Other opacity sources are not included.

There are problems involved with the use of the H_2 dipole opacity. The lowest temperature at which laboratory measurements of H_2 absorption have been made is 77 K (Birnbaum, 1978); more work must be done to understand the behavior of the absorption at lower temperatures (relevant to Uranus and Neptune). Furthermore, atmospheric models have thus far assumed a ratio of para-H_2 to ortho-H_2 which is always in equilibrium at the local temperature; this will not be true if the characteristic time scale for convection is faster than that for ortho-para conversion. Even if the correct ortho-H_2 to para-H_2 ratio were known everywhere, no models have been developed to express the opacity of different state ratios at all temperatures of interest. Finally, the broad features may be confused with other opacity sources, such as the distant wings of strong NH_3 rotational lines (near 100 cm^{-1}) or absorption by NH_3 ice particles. Figure 2 demonstrates the effect of an NH_3 ice haze in the atmosphere of Jupiter on the infrared spectrum. The effect is substantial, but for sufficiently large particle sizes, no distinctive spectral signature is seen. Approaches to this problem must involve simultaneous examination of several spectral regions where the particle optical properties differ substantially (e.g. in single scattering albedo), or simultaneous analysis of reflected solar radiation at shorter wavelengths.

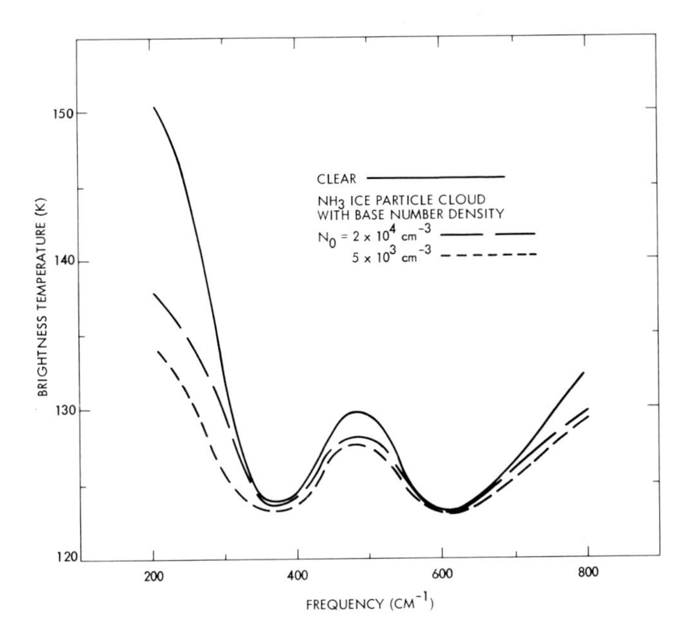

Figure 2. Effect of an NH_3 ice haze in the atmosphere of Jupiter on the thermal spectrum. The NH_3 particles are treated as spherical with a mode radius of 3 μm (and a 10% variance). The base of the cloud is at 670 mb (147 K) with the particle number density given, and the scale height is 20 times smaller than the gas scale height. (Note that the number density values are reversed.)

In order to extend vertical coverage to regions where the pressure is substantially less than 100 mb, where most of the outer planets have thermally inverted stratospheres, sufficient opacity is provided by the ν_4 fundamental band of CH_4 near 7.6 μm (1300 cm^{-1}). For Jupiter and Saturn, CH_4 is expected to be uniformly mixed throughout the atmosphere. Figure 3 shows the pressures associated with unit optical depth for Jupiter as a function of frequency in the region of the CH_4 ν_2 and ν_4 spectral region (for a spectral element with a full width at half maximum, FWHM, of 4 cm^{-1}, similar to that of the Voyager IRIS - Infrared Interferometer Spectrometer - experiment). Figure 4 displays sample weighting functions for both H_2 and CH_4 opacity dominated spectral regions. Weighting functions for CH_4 are somewhat dependent on spectral location and resolution, but they are at least twice as wide as for H_2 (except where the wing of an individual Lorentz line may be isolated).

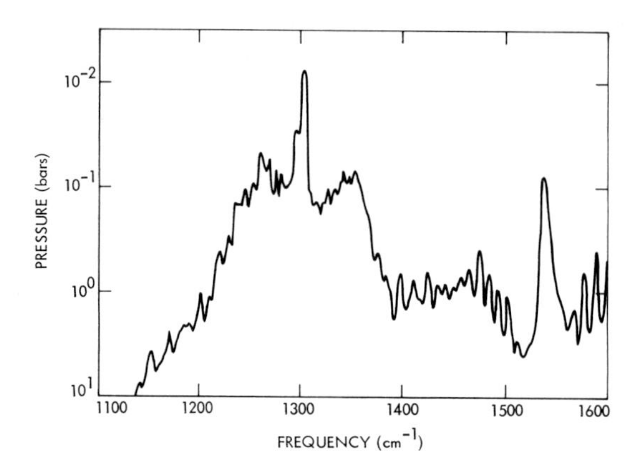

Figure 3. Pressure at which unit optical depth is reached in the atmosphere of Jupiter due to CH_4 opacity in the region of the ν_2 and ν_4 fundamentals. The FWHM of a resolution element is 4 cm^{-1} (from Orton and Robiette, 1980). The CH_4 mixing ratio is 1×10^{-3}.

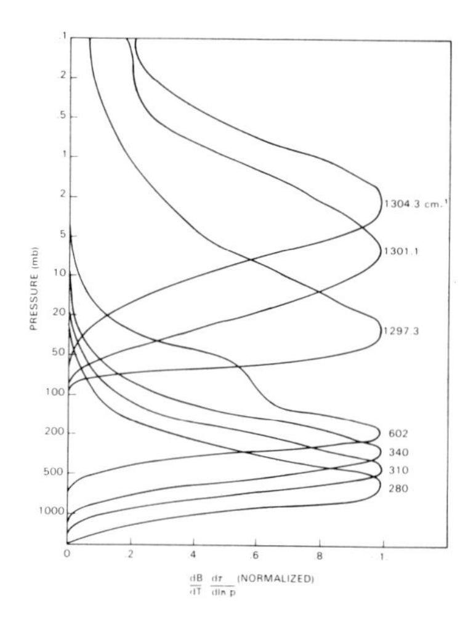

Figure 4. Weighting functions for H_2-dominated and CH_4-dominated regions of opacity in the Jovian spectrum (from Conrath and Gautier, 1980).

Temperature sounding with CH_4 has its own difficulties, beginning with the real problem of determining a good value for the atmospheric mixing ratio. Added to this is the disagreement among various laboratory studies on the absolute strength of ν_4 CH_4 lines, with an overall uncertainty still around 20% as of this writing (Orton and Robiette, 1980). Both Wallace and Smith (1976) and Orton (1977) have demonstrated the problem that thermal sounding at this frequency, where a small change in temperature produces a very large change in brightness, in the presence of a steep temperature gradient is poorly posed. A small change in the assumed temperature or non local thermodynamic equilibrium source function at very low pressures (but relatively high temperatures) may produce substantial changes in the temperatures recovered at higher pressures. Attempts to estimate the non-LTE source function are hampered by the lack of laboratory measurements of the ν_4 relaxation time constant under the influence of H_2 collisions in the relevant temperature range. In the atmospheres of Uranus and Neptune, CH_4 is expected to condense into the solid phase below the respective temperature minima, resulting in a non-uniform vertical distribution and an extremely small mixing ratio in the respective stratospheres. Finally, the ν_4 band is not located in a region of intrinsically high flux; thus, for the low temperatures of relevance, adequate signal is far more difficult to achieve than at longer wavelengths.

Other techniques could be used to extend the vertical sounding range. For example, extremely high resolution measurements of the ν_4 R(0) line could allow the weighting functions shown in Figure 5 to be formed. This could provide the opportunity to sound the stratosphere up to the level of 10^{-7} bars (although the problems associated with the non-LTE function and the extent to which the inversion is well-posed would still be present).

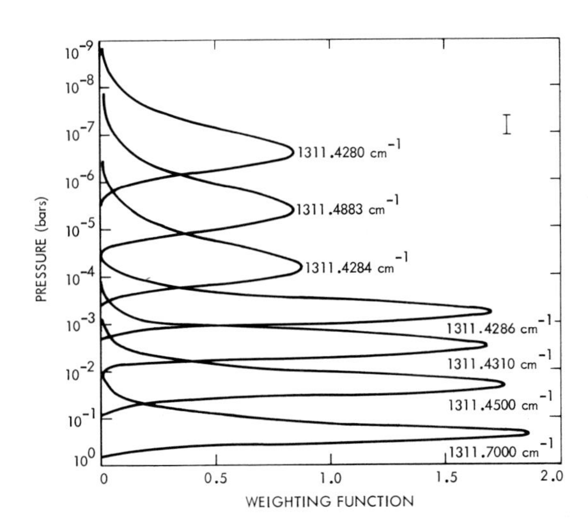

Figure 5. Weighting functions in the Jovian atmosphere, for the same model as in Fig. 3, for monochromatic radiation originating in the vicinity of the R(0) line at 1311.428 cm^{-1} (from Orton and Robiette, 1980).

To extend coverage deeper into the atmosphere, one possibility exists for Jupiter in which the "5-μm window", a spectral region (roughly 1900 - 2300 cm^{-1}) relatively free from strong gaseous absorption, could be used to sound by making use of the weak opacities associated with PH_3, NH_3 or CH_3D. However, not only is there a problem with determining the abundances of these constituents independently, but aerosol effects exert such a substantial influence that the region is best used for cloud and haze sounding. For Saturn and beyond, no substantial radiation is observed from this region, compared with that from Jupiter, and it is likely that a substantial fraction of radiation observed from those planets near 5 μm is, in fact, reflected sunlight. An alternative possibility for deeper atmospheric coverage is at very long wavelengths, where aerosol effects are minimized. Figure 6 illustrates the thermal emission spectra of the outer planets in this region. For Jupiter and Saturn, the gaseous opacity of NH_3 rotation-inversion lines would be useful for temperature sounding if the vertical distribution of NH_3 were well understood. For Uranus and Neptune, the long wavelength (low frequency) wing of the H_2 dipole translational band still dominates the opacity for wavelengths up to about 2 mm. It is a problem, however, to obtain useful spatially resolved data at such long wavelengths, even for spacecraft observations, as relatively large antennas are required to overcome diffraction limitations.

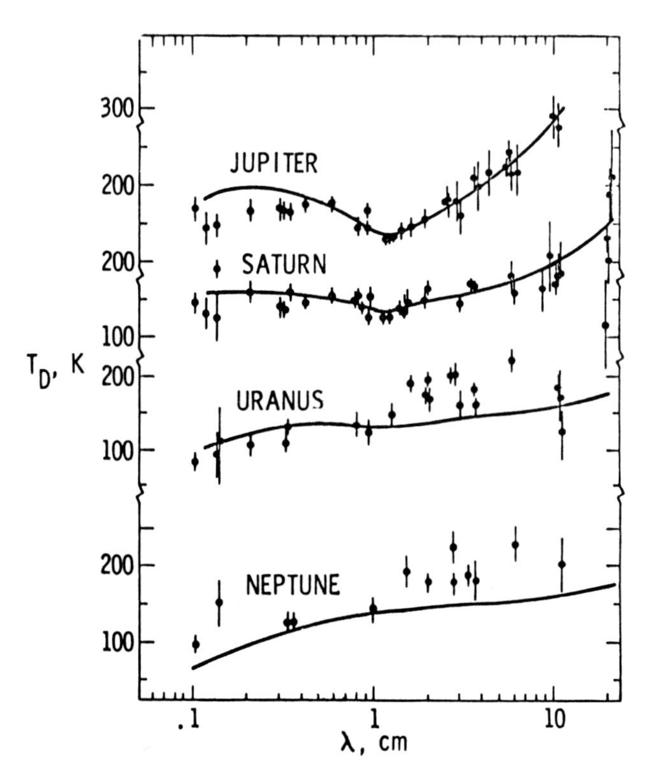

Figure 6. Composite microwave spectra of Jupiter (from Berge and Gulkis, 1976), Saturn (from Klein et al., 1978), Uranus (from Gulkis et al., 1978), and Neptune (from Gulkis and Olsen, 1980). Solid lines represent simple model spectra with H_2 and with NH_3 in saturation equilibrium.

4. RESULTS

Jupiter

Early work on Jovian temperature sounding using earth-based data (Ohring, 1973; Orton, 1975b, 1977; Gautier et al., 1977a, 1979) and using Pioneer Infrared Radiometer results (Orton, 1975a; Orton and Ingersoll, 1976) has largely been superceded by the more recent and complete results of the Voyager IRIS experiment. Some of the first IRIS results (Hanel et al., 1979) are summarized in Figures 7 and 8.

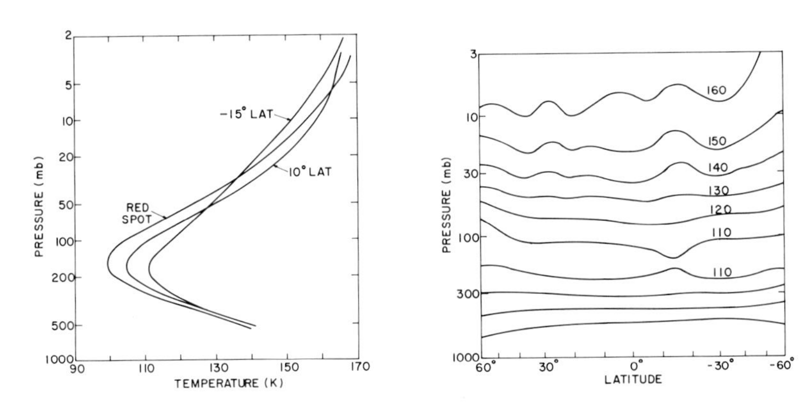

Figure 7. (Left) Examples of temperature profiles obtained by inversion of spectral radiances observed by the Voyager IRIS experiment (from Hanel et al., 1979). Figure 8. (Right) Zonally averaged meridional cross section of Jovian temperature structure recovered from Voyager IRIS experiment data (from Hanel et al., 1979).

One should note the relative steepness of the temperature structure both above and below the temperature minimum. Of further note are features which vary with position, such as the gradual increase in stratospheric temperatures from south to north, noted also in earth-based observations by Sinton et al. (1980). Presuming that the temperature inversions in the outer planets are due to thermalization of absorbed insolation, this temperature increase from south to north is most likely due to the slight tilt of the north pole of Jupiter sunward. Future monitoring of this feature with time as a function of changing insolation geometry will allow some conclusions to be drawn about the radiative relaxation time characterizing the Jovian stratosphere. It should also be noted that there is unexpected detail in the latitudinal structure of stratospheric temperatures, implying a non-uniform horizontal distribution of insolation-absorbing aerosols. One further feature of note (not illustrated in Figs. 7 or 8) is the strong drop (more than 5K) in temperatures over the Great Red Spot compared with the surrounding area near the 100 mb level. These may result from radiative equilibrium with reduced planetary flux emerging locally from below (Orton, 1975a) or from dynamical considerations (Flasar et al., 1980). It should be noted that analysis to date has not attempted to account systematically for the effects of

(NH_3 ice) clouds around the 150 K level on outgoing emission. Latitudinal variations of temperatures recovered near 500 mb may therefore only reflect the variation of cloud thickness or cloud height from region to region.

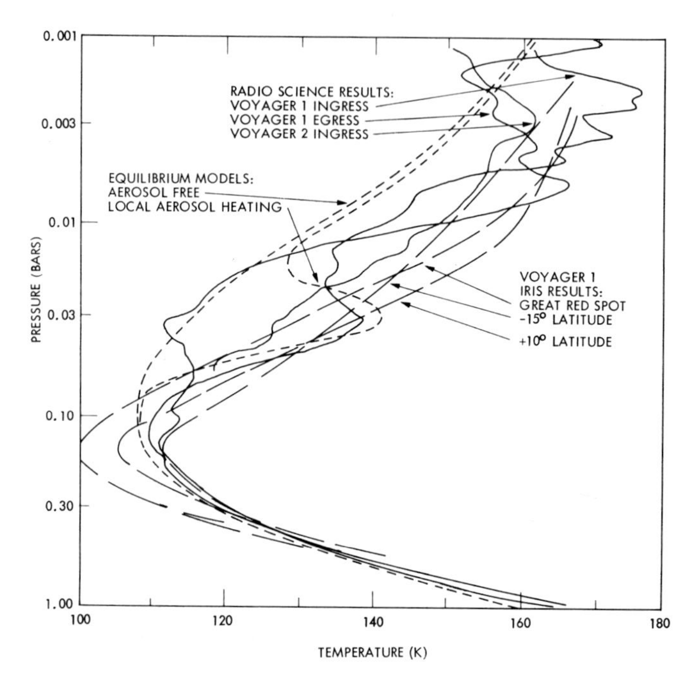

Figure 9. Comparison of Jovian temperature structures derived from Voyager radio occultation results (Lindal et al., 1980; solid line), IRIS results (Hanel et al., 1979; long dashed lines), and radiative-convective equilibrium models (Appleby and Hogan, 1980; short dashed lines). Equilibrium models are for a uniform distribution of insolation-absorbing aerosols and for a "local" distribution absorbing 4% of the total insolation.

Figure 9 demonstrates a visual comparison of the results of IRIS, the radio occultation results, and equilibrium models. The IRIS and radio occultation results are largely in agreement; for the tropopause and below the agreement has been optimized by an appropriate assumption for the bulk composition (Gautier et al., 1980). The radio results do, however, give an independent confirmation of the assumption of an adiabatic lapse rate for pressures near and greater than 1 bar. The detailed vertical structure of the occultation results are beyond the vertical resolution of infrared experiment retrieval. An unresolved question is whether the differences between the relatively cold Voyager 1 egress profile for the lower stratosphere and the relevant infrared data for this region are totally irreconcilable and whether the consistent difference in location of the temperature minimum is significant. The variable radio occultation results for the stratosphere are consistent with IRIS results implying a variation of insolation-absorbing aerosols with location of the planet. The Appleby and Hogan (1980) equilibrium models demonstrate that such an explanation is physically viable. The "cold" Voyager 1 occultation profile is never much lower than the thermal inversion supported by CH_4 absorption alone, whereas the Voyager 1 ingress profile vertical variation could easily be the result of heating by a "local" absorption of 4% of the total absorbed solar flux.

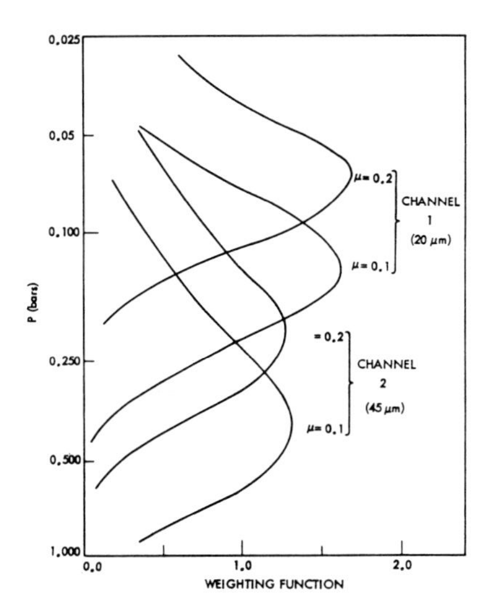

Figure 10. Weighting functions for the flux observed from Saturn in the two channels of the Pioneer 11 IRR (from Orton and Ingersoll, 1980). In this figure, μ is the emission angle cosine.

Saturn

Earth-based spectral observations formed the data base for early Saturn temperature sounding (Ohring, 1975; Gautier et al., 1977b) as well as semi-empirical radiative-convective equilibrium models by Tokunaga and Cess (1977). More recently, detailed information on Saturn has been obtained by the Pioneer 11 Infrared Radiometer in the same manner as the Pioneer 10 and 11 IRR data to determine the Jovian temperature structure (Orton, 1975a; Orton and Ingersoll, 1976). Weighting functions for the two broadband channels of the IRR, centered near 20 μm and 45 μm, are shown in Figure 10. Note that the weighting functions form a useful set only when a given area is observed over a range of emission angles. However, since the IRR observed no planetary position more than once, it was necessary to assume atmospheric longitudinal homogeneity. Figure 11 shows results for one of the warmest and one of the coolest regions observed. All models assume an adiabatic lapse rate below the 500 mb level and use a linear temperature interpolation in between the discrete levels where temperature is actually recovered. Cases 1 and 2 assume overlying inverted lapse rates consistent, respectively, with (1) the Pioneer 11 radio occultation results (Kliore et al., 1980) and (2) the equatorial model of Tokunaga and Cess (1977). Case 3 assumes that the apparently cooler temperatures at 500 mb near the equator are due to an unmodeled cloud influence. If the 500 mb temperatures are to be recovered with nearly equal values as for warmer poleward latitudes, the presence of a cloud must be invoked, modeled crudely as a uniform opaque blackbody surface emitting at approximately 124 K. If the cloud is composed of NH_3 ice particles, as is strongly suspected on Jupiter,

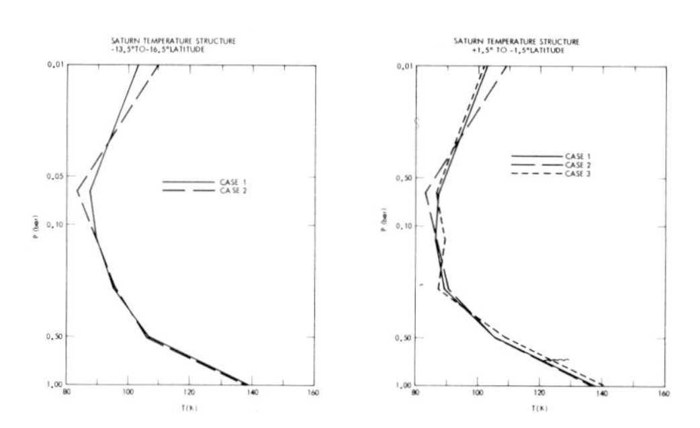

Figure 11. Temperatures recovered by the Pioneer 11 IRR for one of the warmest (left) and one of the coolest (right) regions observed (from Orton and Ingersoll, 1980).

it is substantially thicker on Saturn, since the cloud bottom is near the 150 K level. Figure 12 shows a smoothed plot of recovered temperatures vs. latitude for the region covered by IRR observations. Note that for all latitudes (and all cases) the temperature minimum region is substantially broader than for Jupiter (Figs. 7 and 8). The only major temperature variation with latitude, the cool equatorial (10°N to 10°S) region, correlates strongly with a region which is bright at visual wavelengths.

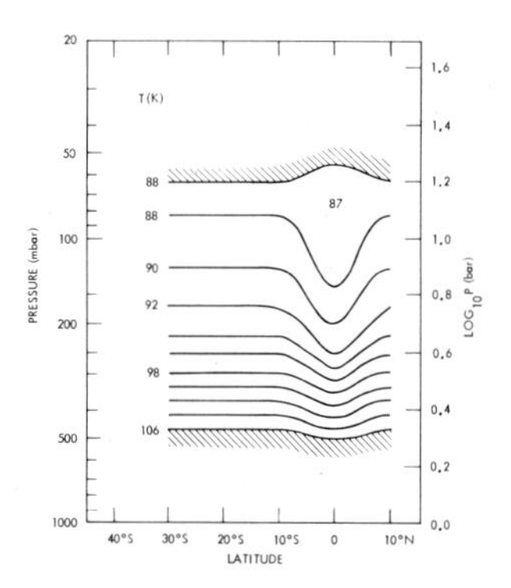

Figure 12. Smoothed contours of meridional cross section of Saturnian temperature structure for the region of the planet observed by the Pioneer 11 IRR (from Orton and Ingersoll, 1980).

Figure 13 displays a comparison of Pioneer 11 IRR and radio occultation results with a radiative-convective equilibrium model (Appleby and Hogan, 1980). The models correspond reasonably well with one another below the temperature minimum, although the correlation between the infrared and radio occultation profiles has been optimized by an appropriate choice for the bulk composition (Kliore et al., 1980; Orton and Ingersoll, 1980). The temperature structures of Kliore et al. and those of Tokunaga and Cess (1977) above the temperature minimum are not reconcilable. In fact, the radio occultation temperatures fall below those of the equilibrium model.

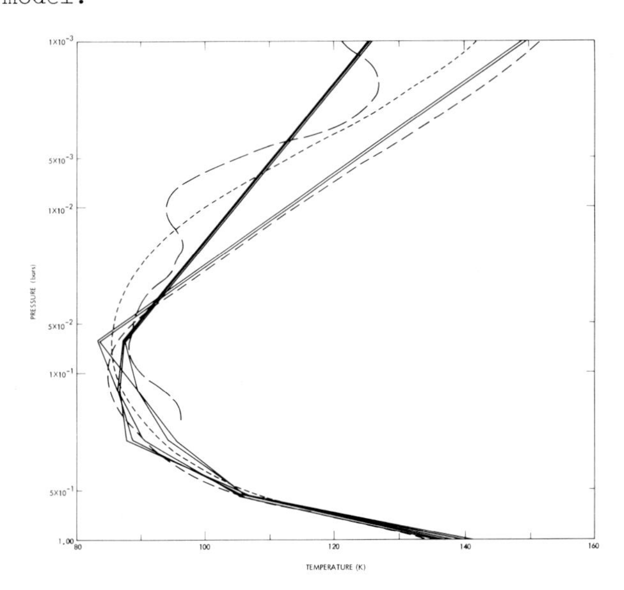

Figure 13. Comparison (from Orton and Ingersoll, 1980) of Saturnian temperature structures derived from Pioneer 11 IRR results for different regions and different assumptions (Orton and Ingersoll, 1980; solid lines), radio occultation results (Kliore et al., 1980; long dashed line), a model derived from earth-based data (Tokunaga and Cess, 1977; medium dashed line), and an equilibrium model (Appleby and Hogan, 1980; short dashed line). The equilibrium model has a uniform vertical distribution of aerosols.

Another noteworthy phenomenon is the substantial latitudinal dependence of the stratospheric temperatures observed from the earth, as illustrated in Figure 14. Elevated temperatures at the south pole are presumed to be due to the substantially greater amount of sunlight incident on the south pole at the time of the measurements. As for Jupiter, continued monitoring of the temperatures at both poles as the north goes into summer and the south into winter will provide useful information on the characteristic time constant for radiative thermal adjustment.

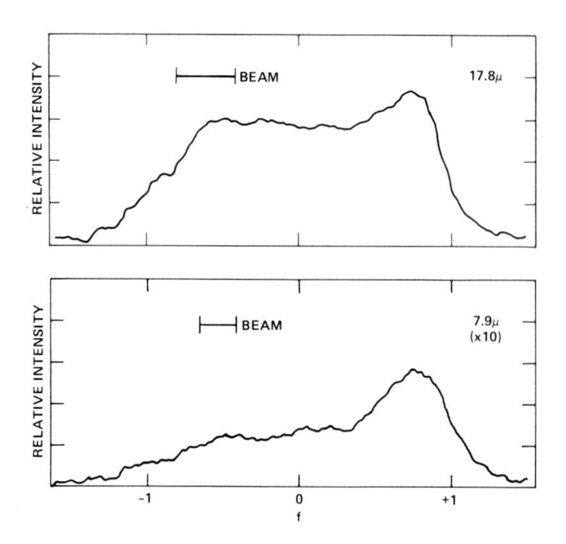

Figure 14. Spatially resolved meridional scans of Saturn obtained in March 1977, showing the south polar brightening in two spectral regions sensitive to stratospheric temperatures (from Tokunaga et al., 1978). Position on disk, relative to the polar radius, is given by f, with positive values to the south.

Finally, not only continued earth-based spatially resolved observations but also Voyager IRIS (and radio occultation) results should be able to resolve some of the questions raised by earlier results. Among these are the temperature structure of the stratosphere, the nature and distribution of NH_3 ice particles, the form of temperature or cloud asymmetries about the equator, and properties poleward of 30°S or 10°N.

Uranus and Neptune

The outermost known giant planets represent a whole separate class of observational problems involving dim objects for which no spatial resolution is available. Most of the thermal flux appears in a spectral region which is obscured from the ground by strong water vapor rotational line absorption. The inference from brightness temperature measurements and equilibrium models is that CH_4 will condense from gaseous to solid form with negligible amounts existing in the stratosphere. Thus, H_2 is the only useful opacity source which is known to be distributed uniformly, although its mixing ratio in the atmospheres of Uranus and Neptune is uncertain or unknown.

Part of the data base for Uranus and Neptune in the 8 - 14 μm region is shown in Figure 15. As expected for Uranus, no ν_4 CH_4 or ν_9 C_2H_6 emission features are seen, as they are in the spectra of Jupiter and Saturn. However, the spectrum of Neptune displays emission features of both bands. The existence of CH_4 above an effective cold trap is difficult to understand. Three simple and somewhat distinct explanations

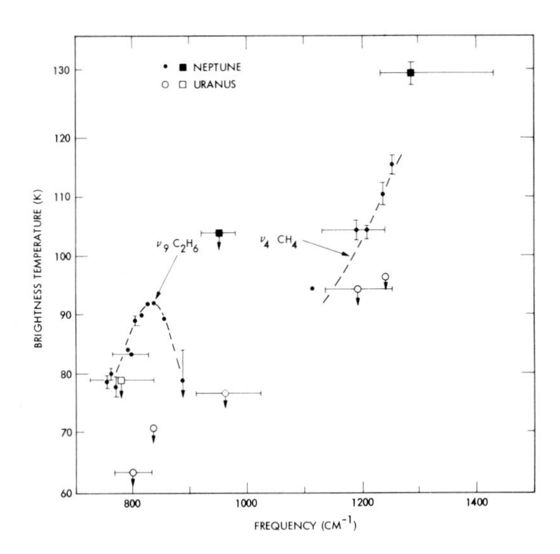

Figure 15. Spectra of Uranus and Neptune for the 8 - 14 μm region (circles from Gillett and Rieke, 1977), squares from Macy and Sinton, 1977). Note the emission features of ethane and methane in the spectrum of Neptune, but not Uranus.

are possible. (A) CH_4 is in saturation equilibrium and strongly depleted in the stratosphere, but is seen because of a very steep temperature inversion supported by aerosol absorption of insolation. (B) CH_4 is in saturation equilibrium below and above the temperature minimum, relatively depleted only near the temperature minimum, although this opens the question as to the origin of the stratospheric CH_4. This model does not require as steep a temperature inversion as model A. (C) CH_4 is uniformly distributed vertically and supersaturated in the stratosphere; presumably this distribution is maintained by a circulation system which acts more rapidly than the time scale for CH_4 freezing. Again, this model does not require as steep a temperature inversion as model A. Why these models (or others) do not appear to operate in the atmosphere of Uranus is unknown, although the answer is probably associated with the peculiar aspect geometry of that planet with respect to the sun and the earth. With the cold south pole facing the earth, all conclusions based on current observational evidence may be characteristic of a colder environment than is true for the global average. The existence of C_2H_6 in the presence of CH_4 is consistent with models of hydrocarbon photochemistry for the outer planets (e.g., Strobel, 1975).

Longer wavelength data for Uranus and for Neptune are displayed in Figures 16 and 17, respectively. Also shown in these figures are spectra resulting from the direct inversion models of Courtin et al. (1978) and Courtin et al. (1979), and the equilibrium models of Appleby and Hogan (1980). The temperature structures corresponding to these models are shown in Figure 18 for Uranus and Neptune. The data are relatively sparse and often broadbanded, attesting to the difficulty of making observations of these objects in this spectral region. Figures 16 and 17 show that the temperature structure models cannot fit all the available data within quoted errors (although some data in the figures were not considered in the models when published). In addition, the data are occasionally not consistent with each other. Work is clearly called for to remeasure parts of the spectrum, especially with higher spectral resolution where possible.

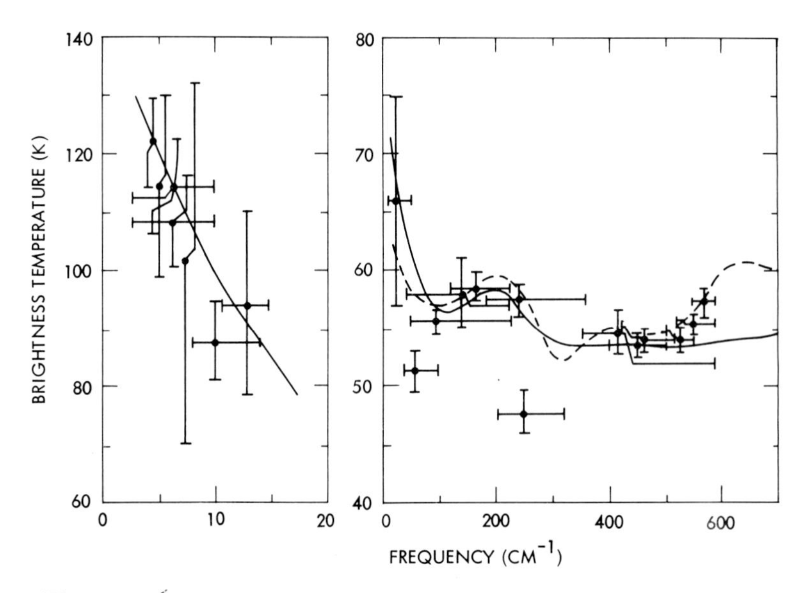

Figure 16. Long wavelength data for Uranus, representing the work of Courtin et al. (1977), Fazio et al. (1976), Gillett (1979), Gillett and Rieke (1977), Hildebrand et al. (1980), Kostenko et al. (1971), Morrison and Cruikshank (1973), Lowenstein et al. (1977a), Rieke and Low (1974), Rowan-Robinson et al. (1978), Stier et al. (1978), Ulich (1974), Ulich and Conklin (1976), Werner et al. (1978) and Whitcomb et al. (1978). Data are plotted at effective spectral positions with FWHM bandpasses (or best estimates where not given explicitly) represented by horizontal bars. No data are shown for frequencies less than 4.5 cm^{-1}. The solid line represents the model spectrum of Courtin et al. (1978) and the dashed line the model spectrum of Appleby and Hogan (1980) (see Fig. 18).

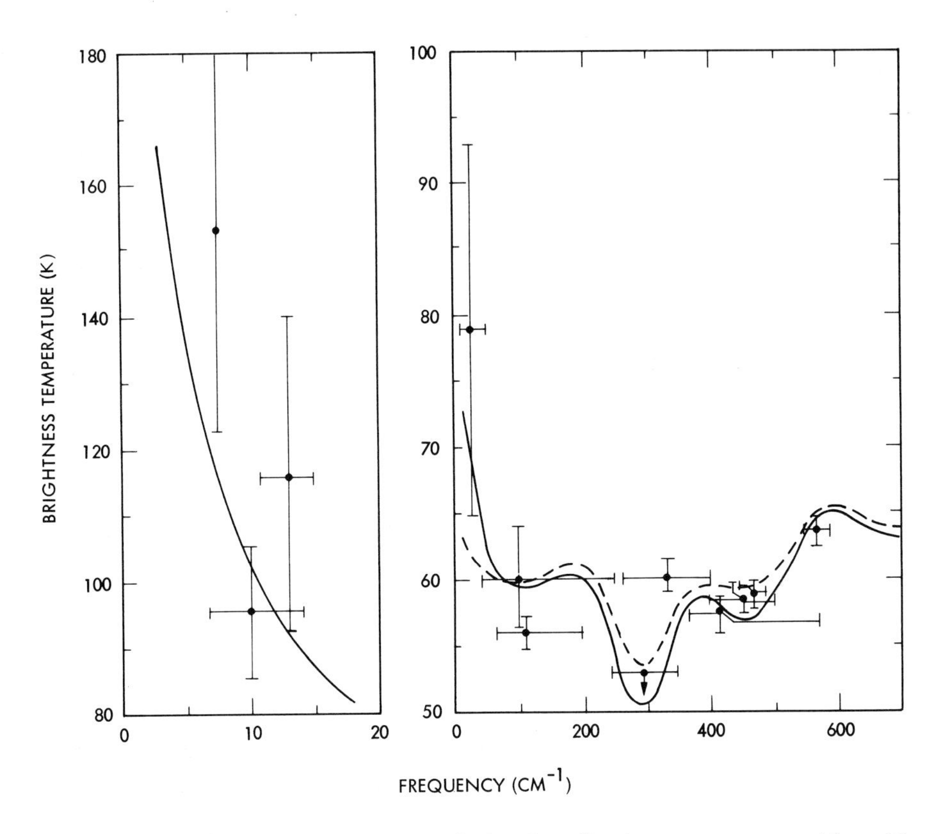

Figure 17. Long wavelength data for Neptune, representing th work of Courtin et al. (1977), Gillett and Rieke (1977), Hild brand et al. (1980), Low et al. (1973), Lowenstein et al. (1977b), Morrison and Cruikshank (1973), Rieke and Low (1974) Steir et al. (1978), Ulich and Conklin (1976), Werner et al. (1978) and Whitcomb et al. (1979). Symbols and spectral ran are the same as for Figure 16. The solid line represents th model spectrum of Courtin et al. (1979) and the dashed line the model spectrum of Appleby and Hogan (1980) (see Fig. 18)

The equilibrium models and direct inversion models are not extremely different from each other (Fig. 18). Differences in the assumed bulk composition for the models shown account for the somewhat different adiabatic lapse rates in the convective region (pressures near or gre than 1 bar). For Uranus, Courtin et al. (1978) assumed an isotherm o lying the highest level sounded in their model I which is shown here (although they did also present results from a model with a hypotheti inverted stratosphere). The shallow thermal inversion in the model c Appleby and Hogan (1980) is a result of insolation absorbed by the existing CH_4 in the stratosphere and a uniform vertical distribution insolation absorbing aerosols. The temperature inversion is, in fac required to match the temperatures deduced by Dunham et al. (1980) f the occultation of SA0158687 by Uranus. For Neptune the models of

Courtin et al. (1979) and Appleby and Hogan (1980) best fitting the data are shown in Fig. 16, chosen from the models for stratospheric CH_4 distribution discussed earlier. Models do not appear to be able to fit all the data well with a "model A type" CH_4 distribution, as discussed above. The Courtin et al. (1979) model corresponds to a "model C type" CH_4 distribution and the Appleby and Hogan (1980) model to a "model B type" distribution.

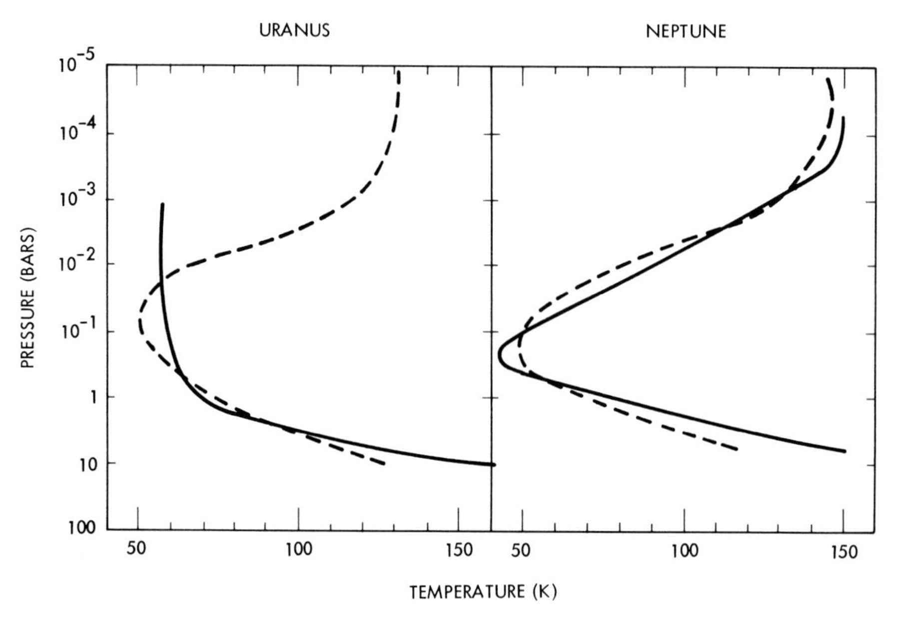

Figure 18. Model temperature structures for Uranus (left) and Neptune (right). Direct inversion models (model I of Courtin et al., 1978) and for Neptune (model N of Courtin et al., 1979) are shown as solid lines. Equilibrium models of Appleby and Hogan (1980) are shown as dashed lines.

Undoubtedly, the models will improve with the addition of better data, ultimately whose spectral resolution will be sufficient to confirm that the collision induced H_2 dipole is, in fact, the dominant opacity. Other substantial improvements in the earth-based data record could result from self-consistent calibration sources from short to long wavelengths. Where broad-band measurements are used in temperature sounding, the full filter response (including telluric extinction) should be convolved with a detailed spectrum, instead of using the monochromatic approximation represented by an effective wavelength or frequency. This would allow a truer calculation of the real weighting function for the filter. As for Jupiter and Saturn, consideration should be given to the possible effect of aerosols (NH_3 and CH_4 ice particles, at least) on the outgoing thermal radiance. Spacecraft experiments to measure thermal flux from Uranus or Neptune with spatial resolution will be difficult due to the low flux levels involved and the associated instrument cooling requirements.

Nevertheless, substantial information is latent in such measurements which can be a key to understanding the climatology and meteorology of these planets, especially for the peculiar insolation conditions of Uranus.

5. CLOSING REMARKS

The most complete kind of planetary information involving high spatial resolution and global coverage can only be provided by spacecraft instrumentation. However, the role of earth-based data and the potential role of infrared observations from earth-orbiting platforms is substantial. Such observations provide the only realistic basis for providing long-term monitoring of time dependent phenomena, and the opportunity for new discovery using spectral regions or resolutions which are, for a variety of reasons, outside the scope of spacecraft experiments.

It is apparent that, even with often very sparse available data, the outer planets demonstrate a remarkable variety of atmospheric properties. Ultimately the results of temperature sounding and energy transport equilibrium models must be combined in order to provide a useful interpretation of the phenomena observed. Future work must address not only the details of the differences among the physical properties of the planets, but underlying reasons and the extent to which they are based on differences of solar distance, obliquity, rotation period and chemical composition.

Acknowledgements are gratefully given to several persons for valuable discussion and support: J. Appleby, J. Bergstralh, J. Caldwell, B. Conrath, D. Gautier, S. Gulkis, R. Hanel, T. Jones, M. Klein, V. Kunde, J. Martonchik, E. Olsen, A. Robiette, and A. Tokunaga; also to D. Gautier and R. Courtin for their earlier review of outer planet thermal structures (1979). This is JPL Atmospheres publication No. 980-13 (internal number) and presents one phase of research carried out at the Jet Propulsion Laboratory, California Institute of Technology, under contract NAS 7-100, sponsored by the Planetary Atmospheres Program Office, Office of Space Science, National Aeronautics and Space Administration.

REFERENCES

APPLEBY, J.F., HOGAN, J., 1980 (in preparation).
ATREYA, S., 1980, communication presented at 23rd Plenary Session of COSPAR, Budapest.
BERGE, G.L., GULKIS, S., 1976, In "Jupiter", T. Gehrels ed., U. of Arizona Press, p. 621.
BIRNBAUM, G., 1978, J. Quant. Spectrosc. Rad. Transf. 19, 51.
CESS, R.D., KHETAN, S., 1973, J. Quant. Spectrosc. Rad. Transf. 13, 995.
CESS, R.D., CHEN, S.C., 1975, Icarus 26, 444.
CHAHINE, M.T., 1972, J. Atmos. Sci. 29, 741.
CHAHINE, M.T., 1975, J. Atmos. Sci. 32, 1946.
CONRATH, B.J., 1972, J. Atmos. Sci. 29, 1262.
CONRATH, B.J., GAUTIER, D., 1980, In "Interpretation of Remotely Sensed Data", A. Deepak ed., Academic Press.

COURTIN, R., CORON, N., ENCRENAZ, T., GISPERT, R., BRUSTON, P., LEBLANC, J., DAMBIER, G., VIDAL-MADJAR, A., 1977, Astron. Astrophys. 60, 115.
COURTIN, R., GAUTIER, D., LACOMBE, A., 1978, Astron. Astrophys. 63, 97.
COURTIN, R., GAUTIER, D., LACOMBE, A., 1979, Icarus 37, 236.
DANIELSON, R.E., 1977, Icarus 30, 462.
DEEPAK, A., 1977, (ed.) "Inversion Methods in Atmospheric Remote Sensing", Academic Press.
DUNHAM, E., ELLIOT, J.L., GIERASCH, P.J., 1980, Astrophys. J. 235, 274.
FAZIO, G.G., TRAUB, W.A., WRIGHT, E.L., LOW, F.J., TRAFTON, L.M., 1976, Astrophys. J. 209, 633.
FLASAR, F.M., CONRATH, B.J., PIRRAGLIA, J.A., CLARK, P.C., FRENCH, R.G., GIERASCH, P.J., 1980, J. Geophys. Res. (in press).
GAUTIER, D., LACOMBE, A., REVAH, I., 1977a, J. Atmos. Sci., 34, 1130
GAUTIER, D., LACOMBE, A., REVAH, I., 1977b, Astron. Astrophys. 61, 149.
GAUTIER, D., MARTEN, A., BALUTEAU, J.P., LACOMBE, A., 1979 Icarus 37, 214.
GAUTIER, D., COURTIN, R., 1979, Icarus 39, 28.
GAUTIER, D., CONRATH, B., FLASAR, M., HANEL, R., KUNDE, J., CHEDIN, A., SCOTT, N., 1980, J. Geophys. Res. (in press).
GILLETT, F.C., RIEKE, G.H., 1977, Astrophys. J 218, L141.
GILLETT, F.C., 1979, unpublished communication.
GULKIS, S., JANSSEN, M.A., OLSEN, E.T., 1978, Icarus 34, 10.
GULKIS, S., OLSEN, E.T., 1980, unpublished communication.
HANEL, R., CONRATH, B., FLASAR, M., KUNDE, V., LOWMAN, P., MAGUIRE, W., PEARL, J., PIRRAGLIA, J., SAMUELSON, R., GAUTIER, D., GIERASCH, P., KUMAR, S., PONNAMPERUMA, C., 1979, Science 204, 972.
HILDEBRAND, R.H., KEENE, J., WHITCOMB, S.E., 1980 (in preparation).
HOGAN, J., RASOOL, I., ENCRENAZ, T., 1969, J. Atmos. Sci. 26, 898.
KLEIN, M.J., JANSSEN, M.A., GULKIS, S., OLSEN, E.T., 1978, In "The Saturn System", NASA Conference Publication CP-2089, p. 195.
KLIORE, A.J., PATEL, I.R., LINDAL, G.F., WAITE, J.H., MCDONOUGH, T.R., 1980, J. Geophys. Res. (in press).
KOSTENKO, V.J., PAVLOV, A.V., SCHOLOMITSKY, G.B., SLYSH, V.I., SOGLASNOVA, V.A., ZABOLOTNY, V.F., 1971, Astrophys. Lett. 8, 41.
LINDAL, G.F., WOOD, G.E., LEVY, G.S., ANDERSON, J.D., SWEETNAM, D.N., HOTZ, H.B. BUCKLES, B.J., HOLMES, D.P., DOMS, P.E., ESHLEMAN, V.R., TYLER, G.L., CROFT, T.A., 1980, J. Geophys. Res. (in press).
LOEWENSTEIN, R.F., HARPER, D.A., MOSELEY, S.H., TELESCO, C.M., THRONSON, H.A., HILDEBRAND, R.H., WHITCOMB, S.E., WINSTON, R., STIENING, R.F., 1977a, Icarus 31, 315.
LOEWENSTEIN, R.F., HARPER, D.A., MOSELEY, H., 1977b, Astrophys. J. 218, L145.
LOW, F.J., RIEKE, G.H., ARMSTRONG, K.R., 1973, Astrophys. J. 183, L105.
MACY, W., SINTON, W.M., 1977, Astrophys. J. 218, L79.
MORRISON, D., CRUIKSHANK, D.P., 1973, Astrophys. J. 179, 329.
OHRING, G., 1973, Astrophys. J. 184, 1027.
OHRING, G., 1975, Astrophys. J. 195, 223.
ORTON, G.S., 1975a, Icarus 26, 125.
ORTON, G.S., 1975b, Icarus 26, 142.

ORTON, G.S., INGERSOLL, A.P., 1976, In "Jupiter", T. Gehrels ed., U. of Arizona Press, p. 206.
ORTON, G.S., 1977, Icarus 32, 41.
ORTON, G.S., ROBIETTE, A.G., 1980, J. Quant. Spectrosc. Rad. Transf. (in press).
ORTON, G.S., INGERSOLL, A.P., 1980, J. Geophys. Res. (in press).
RIEKE, G.H., LOW, F.J., 1974, Astrophys. J. 193, L147.
ROWAN-ROBINSON, M., ADE, P.A.R., ROBSON, E.I., CLEGG, P.E., 1978, Astron. Astrophys. 62, 249.
SINTON, W.M., MACY, W.W., ORTON, G.S., 1980 Icarus 42, 86.
STIER, M.T., TRAUB, W.A., FAZIO, G.G., WRIGHT, E.L., LOW, F.J., 1978, Astrophys. J. 226, 347.
STROBEL, D.F., 1975, Rev. Geophys. Space Phys. 13, 372.
TOKUNAGA, A., CESS, R.D., 1977, Icarus 32, 321.
TOKUNAGA, A.T., CALDWELL, J., GILLETT, F.C., NOLT, I.G., 1978, Icarus 36, 216.
TRAFTON, L.M., 1967, Astrophys. J. 147, 765.
TRAFTON, L.M., STONE, P.H., 1974, Astrophys. J. 188, 649.
ULICH, B.L., 1974, Icarus 21, 254.
ULICH, B.L., CONKLIN, E.K., 1976, Icarus 27, 183.
WALLACE, L., PRATHER, M., BELTON, M.J.S., 1974, Astrophys. J. 193, 481.
WALLACE, L., 1975, Icarus 25, 538.
WALLACE, L., SMITH, G.R., 1976, Astrophys. J. 203, 760.
WERNER, M.W., NEUGEBAUER, G., HOUCK, J.R., HAUSER, M.G., 1978, Icarus 35, 289.
WHITCOMB, S.E., HILDEBRAND, R.H., KEENE, J., STIENING, R.F., HARPER, D.A., 1979, Icarus 38, 75.

DISCUSSION FOLLOWING PAPER DELIVERED BY G. S. ORTON

T. JONES: Occultation data shows a temperature inversion in the atmosphere of Uranus. What can cause the heating?

ORTON: The models derived by Wallace and by Appleby and Hogan have no difficulty in achieving the "warm" temperatures near 10^{-5} bar derived from stellar occultation data. Heating via insolation absorbed by the residual CH_4 above the temperature minimum is sufficient, although stratospheric aerosols may also play an additional role.

WRIGHT: Has He-H_2 collision-induced absorption been measured to 77 K?

ORTON: Yes; the measurements can be found in Figure 6 of Birnbaum (1978, J. Quant. Spectrosc. Rad. Transf. 19, p. 51). Birnbaum has improved these measurements recently, although they are not yet in publication.

WRIGHT: What do you assume for the ortho-para ratio for H_2 in Uranus and Neptune?

ORTON: Equilibrium at the ambient temperature.

WRIGHT: Is there any constituent that could establish equilibrium?

ORTON: We don't know of any. Observations of the H_2 quadrupole absorptions by Smith (1978, Icarus 33, p. 210) tend to indicate equilibrium in Jupiter and Saturn and some evidence for non-equilibrium in Uranus and Neptune.

WRIGHT: There is no translational collision-induced absorption when two $J = 0$ H_2 molecules collide, so ortho-para equilibrium gives less long wavelength opacity than a 3:1 ortho-para ratio.

ORTON: At this time, an equal or greater source of uncertainty in the infrared absorption is due to the uncertainty in the bulk compositions of Uranus and Neptune.

GULKIS: Your equation of radiative transfer did not include scattering. If scattering is present, how do you differentiate between effects due to thermal structure and effects due to scattering?

ORTON: We cannot differentiate initially between the two effects. We have to look at spectral regions where we think there is evidence for aerosols on some other basis in order to evaluate the scattering. Specific candidates for scatterers can be modeled, for example, but to date we have modeled only a few possible substances.

SPECTROPHOTOMETRIC REMOTE SENSING OF PLANETS AND SATELLITES

T. B. McCord and D. P. Cruikshank
University of Hawaii
Honolulu, Hawaii 96822

1. INTRODUCTION

Near infrared spectrophotometry has vastly increased our knowledge of the composition and structure of asteroids, satellites and planetary surfaces over the past ten years. In this article we will attempt to summarize the most recent comprehensive results. We will emphasize the interpretations and present only examples of the data.

At visible and near infrared wavelengths the radiation received from solar system objects is passively scattered solar radiation. At longer wavelengths thermal emitted radiation dominates. (Figure 1 shows the two radiation components for the lunar case.) The wavelength at which the two components contribute equally is about 2.5 μm for the moon's sub-solar point. For Mercury this wavelength is near 1.9 μm and for the colder outer solar system objects the wavelength is farther into the infrared.

At certain energies associated with the visible and near-infrared wavelengths, electronic and molecular transitions occur which result in absorption of solar radiation by surface material. Spectra of solar radiation reflected from surfaces of solar system objects show these absorptions. At visible and near infrared wavelengths longward to about 2.5 μm, transitions of d-shell electrons in transition metal ions and electron exchange between ions are mainly responsible; iron ions are particularly important. Molecular oscillations result in absorptions longward of about 1 μm, with H_2O and OH being most important.

The electronic absorptions occur at energies controlled by the particular ion present and by the electric field the ion experiences. These properties generally define mineralogy. Similar interpretations can be made of certain molecular absorptions; ice, for example, is a mineral.

In this review we first consider the relatively nearby objects which can be resolved into several individual surface elements by

C. G. Wynn-Williams and D. P. Cruikshank (eds.), Infrared Astronomy, 57–87.

present techniques of remote sensing. Progress has been made in mapping the surface compositions of some of these bodies, and the work continues with large telescopes at high-quality ground-based observatories. In a later section we consider the smaller and more distant solar system objects which cannot be resolved into surface elements and for which global, or hemispheric, averages of surface composition are revealed by the techniques of remote sensing.

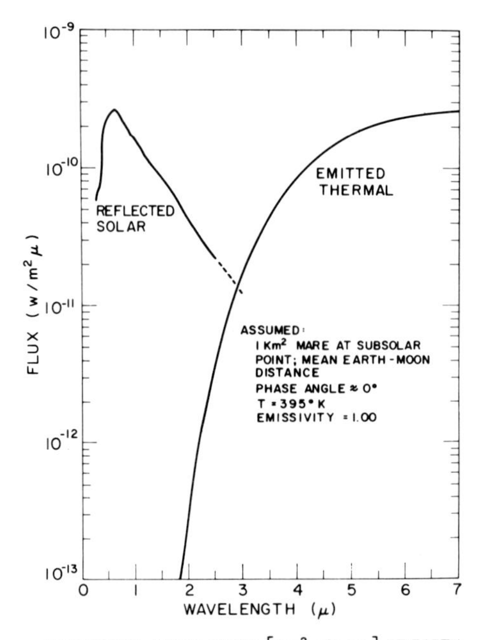

Figure 1. The radiation received from the moon is shown for the extended visible and near- to mid-infrared spectral regions. The two radiation components, reflected and emitted, are obvious.

2. THE MOON AND TERRESTRIAL PLANETS

Mercury

Ground-based telescopic observations of Mercury are difficult to make because Mercury is close to the sun and only appears in the sky during the day or near the horizon at twilight, when the sky is still bright and line-of-sight air mass is great. McCord and Adams (1972a,b) reported a possible weak absorption in the Mercury spectrum near 0.95 μm, but the data were of insufficient quality to be certain. Vilas and McCord (1976) later found no feature with greater contrast than a few percent. Tepper and Hapke (1977) reported a weak absorption band near 1 μm, but as in the earlier measurements, the feature was about the same strength as the uncertainty in the measurement.

More recently, a continuously spinning CVF spectrometer (McCord et al., 1978) was used to obtain the Mercury spectrum between 0.65 μm and 2.5 μm (McCord and Clark, 1979). The reflectance was calculated and is shown in Figure 2 for three different observing sessions.

The Mercury reflectance spectrum shows a continuous increase in reflectance from visible to infrared wavelengths. A very similar continuum is evident for lunar surface material, as measured through the telescope (Figure 2). With the possible exception of the dark side of Saturn's satellite Iapetus, no other objects in the solar system so far observed have similar spectral reflectance.

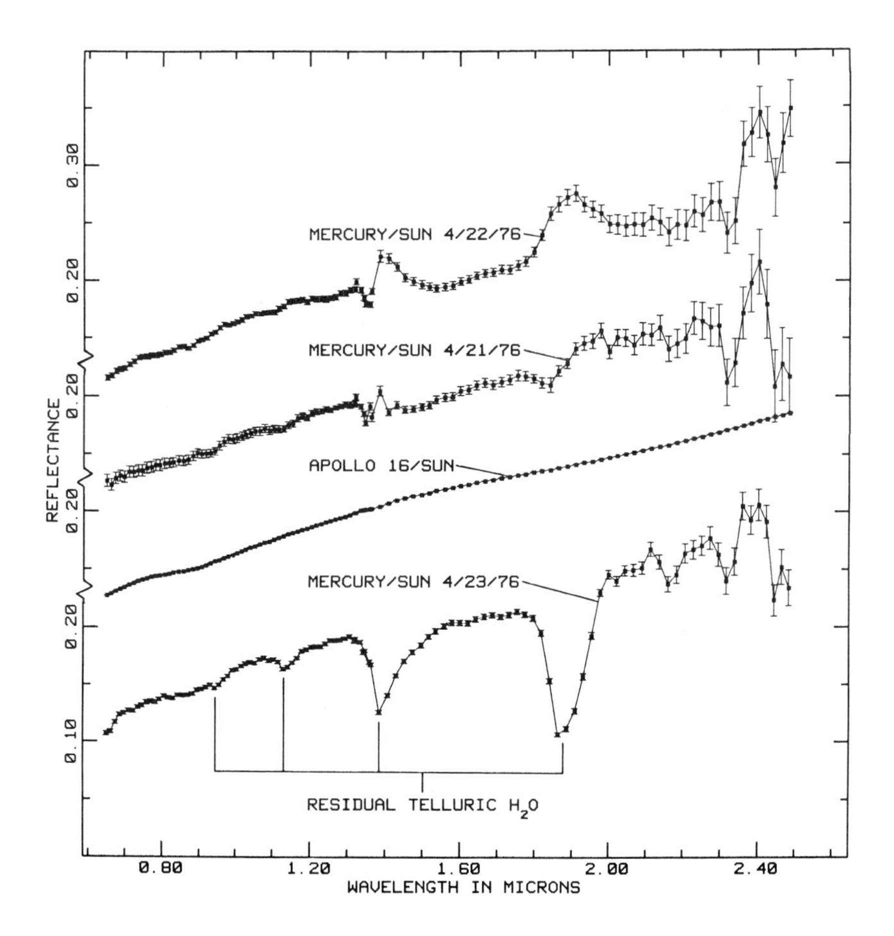

Figure 2. Spectral reflectance of Mercury for the three nights on which observations were made. The stronger telluric H_2O bands, which were not completely removed by extinction corrections, are noted. The spectral reflectance of Apollo 16 site soil as measured in the laboratory is shown for comparison. This and Figures 3 and 4 from McCord and Clark (1979).

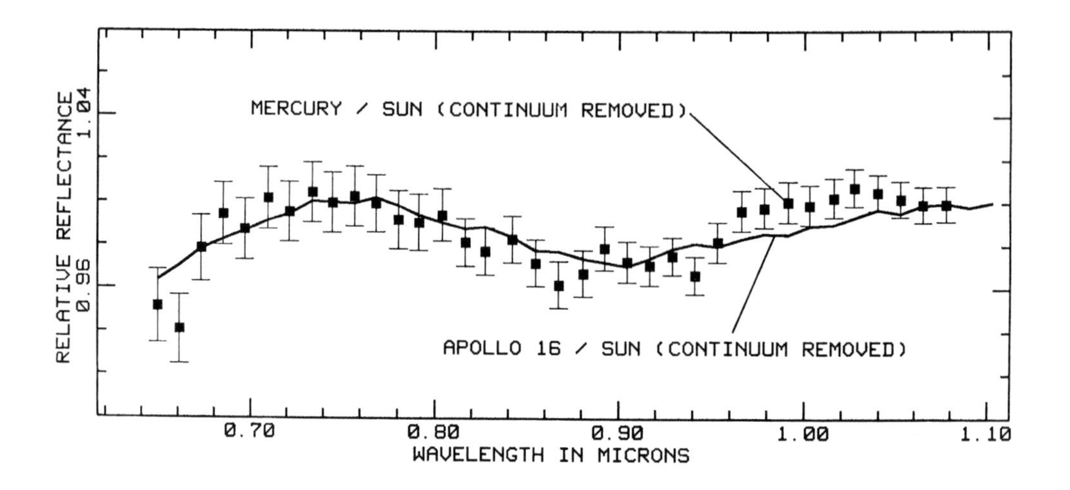

Figure 3. Spectral reflectance of Mercury (points) and of Apollo 16 soil measured in the laboratory [Adams and McCord, 1973] (continuous line) are shown overlayed. A straight line continuum slope fitted at 0.745 and 1.078 μm has been removed from each spectrum.

The weak reduction in reflectance between 0.75 and 0.95 μm is of particular interest, and it is more clearly shown in Figure 3. The continuum of the reflectance spectrum in this spectral region has been removed from both the Mercury and Apollo 16 site spectra by dividing by a straight line fitted to the spectra at 0.745 μm and 1.078 μm.

As originally proposed by McCord and Adams (1972a, b), this very weak and broad absorption is probably due to d-shell electronic transitions in Fe^{2+} ions either in a mafic mineral, probably pyroxene, in glass, or in both. The band center spectrum is located at 0.89 μm, as seen in Figure 4, according to a Gaussian band-fitting calculation. This band position suggests an orthopyroxene as the major mafic mineral affecting the spectrum and tends to rule out glass.

The absorption feature between 0.75 and 0.95 μm in the Mercury spectra is nearly the same strength, but the band center is at slightly shorter wavelengths in comparison to that in the spectrum of lunar highlands Apollo 16 site soil shown in Figures 2 and 3. The absorption band is weaker than for most lunar maria and all lunar fresh crater material. The weak band strength could be due to the amount of Fe^{2+} ions present or to an opaque phase reducing the spectral contrast. The latter case would result in a very low albedo soil, much lower in albedo than for Mercury (Hapke et al., 1975). Thus it was argued (McCord and Clark, 1979) that the average Mercury soil has about the same amount of Fe^{2+} as Apollo 16 site lunar highland soil, which averages about 5.5% FeO (Taylor, 1975; Rose et al., 1975).

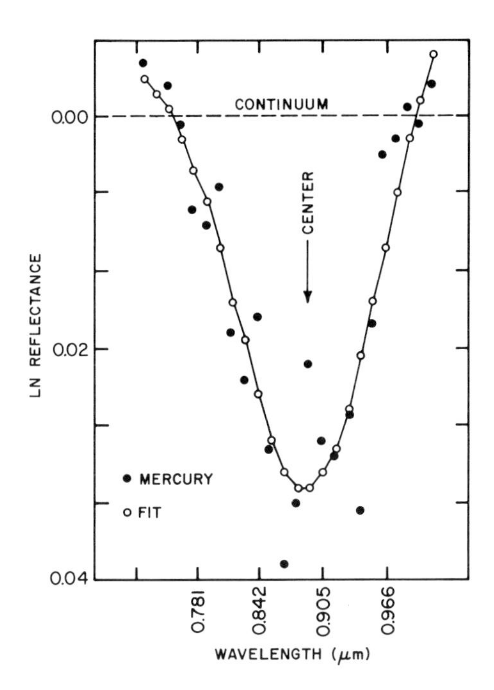

Figure 4. The Mercury reflectance spectrum segment shown in Figure 3 (solid circles) is shown computer-fitted by a single Gaussian curve (open circles). Band depth is 4%, band center is 0.890 μm, and bandwidth is 0.165 μm using the continuum shown.

The presence of Fe^{2+} in the soil of Mercury is important, since it suggests that Mercury surface soil is not reduced completely and all the iron is not in metallic form. In the case of Mercury, it is difficult to attribute a soil rich in oxidized Fe to primary condensation processes, because oxidized iron does not appear in quantity in the projected condensation sequence until temperatures far below those envisioned for condensation of average Mercury material. A more likely explanation is that a Mercury crust rich in iron-bearing pyroxene resulted from familiar processes leading to basaltic surface volcanism.

Hapke et al. (1975) used Mariner 10 images in two spectral bands centered at 0.355 μm and 0.575 μm to map albedos and colors over portions of the Mercury surface. The higher albedos are similar to those for lunar highlands, and the lower albedos are somewhat higher than for lunar maria. The color maps showed differences and boundaries. Apparently fresher material exposed near craters is bluer in color as it is in the lunar case.

The Moon

Spectrophotometry of the lunar surface has been an area of active research since early in this century, when it was realized that the color of a material was related to its composition. Until 1965 the observations were confined to the visible spectral region, where spectral coverage was not sufficient to define electronic absorption features. Consistency of the measurements with volcanic materials was often mentioned, but conclusive evidence was missing. Perhaps most important was considerable evidence that there are surface units of differing composition with sharp boundaries. A review of this early work is given by McCord (1968).

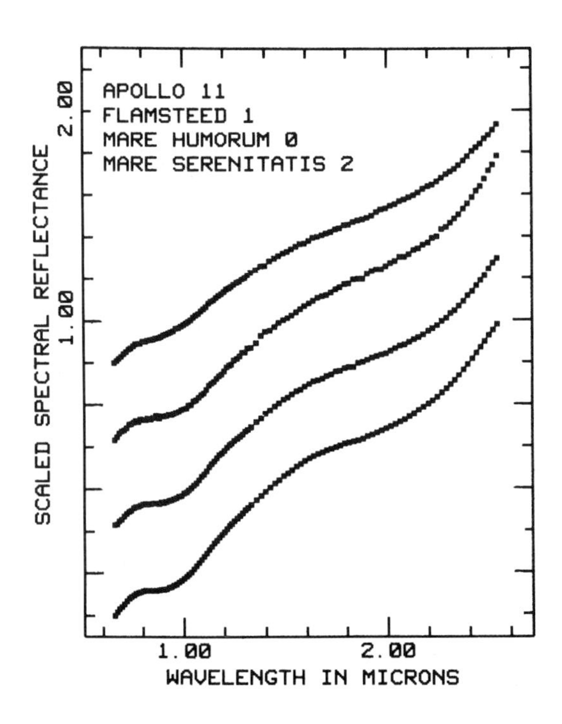

Figure 5. The spectral reflectance for several lunar areas is shown (McCord et al. 1980). Note the variety of spectrum slopes and the different intensities of the Fe^{2+} bands near 1 μm and 2 μm.

Near infrared observations of lunar areas began in the 1960s as the technology developed. Early measurements revealed little compositional information because of insufficient photometric precision (see McCord et al. 1980, Figure 1). Later differential measurements of lunar reflectance for area-pairs in the 0.7 μm to 2.5 μm spectral region showed structure near 1.0 μm indicating the presence of Fe^{2+} electronic transition bands in mafic minerals.

The spectral structure varied from area to area suggesting mineralogic variations. Adams and McCord (1970) compared new telescopic reflectance spectra of lunar areas with laboratory spectra of lunar samples from Apollo 11 to show Fe^{2+} absorption features near 0.95 μm in both spectra. The 0.95 μm absorption was attributed to Fe^{2+} in the mafic mineral pyroxene, a major constituent of basalt. McCord et al., (1972) reported a number of lunar reflectance spectra to 1.1 μm defining the Fe^{2+} absorption feature and showing the variation of it and the spectra continuum with lunar terrain type. Since then a large number of visible and very near infrared reflectance spectra of lunar areas have been collected and used (e.g. Pieters and McCord, 1976; Pieters, 1978; Pieters et al., 1980) to derive information on the composition and structure of the lunar surface. But it was clear that measurements beyond 1.0 μm would provide additional information.

Recently, McCord et al. (1980) obtained reflectance spectra of high photometric precision of small lunar areas between 0.65 μm and 2.5 μm, including through the regions of the terrestrial water bands near 1.4 μm and 1.9 μm (Figure 5). Several electronic absorption features are revealed in the spectra; in particular, bands due to pyroxene and plagioclase are present. Computer analysis of the bands (in these very precise data) provides quantitative mineralogical information. Figure 6 demonstrates the removal of the reflectance continuum and the quantitative specification of the bands.

The quantitative information on absorption bands can be used with laboratory studies of the optical properties of lunar and terrestrial materials to develop quantitative mineralogical information about small areas on the lunar surface (McCord et al., 1980). For example, consider the spectrum for a ten-kilometer area in the crater Aristarchus (Figure 7). The band positions are 0.96 μm and 2.26 μm for pyroxene and 1.28 μm for the mineral plagioclase.

Adams (1974) has explored in the laboratory the optical properties of pyroxene and has shown that the positions of the two bands change in a regular way with the mineral composition (Figure 8). The bands move to longer wavelengths (lower energies) as the iron and calcium content increases and the crystal structure expands. The position of the Aristarchus bands on the plot indicates the presence of a pyroxene of augite composition. Plots of Fe and Ca content versus band positions have been made by Adams (1974) and they can be used to show that the Aristarchus augite has 25 ± 5% Ca/(Ca + Mg + Fe). This composition is consistent with a mare basalt. Since the crater is located in a region of apparently only thin mare covering over underlying terra material, one would expect to sense terra material in the crater. Perhaps the mare is deeper than thought at this place or a unit of mare material became incorporated in the crater so that it composed the floor of the crater. More spectra are being obtained around the crater to map the ejecta blanket composition and work out the structure of the crater and the lunar surface in the crater region.

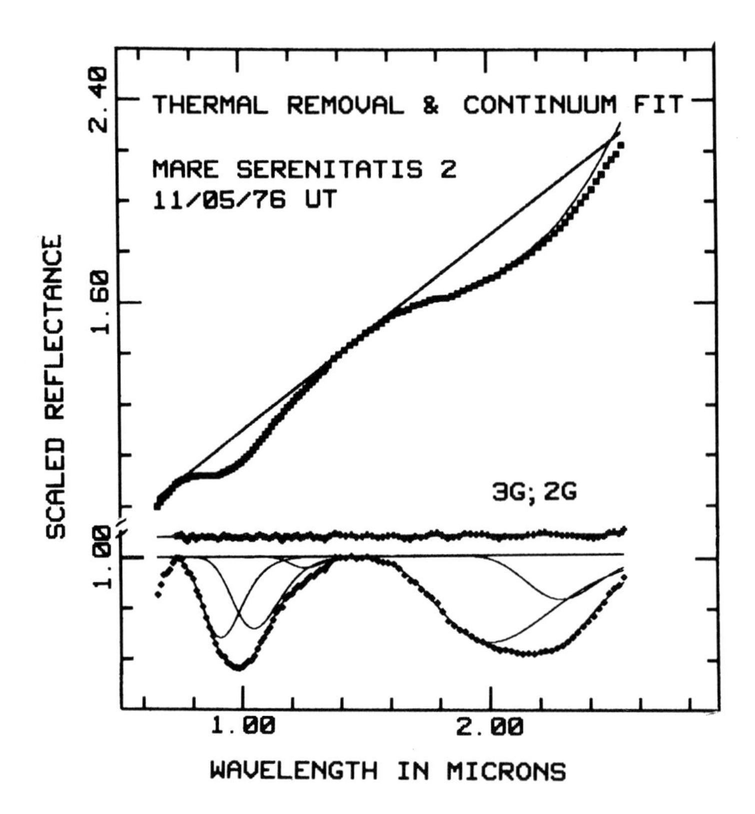

Figure 6. The method of continuum removal and absorption band analysis is illustrated here using the spectrum for an area in Mare Serenitatis (McCord et al. 1980). The line spectrum in the upper plot is the reflectance with the thermal emission component of radiation; the point spectrum has the thermal contamination removed.

From the spectroscopic studies it has been possible to determine relationships between measurements in a few spectral bands and certain compositional properties. These optical properties and thus compositional properties can then be mapped in two dimensions. An example is the map of basalt types developed for the geological study of the

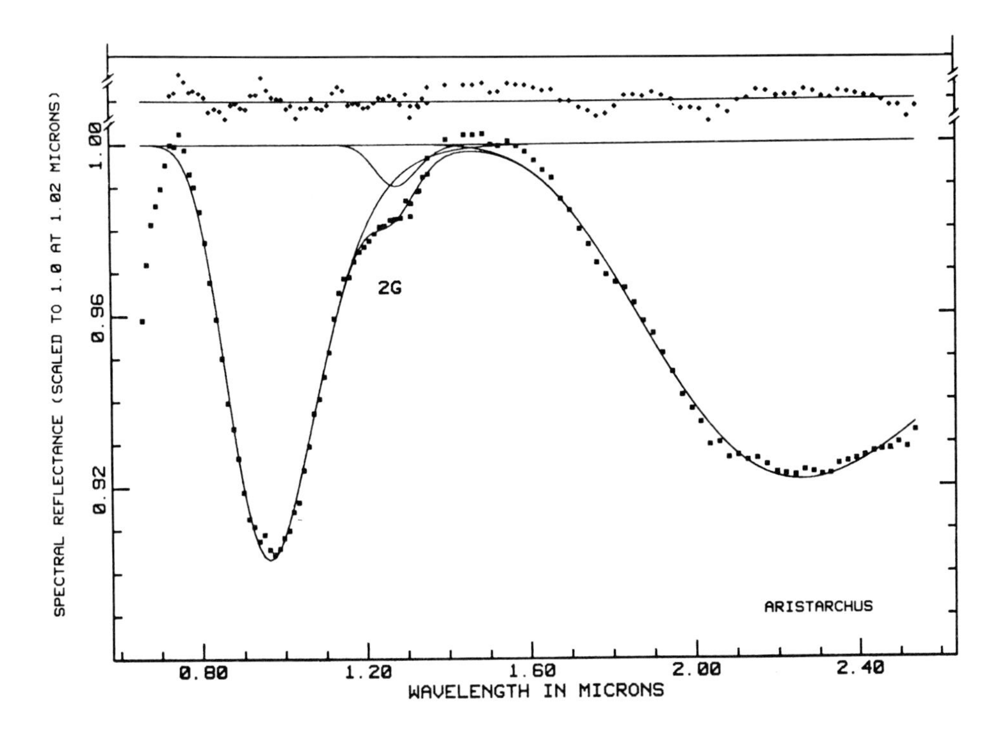

Figure 7. The spectrum for the crater Aristarchus is shown with the continuum removed and three gaussian functions fitted (McCord et al., 1980). The bands at 0.96 μm and 2.26 μm are due to pyroxene and augite; the band at 1.28 μm is due to plagioclase.

Flamsteed region of Oceanus Procellarum (Figure 9) (Pieters et al., 1980). This map was developed using a relationship between titanium content in the lunar surface material and the slope of the reflectance spectrum between 0.40 μm and 0.56 μm (Charette et al., 1974; Pieters, 1978). Titanium content is a good basis on which to distinguish lunar maria basalt types. The map in Figure 9 was produced by obtaining digital photometric images of lunar regions through filters with bandpasses centered at 0.40 μm and 0.56 μm (McCord et al., 1976, 1979). The calibrated images are divided, pixel by pixel, to produce a map of reflectance spectrum slope and thus of titanium. Other regions of the moon have been studied this way (e.g. Johnson et al., 1977a,b; Head et al., 1978) and the number of studies and compositional-optical property relationships is expanding.

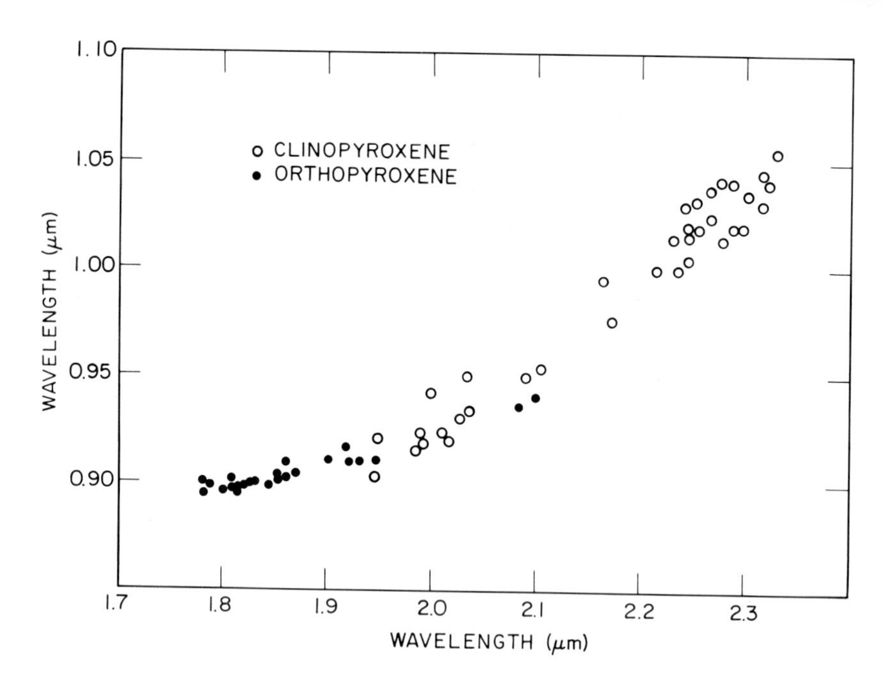

Figure 8. The mineral pyroxene has two strong Fe^{2+} bands in its diffuse reflectance spectrum. This plot shows the shift in wavelength of these two bands as the pyroxene composition changes (Adams, 1974).

Mars

To a visual observer Mars is red with spots of high and low albedo. The distinctive red color led early observers to suggest that iron oxide is a constituent of the surface material. This has been repeatedly confirmed by detailed studies of the Mars spectrum and considerably more compositional information also has been derived. Spectroscopic studies and compositional interpretation of Mars have been reviewed recently (Singer et al., 1979). The reader is refered to that review for a current, detailed discussion.

Visible and near-infrared (0.3 - 2.6 μm) reflectance spectra of the martian surface have been obtained primarily from earth-based telescopic observations, and multispectral images have been obtained both from spacecraft and earth-based observations. Observations in this wavelength region have confirmed the bimodal albedo distribution of surface materials first observed visually. All spectra of Mars are characterized by strong Fe^{3+} absorptions from the near-UV to about 0.75 μm. Darker regions show this effect to a lesser degree, and are interpreted to be less oxidized materials. In addition, dark areas have Fe^{2+} absorptions near 1.0 μm, attributed primarily to olivines and pyroxene. There is evidence at infrared wavelengths for highly

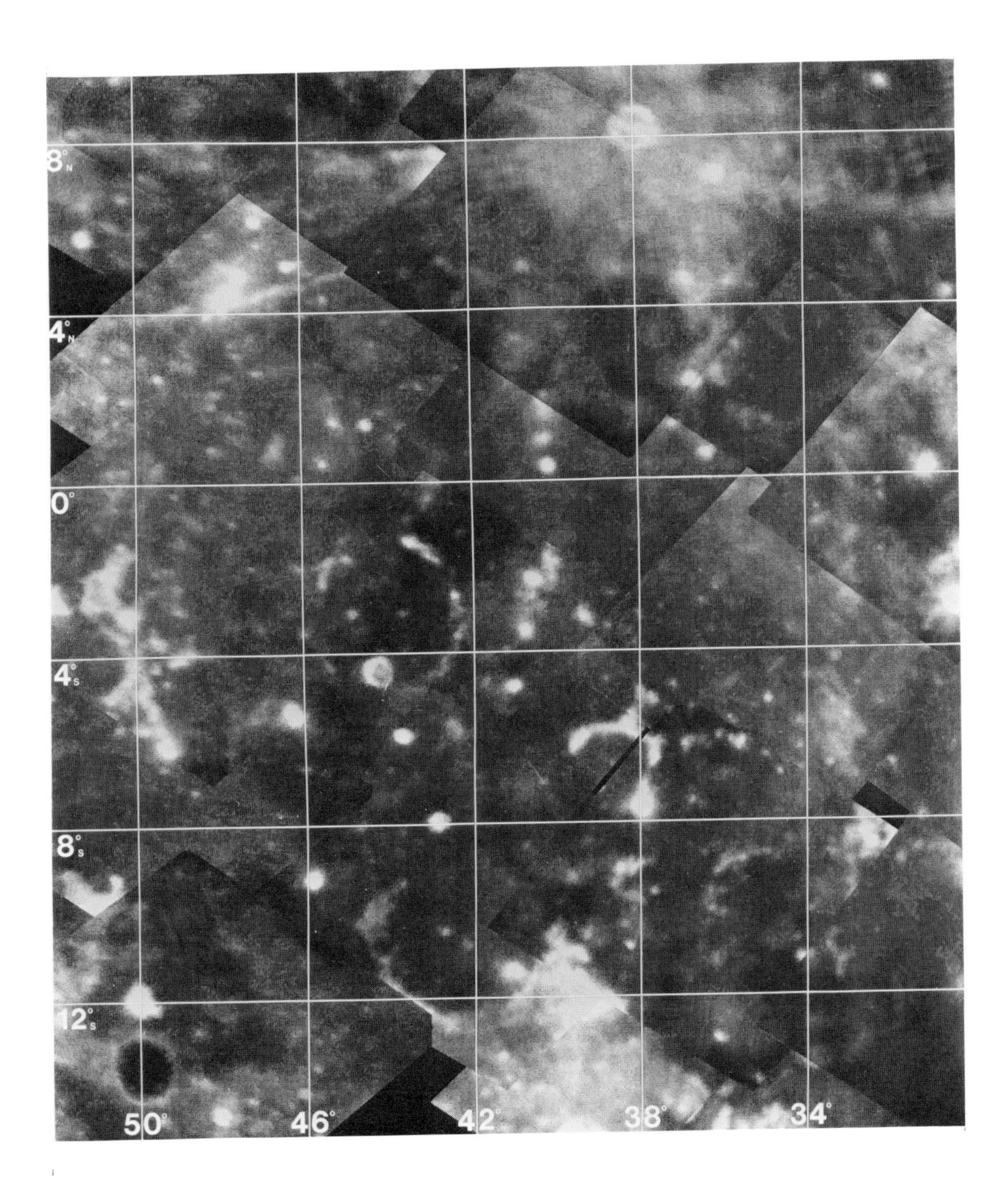

Figure 9a. Multispectral maps of the Flamsteed region of the Moon. These maps are mosaics of digital images acquired using groundbased telescopes. This is an image mosaic of the Flamsteed region made through a 0.56 μm filter; grey tones correspond to albedo. This and Figure 9b from Pieters et al. (1980).

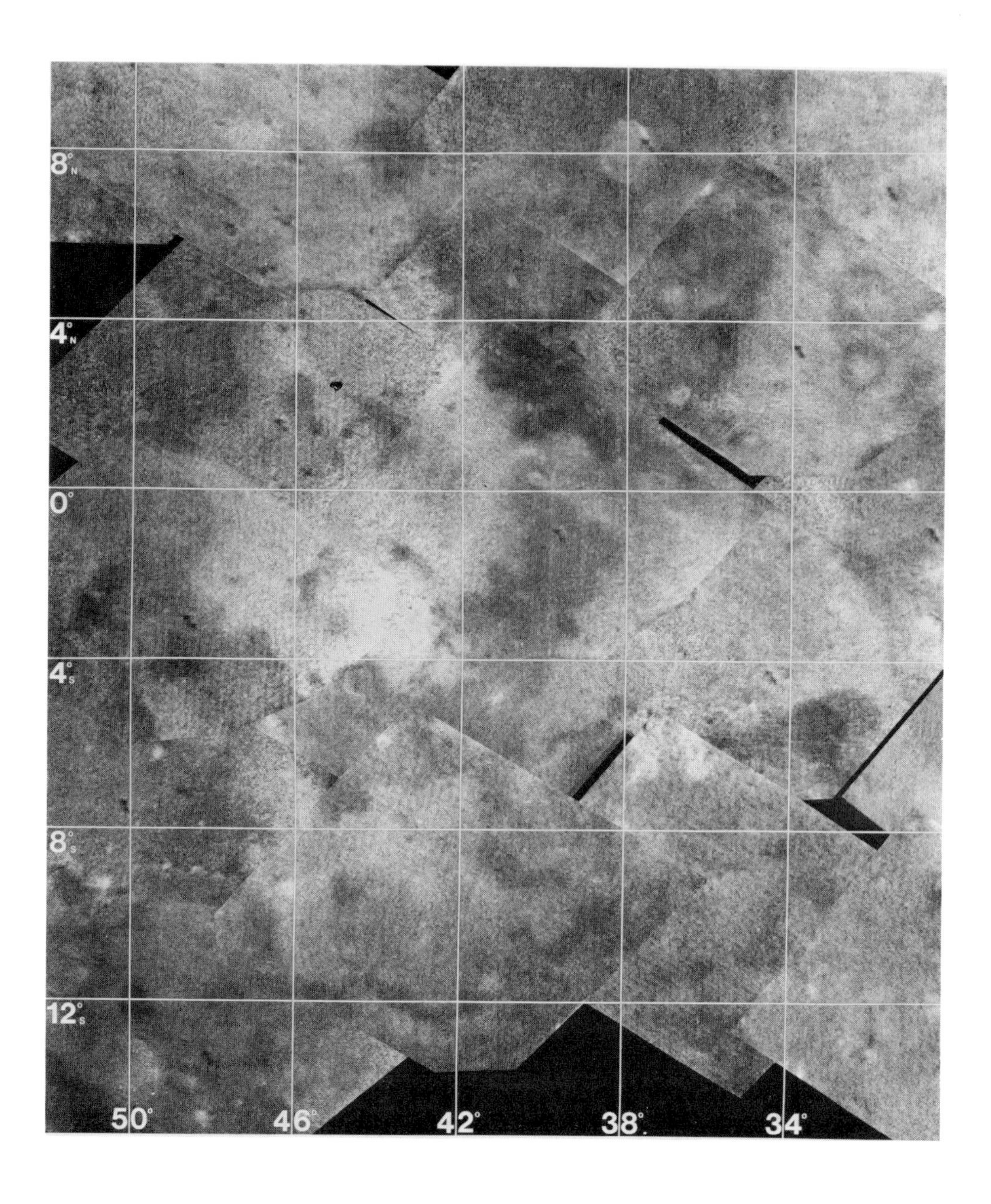

Figure 9b. Grey tone representation of the areal distribution of the reflectance ratio 0.40/0.56 μm. This figure is produced by dividing images made at 0.40 by images made at 0.56 μm and contrast enhancing the result. Bright indicates a high UV/VIS ratio.

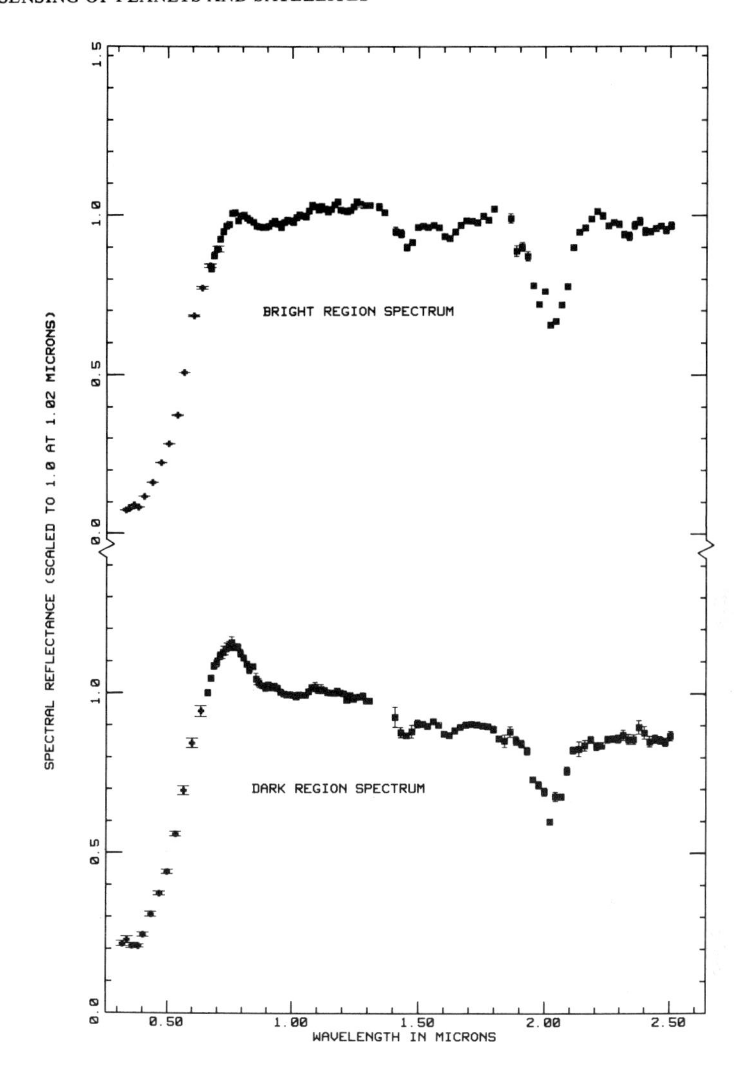

Figure 10. Representative bright and dark region reflectance spectra, scaled to unity at 1.02 μm. The bright region spectrum (top) is composed of an average of the brightest areas observed. The dark region spectrum (bottom) is a composite of data from two nearby locations in Iapygia. This and Figure 11 from Singer et al. (1979).

dessicated mineral hydrates and for H_2O-ice and/or adsorbed H_2O. Observations of the north polar cap show a strong H_2O-ice spectral signature but no spectral evidence for CO_2-ice, while only CO_2-ice has been identified in spectra of the south polar cap. The brightest materials on Mars show greater mineralogic variability and are thought to be closer in petrology and physical location to their parent rock. At present the best model for the dark materials is a somewhat

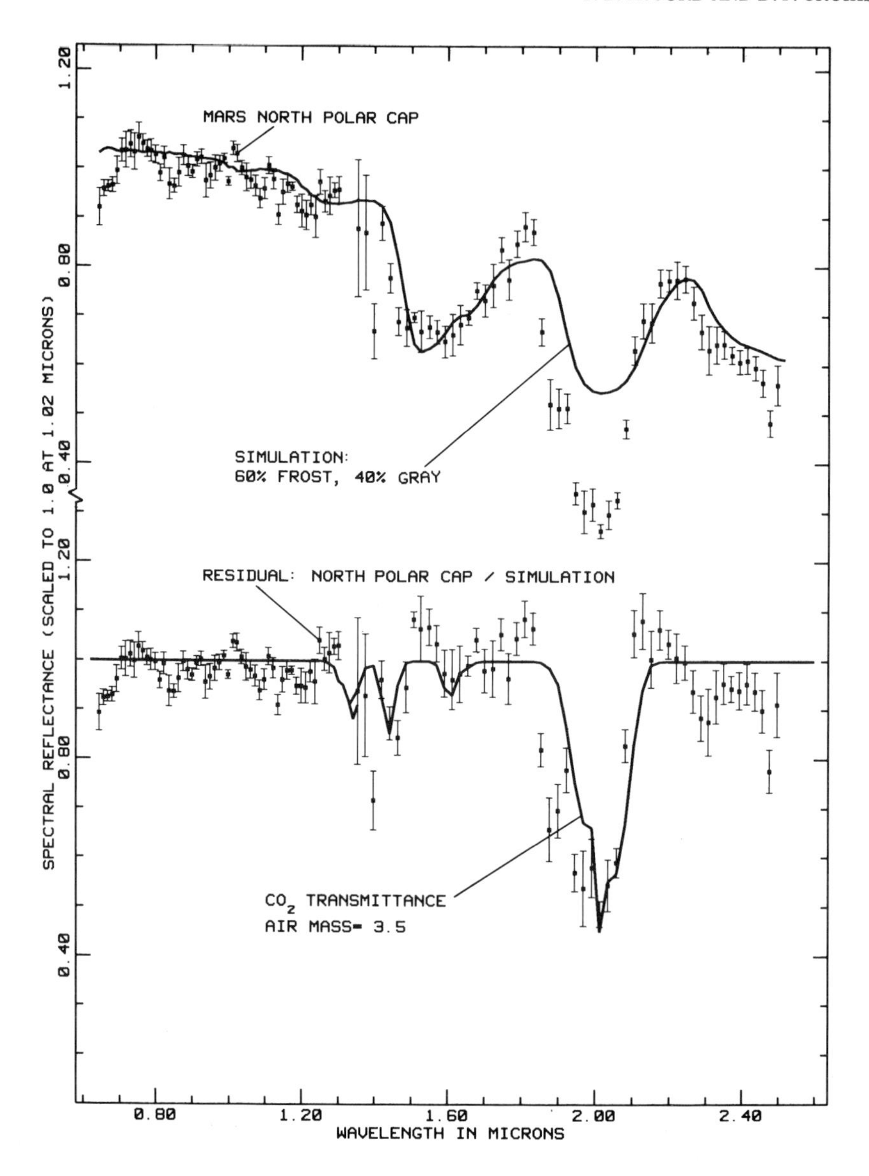

Figure 11. The martian north polar cap spectrum compared with an additive simulation of ice and a grey material (top). The ratio of the polar cap spectrum to the simulation is then compared with the expected CO_2 martian transmittance (bottom) [from McCord et al., 1979].

oxidized basaltic or ultramafic rock, regionally variable in composition and details of oxidation. The bright materials appear to be finer-grained assemblages of primarily highly oxygen-sharing dessicated mineral hydrate, some ferric oxides, and other less major constituents, including a small amount of relatively unaltered mafic material. The bright materials seem likely to be primary and/or secondary alteration products of the basaltic or ultramafic dark materials.

Representative Mars spectra are shown in Figure 10 and 11. Examples of visible spectra are given in Singer et al. (1979).

Reflectance observations from spacecraft consist mainly of multispectral images from Viking orbiters and landers. Soderblom et al. (1978) have prepared three-color photometric maps for a large portion of the planet between latitudes 30° N and 60° S from VO 2 approach images (L_S = 105°). These have good spatial resolution (10-20 km) but limited spectral coverage and resolution (three broad bands: 0.45 ± 0.03 μm, 0.53 ± 0.05 μm, and 0.59 ± 0.05 μm). Soderblom and others are preparing additional multispectral maps using Viking orbital images of selected regions and at higher spatial resolution (L. A. Soderblom, personal communication, 1979). Viking lander cameras are capable of taking images in six spectral bandpasses from 0.4 to 1.0 μm. Huck et al. (1977) developed a technique for transforming these six brightness values into an estimate of spectral reflectance. These data are being used successfully for determining color differences and properties of the surface at the two landing sites (Evans and Adams 1979; Strickland 1979). As with orbital data, repeat coverage is available throughout a martian year, permitting monitoring of variations in surface optical properties (Guinness 1979).

3. DISTANT, UNRESOLVED OBJECTS

The remote sensing of the compositions of solar system objects which are too small or distant to be resolved into a significant number of surface elements has progressed along several lines. Photometry and polarimetry of these objects have given information on their surface microstructures, and virtually all of those bodies not having atmospheres are found to have surface microtextures indicative of a regolith of finely divided material, whether composed of rocky/dusty powder or frozen volatiles (water or methane ice). Spectrophotometric investigations in the photovisual region (0.2 - 1.0 μm) have revealed specific absorptions in the iron-bearing minerals, as already described, and slopes, usually upward toward the red, of varying degrees and indicative of mineral absorptions in the violet and blue spectral region. This work, when extended further to the near infrared (1.0 - ∿5 μm), where the light received from the objects is reflected sunlight, shows additional stronger absorption bands characteristic of various frozen volatiles and minerals. It is in this spectral region where the greatest amount of diagnostic compositional information has been obtained in recent years, both from spectrophotometry, spectroscopy, and filter photometry. In this section we will review the major results of near infrared studies of the outer planetary satellites and Pluto.

Asteroids

Almost all of the available information on the chemistry of the asteroids, as well as that on their physical states, has been developed in the past decade, and this knowledge has been summarized in two books (Gehrels, 1971, 1979). Photoelectric photometry (UBV) has produced colors for over 700 asteroids (cf. Bowell and Lumme, 1979). These colors have been plotted on a B-V, U-B diagram to show clumpings. These clumpings, when used with albedo and spectra, have been used to infer statistical properties of the asteroids. A major advantage of the UBV measurements is that they can be made for fainter asteroids than those for which full spectra can be obtained.

The reflectance of an asteroid surface as a function of wavelength from UV to the IR (where thermal emission becomes important near 5 μm) is the property found to be most directly indicative of surface composition. Therefore a great deal of effort has gone into the acquisition and interpretation of such data during the past decade. Since the first observations of the spectral reflectance of Vesta and the discovery that its surface contained basaltic material (McCord et al., 1970), reflectance spectra between 0.35 μm and about 1.1 μm have been obtained for over 277 asteroids (cf. Chapman and Gaffey, 1979).

Inspection of the representative spectra in Figure 12 shows that a wide variety of spectral features is evident, indicating heterogeneous composition. The mineral assemblages detected on asteroids are generally similar to meteoritic minerals. Mafic silicate minerals (pyroxene and olivines), opaques and metals are the major constituents (cf. Gaffey and McCord, 1978, 1979).

Spectra in the near infrared spectral region to 2.5 μm (reviewed by Larson and Veeder, 1979) of a few asteroids have confirmed the presence of pyroxene and have demonstrated the existence of plagioclase. Spectrophotometry of Ceres in the 3.0 μm region revealed H_2O absorptions suggesting hydrated minerals (Lebofsky, 1978). A few other asteroids appear to show a band of adsorbed H_2O, but these measurements are still preliminary. The continued exploration of the asteroids by spectrophotometry will be a major endeavor for the next decade. As the statistical sample improves, more detailed studies of compositional families and their possible relationship to dynamical families of asteroids can procede. One major goal of this work is the search for parent bodies of the meteorites found on Earth, and another is the search for asteroidal sources of natural materials of potential economic importance (Gaffey and McCord, 1977a,b).

The Galilean Satellites

The surface compositions of the Galilean satellites have been thoroughly reviewed by Sill and Clark (1981) in the context of ground-based observations and the results of Voyager 1 and 2. Before the Voyager discovery of active volcanism on Io, absorption bands at 4.08 μm

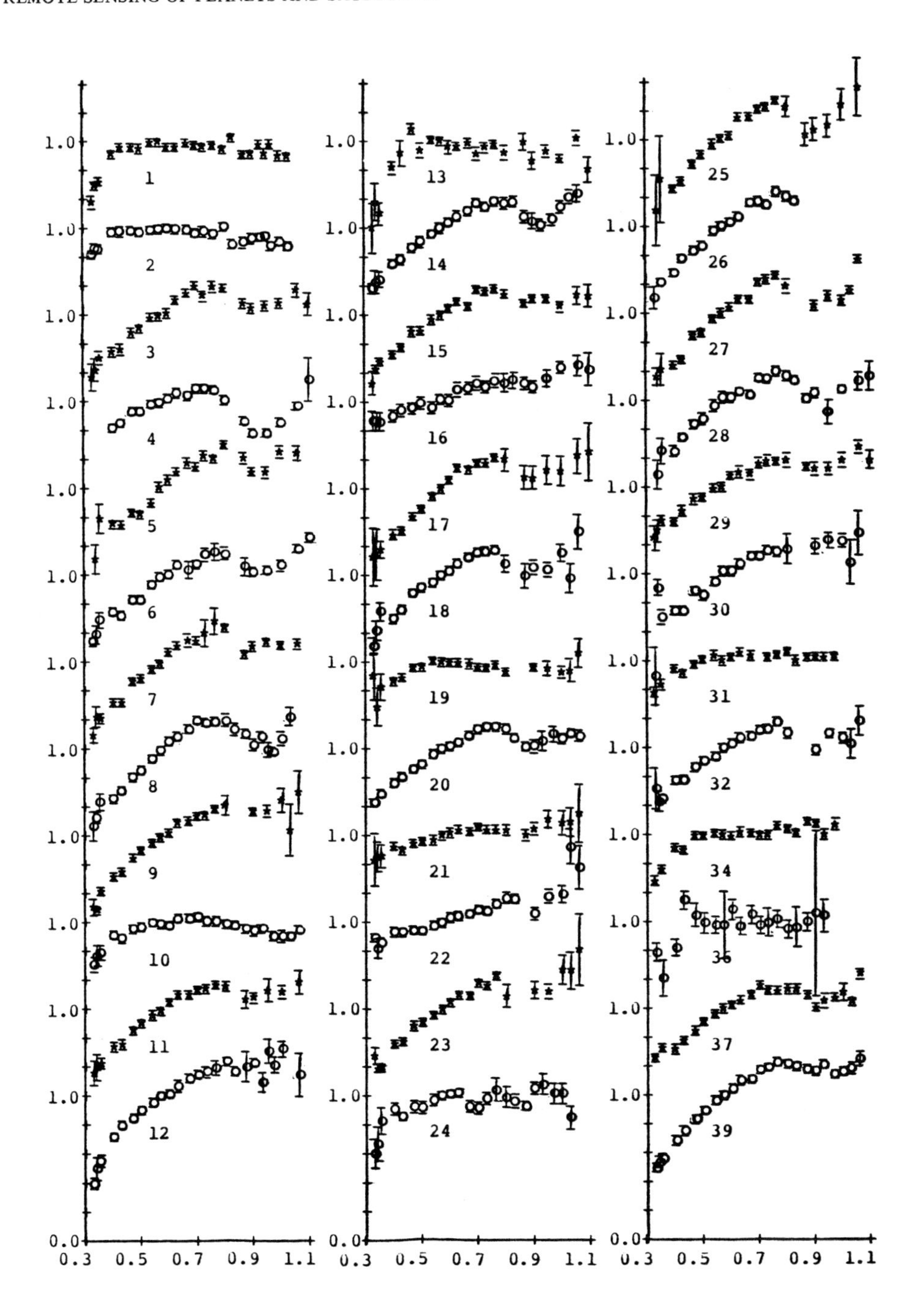

Figure 12. The spectral reflectance for a selection of the brighter asteroids is shown here to illustrate the variety of curve types (Chapman and Gaffey 1979).

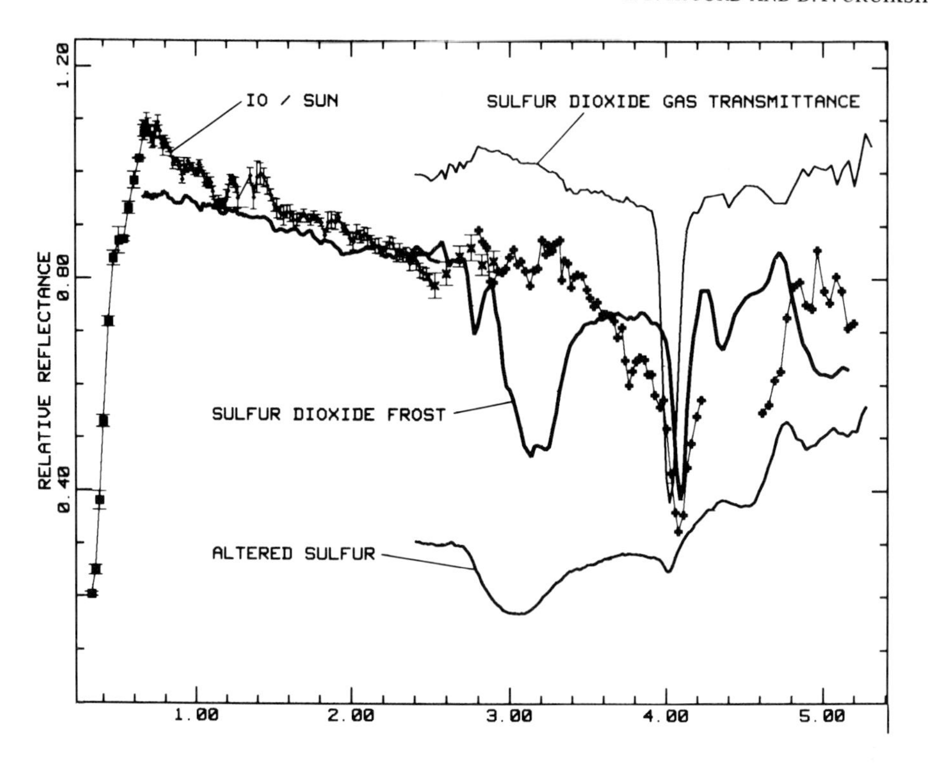

Figure 13. The reflectance spectrum of Io, with laboratory spectra of SO_2 gas and frost and a sample of a sulfur allotrope. The absorption at 4.08 μm in the Io spectrum is attributed to SO_2 frost on the basis of this comparison. From Fanale et al. (1979).

and nearby were found from ground-based spectra (Cruikshank et al. 1978, Fink et al. 1978, Cruikshank 1980a) and identified as SO_2 frost (Fanale et al. 1979). The discovery of active volcanism that quickly followed this identification explained the origin and persistence of the frost in the intense radiation field characteristic of the surface of Io. Figure 13 shows the spectrum of Io with SO_2 gas and frost, plus a solid allotrope of sulfur. Additional studies of Io from Voyager data and from spectra in the short wavelength region of the spectrum have given evidence for extensive surface deposits of numerous temperature-sensitive allotropes of sulfur (Sill and Clark 1981) as well as the SO_2 frost. In addition to frost, gaseous SO_2 was found in localized regions on the satellite with the Voyager spectrometer which revealed an absorption band at 7.35 μm (Pearl et al. 1979).

Infrared reflectance studies of the Galilean satellites began with Kuiper (1957) whose initial results in the 1-2.5 μm region led him to propose that water frost occurs on Europa and Ganymede. Moroz (1965) obtained better data in the same spectral region and reached the same conclusion, but the identification gained acceptance only when telescopic

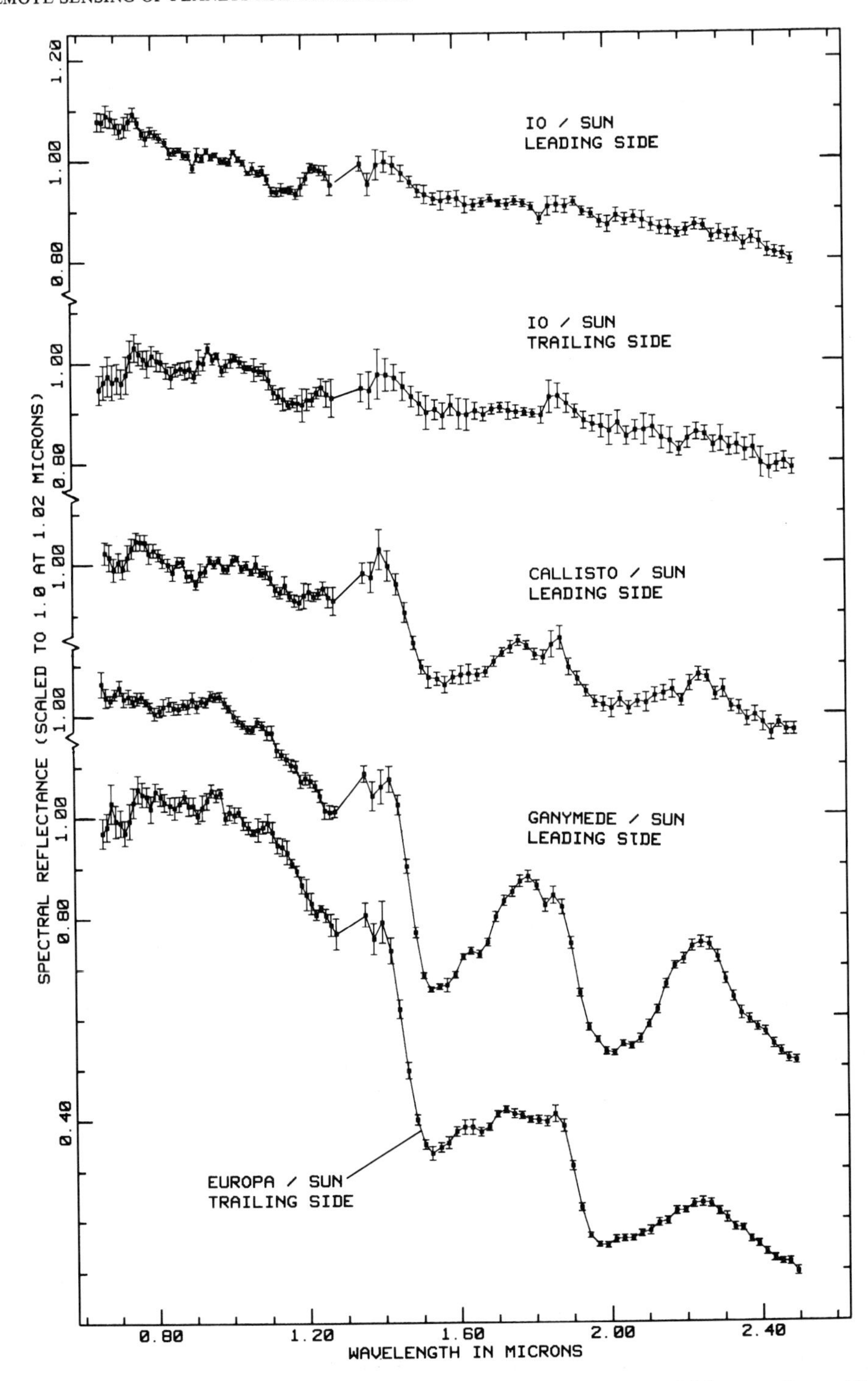

Figure 14. Reflectance spectra of the Galilean satellites, from Clark, and McCord (1980).

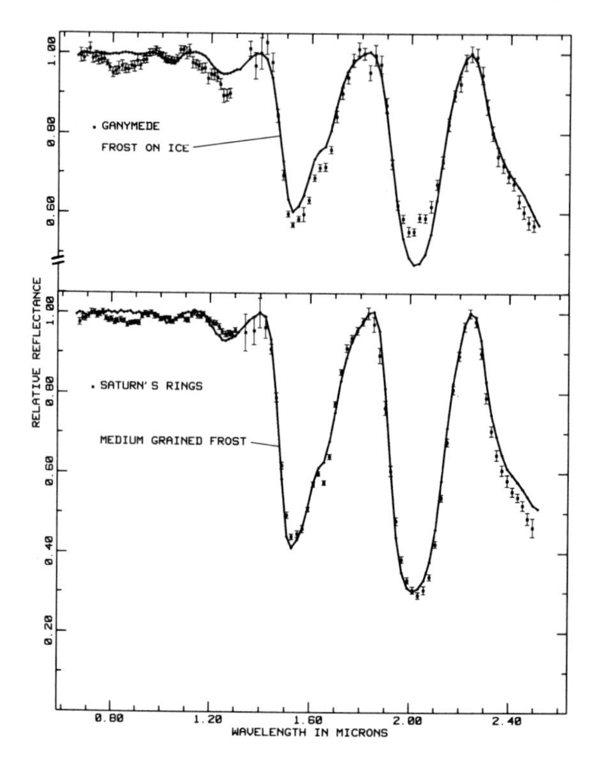

Figure 15. Infrared spectra of Ganymede and Saturn's rings with continua removed for comparison of apparent band depths. From Clark (1980).

data of high quality and better frost data than had previously appeared were presented by Pilcher et al. (1972) and Fink et al. (1973). These early results were of particular importance for Europa and Ganymede, which showed strong absorptions due to water frost in the 1-2.6 μm region, and gave only hints that some frost might be present on Callisto. The work of Clark and McCord (1980) revealed with certainty the weaker frost bands in the spectrum of Callisto, and in his analysis of all three water-frost bearing satellites, Clark (1980) showed that Ganymede's surface is about 90 percent covered with water frost, and that of Callisto about 30 to 60 percent. He notes that "The surface of Europa has a vast frozen water surface with only a few percent impurities." Those impurities, more abundant on Callisto and Ganymede, appear to consist of silicate material having reflectance spectra typical of minerals containing Fe^{3+}, and the material is intimately mixed with the frost. While the interpretation by Clark (1980) is based primarily upon ground-based spectroscopic observations in the near infrared, it is consistent with the morphological structures and distribution of albedo features observed on the Voyager images of the satellites. The best spectrophotometric data for Europa, Ganymede, and Callisto, together with water frost and ice comparisons, are shown in Figures 14, 15, and 16.

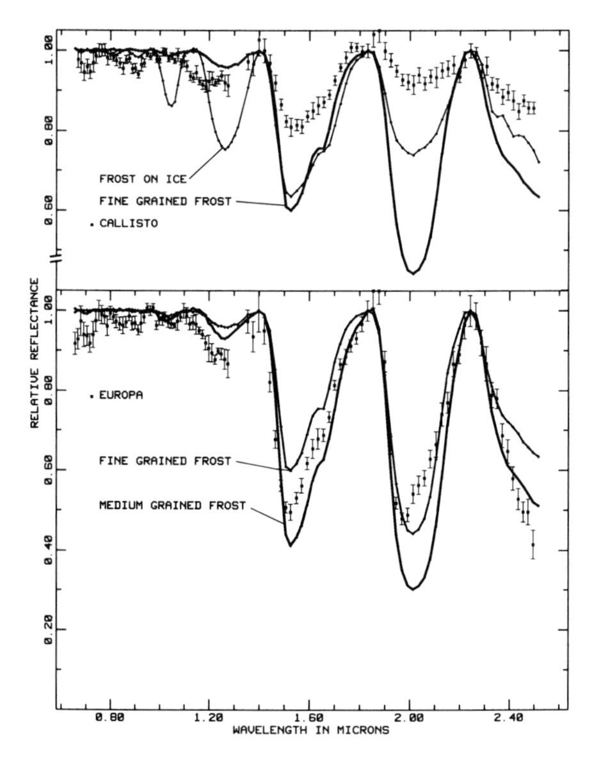

Figure 16. Reflectance spectra of Callisto and Europa with continuua removed for comparison of apparent band depths. From Clark (1980).

The Small Satellites of Jupiter

There are three known satellites interior to Io. J5, Amalthea, was discovered in the last century, but only with the Voyager studies of the Jupiter system has much been learned about it. Two other small satellites, J14 and J15, were discovered with the imaging system of the Voyagers. Rieke (1975) measured thermal radiation from Amalthea, and with his data and photoelectric photometry it was possible to ascertain that the surface geometric albedo of the satellite is on the order of 5 percent, making it a member of the large family of dark objects in the outer solar system. Studies of the Voyager images and photometry by Thomas and Veverka (1981) show that Amalthea is a very irregular object in shape, about 270 x 165 x 150 km, and that its surface apparently has been contaminated by material from Io, particularly sulfur. Its bulk composition cannot, therefore, be determined from studies of its altered surface. There is no compositional information on J14 and J15, nor are there any infrared observations of any kind relevant to these objects.

In the context of the small interior satellites of Jupiter we note that infrared reflectance observations of the ring of Jupiter at 2.2 μm have been reported by Becklin and Wynn-Williams (1979); these data

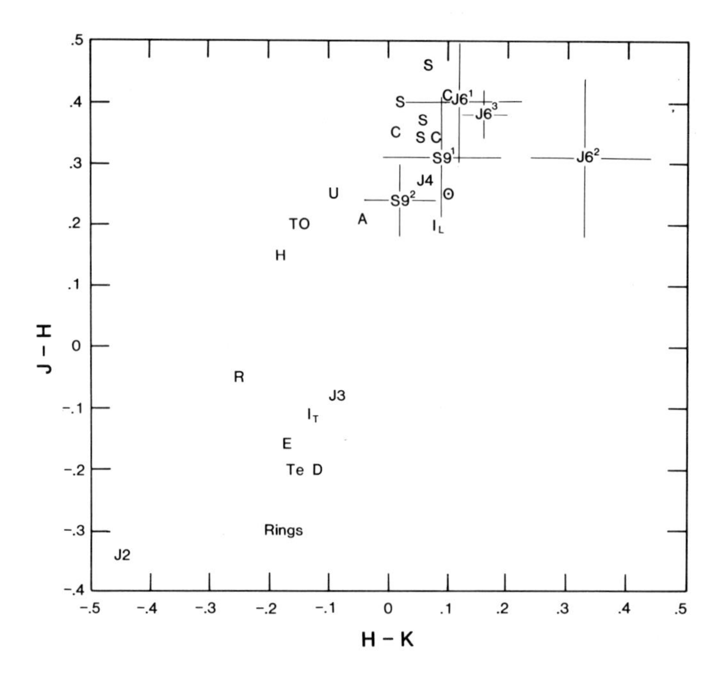

Figure 17. JHK color diagram for small solar system bodies. Key: J2 = Europa, J3 = Ganymede, J4 = Callisto, R = Rhea, E = Enceladus, Te = Tethys, D = Dione, I_T = Iapetus (trailing hemisphere), I_L = Iapetus (leading hemisphere), Rings = Saturn's rings, H = Hyperion, S9 (with superscripts) = independent observations of Phoebe, J6 (with superscripts) = independent observations of Himalia, A = Ariel, U = Umbriel, T = Titania, O = Oberon, C = C-type asteroids, S = S-type asteroids. From Degewij et al. (1980b).

together with crude spectral data in the 2-μm region by Becklin and Neugebauer (in preparation) suggest that the ring particles are of low albedo and that the reflectance is relatively flat in the narrow spectral region studied. Particles of ice or ice-covered dust grains are unlikely because of the high density of energetic particles in the region close to Jupiter; the outer edge of the ring lies at 1.81 R_J.

The satellites exterior to the Galilean satellites, of which eight are presently known with certainty, comprise two dynamical groups. One, at $\sim$164 R_J has orbital inclinations on the order of 27°, and the other at $\sim$322 R_J with i $\sim$ 150°; the orbits of the Galilean and interior satellites are nearly circular with negligible inclination to the planet's equatorial plane. UBVRI photometry of J6 and J7 (in the first group) and J8 (second group) and infrared radiometry of J6 and J7 indicate low geometric albedos and reflectance similar to C-type asteroids (Degewij and van Houten 1979; Cruikshank et al. 1981; Degewij et al.

1980a,b). There is spectrophotometric evidence (Smith et al. 1981) that J9 has a steep red slope in the photovisual spectral region, as do J6, J7, and J8, suggesting that these objects similarly have low albedos with dusty (as opposed to icy) surfaces, but that some differences in the surface mineralogies may occur.

Infrared observations, either of reflected sunlight or intrinsic thermal emission, of the outer satellites of Jupiter are difficult to obtain because of the small size and great distance of these bodies. Probable errors in the data are consequently rather large. In spite of the intrinsic imprecision of infrared broadband color observations, such as JHKL, filter data are useful in determining colorimetric similarities among types of objects in the solar system. Figure 17 shows a JHK color plot of several small solar system objects. Asteroids of all types, C, S, and U, are clustered toward the top of the diagram, while icy satellites of Saturn and Jupiter occupy the lower left-hand region. The Uranian satellites, known to have icy surfaces (see below), cluster in a region between the Saturn satellites and the asteroids. While the data for J6 are crude (two conflicting observations are shown on the diagram, it seems clear that this object is more similar to asteroids than to icy satellites.

The Rings and Satellites of Saturn

The surface compositions of the particles in the main ring system of Saturn and the satellites have been determined from infrared spectrophotometry, spectroscopy, and photometry. The basic references to the discovery of water frost on the ring particles are Kuiper et al. (1970), Pilcher et al. (1970), and Clark (1980). The subject has been reviewed by Pollack (1975) and Cuzzi (1978). The presence of water frost/ice on the inner satellites was inferred by Johnson et al. (1975) and confirmed by Morrison et al. (1976) and Fink et al. (1976) Cruikshank (1979a) has reviewed the nature of the satellite system of Saturn, but some new information about Enceladus (S2), Iapetus (S8), and Phoebe (S9) has been obtained since that review was published.

The picture of the Saturn system of rings and satellites that emerges from the (largely infrared) data collected in the last decade is as follows: The brightest components of the rings show deep infrared absorption bands due to water frost, and laboratory simulations show that the frost is essentially pure with no significant admixture of silicate material (Clark 1980). The extended E ring is probably similarly composed of ice particles, but the observational data have not yet been completely interpreted. The bodies in the inner system of regular satellites (those in circular orbits of low inclination), of which three new members were discovered or confirmed during the 1980 ring-plane passage and the 1979 survey by the Pioneer 11 flyby, appear to be icy objects with surfaces covered completely with water frost or ice. Near-infrared spectroscopic data for S2 Enceladus, S3 Tethys, S4 Dione, and S5 Rhea show water ice absorptions clearly. Titan has a dense methane atmosphere which shows strong infrared absorptions of

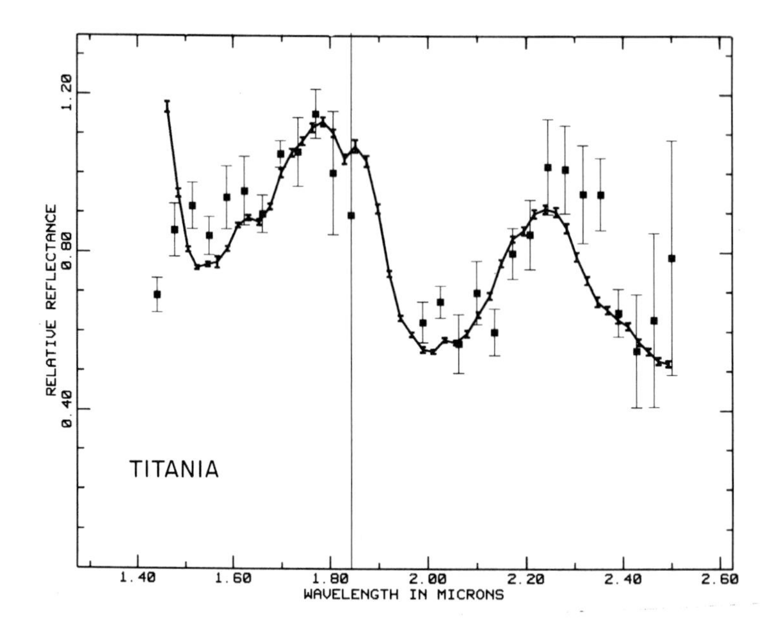

Figure 18. Spectrum of Titania (points with one-sigma error bars) compared with the spectrum of Ganymede (solid line). From Cruikshank (1980b).

variable strength. The observed variability suggests changes in the density of the aerosol haze in the atmosphere on a time scale of several hours. Emissions of methane and ethane are found in the mid-infrared spectrum of Titan, and the composition and mass of the satellite's atmosphere have been widely studied and speculated upon (Caldwell 1978; Hunten 1976,1978). Beyond Titan, three satellites in non-circular and inclined orbits are known. S7 Hyperion appears to be an icy object (Cruikshank 1979b,1980b), while S8 Iapetus is an icy body with its leading hemisphere (in the sense of the satellite's orbital motion around Saturn) covered by dark dust presumably of iron-bearing silicate material (Morrison et al. 1975; Cruikshank 1979a; Cruikshank et al. in preparation). A probable source of the dark material on the leading face of Iapetus is dust spiraling inward toward Saturn from S9 Phoebe, the irregular retrograde satellite exterior to Iapetus (Soter 1974; Cruikshank et al, in preparation; Degewij et al 1980b). Phoebe's surface composition is not yet established with complete certainty, but the best colorimetric evidence in the photovisual region and in the near infrared (shown in Figure 17) suggests that its surface is comparable to that of dark asteroids rather than the icy inner satellites of Saturn. Phoebe is most likely a captured object of asteroidal character.

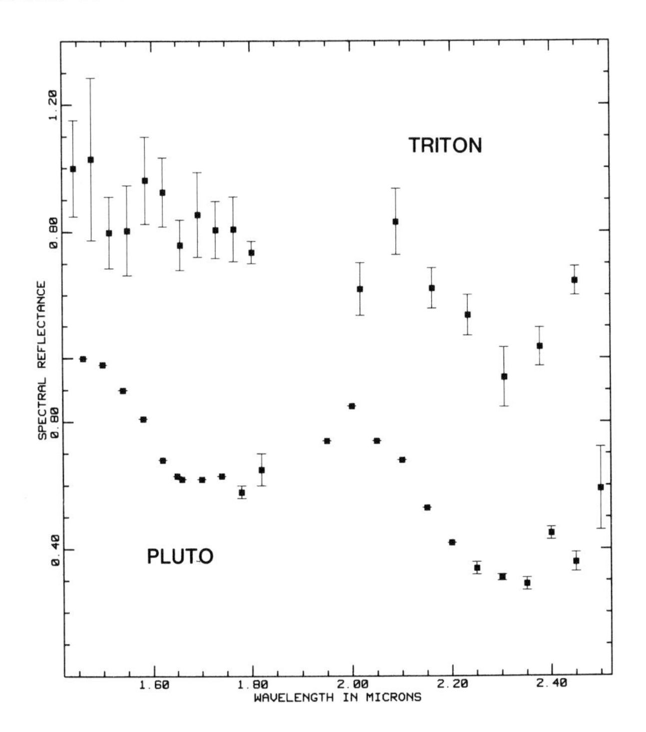

Figure 19. Spectra of Triton and Pluto showing absorptions attributed to methane gas (in the case of Triton) and methane frost (in the case of Pluto). The Triton data are adapted from Cruikshank and Silvaggio (1979) and those for Pluto from Soifer et al. (1980).

The Satellites and Rings of Uranus

Improvements in spectrometer sensitivity have made it possible to study the near-infrared spectra of the Uranian satellites with sufficient signal precision and spectral resolution to reveal the strong absorption bands at 2.0 and 2.4 μm characteristic of water ice/frost (Cruikshank 1980b; Cruikshank and Brown, 1980). JHK photometric observations (Cruikshank 1980b; Nicholson and Jones 1980) are consistent with this interpretation, and are shown in Figure 17. The spectrum of the outermost satellite, U4 Oberon, is shown in Figure 18 in comparison to the spectrum of Jupiter's Ganymede. Ganymede is known to have impurities of iron-bearing silicate material in its surface deposits of water frost, and because the depths of the ice absorption bands on Oberon are comparable to those on the Jovian satellite, Cruikshank (1980b) suggested that Oberon likewise has an impure surface layer of frost. Dynamical studies of the Uranian system (Greenberg 1975, 1976, 1978) together with the new infrared result, suggest that the bulk compositions of the satellites are dominated by water. To date, only the four largest satellites have been studied. The fact that the Uranian satellites lie in a field separate from Ganymede and the icy satellites of Saturn in the JHK color plot (Figure 17) may indicate that the impurities in the frost are of different composition, or that the color of ice at the temperature of the Uranian satellites (~78K) is different from that at the temperature of the Saturn satellites (~110K).

While all four of the outer Uranian satellites show water frost absorption, the strength of the bands is not the same on all objects. In particular, U2 Umbriel has weaker absorption at 2 μm than do the other satellites or Ganymede. Comparison with the spectrum of Callisto shows that the water frost bands are intermediate in strength between this satellite and Ganymede, suggesting that there is a larger fraction of impurities in the surface frost than on U1 Ariel, U3 Titania, and U4 Oberon. Umbriel is the faintest of the four large satellites, and it may be that the larger amount of dark impurities in the ice or frost results in a lower geometric albedo. None of the Uranian satellites can be resolved as disks from the Earth, and inferences as to their dimensions must presently be made on the basis of an assumed geometric albedo. The question of dimensions, mean densities, and masses is reviewed by Cruikshank (1980b).

The innermost known Uranian satellite, Miranda, has not been observed in the infrared because of its faintness (m_V = 16.5) and its proximity to the planet.

Unlike the rings of Saturn, the particles comprising the ring system of Uranus appear to be dark (Sinton 1977; Smith 1977). In the 2-μm spectral region where Uranus is very dark, Matthews et al. (1978) have obtained images of the Uranian ring system, though individual components are not resolved. Nicholson and Jones (1980) observed the reflectance spectrum of the rings in this same region and find marginal evidence for an absorption at 2.1 μm. The distinctive spectral signature of water ice or frost seen in the satellites is apparently not observed in the rings.

Triton and Pluto

Spectroscopy of Triton and Pluto in the photovisual spectral region (0.3 to 1.1 μm) has revealed only hints of absorptions attributable to atmospheres of methane evidenced by a suggestion of an absorption band at 8900-Å (Benner et al. 1978). Until 1976, the infrared region, where most molecular absorptions are stronger than in the photovisual, had been neglected because of the faintness of the two bodies and the insensitivity of infrared detectors. Two-color photometric observations of Pluto (Cruikshank et al. 1976) suggested the presence of frozen methane through the strong absorption band at 1.7 μm compared to 1.5 μm. Later spectrophotometry at 12 wavelengths between 1.44 and 1.84 μm by Cruikshank and Silvaggio (1980) confirmed the presence of an absorption in the region of the strong band of methane, both gaseous and solid, while similar data for Triton in the region 1.44 and 2.52 μm (19 points) showed a moderate absorption at 2.3 μm and very little at 1.7 μm (Cruikshank and Silvaggio 1979).

The presence of methane frost on Pluto was supported by filter photometry obtained by Lebofsky et al. (1979), but a spectrum of high quality in the region 1.46 - 2.46 by Soifer et al. (1980), reproduced in Figure 19, finally confirmed this interpretation by showing the strong frost absorption bands at 1.7 and 2.3 μm.

Cruikshank and Silvaggio (1980) reasoned that gaseous and solid methane cannot be distinguished from one another on Pluto with the existing data because their absorptions overlap. Because the absorptions at 1.7 μm and 2.3 μm are strong on Pluto, the presence of the solid material is assumed, and the presence of gaseous methane is inferred from vapor pressure considerations.

The coverage of solid methane on Pluto is not complete as evidenced by the downward slope toward the violet in the planet's photovisual reflectance spectrum. This behavior is suggestive of exposures or admixtures of silicate material with the ice or frost, as is the case with Europa and Ganymede.

The infrared data shown in Figure 19 for Triton are quite crude, especially in the 1.6-μm region, but show a distinct absorption centered at 2.3 μm attributed to gaseous methane. A new and improved spectrum obtained by Cruikshank (in preparation) in 1980 confirms the 2.3-μm absorption and reveals the 1.7-μm band as well. Cruikshank and Silvaggio (1979) consider the 2.3-μm absorption evidence of gaseous methane because of the relative weakness of the strong 1.7-μm absorption that would be expected for the solid form.

As with Pluto, the photovisual reflectance of Triton slopes down toward the violet, but it may be variable (Bell et al. 1979, Franz 1979). The slope implies the presence of silicate materials (Cruikshank et al. 1979) with no evidence for Rayleigh scattering at the shortest wavelengths.

The 2.3-μm absorption of methane on Triton was synthesized from laboratory data and a random absorption band model developed by Silvaggio (1977) with the result that for pure methane the surface column abundance is $(7 \pm 3) \times 10^2$ cm-agt, corresponding to a surface pressure of about $(1 \pm 0.5) \times 10^{-4}$ bars, a value consistent with the calculated vapor pressure of methane gas above methane ice at a temperature of 57-60 K. Thus, the vapor pressure considerations suggest that some frost is present on Triton, and Golitsyn (1979) has argued that most of the gas should be cold-trapped on the dark part of the satellite or toward the poles.

This work was supported in part by NASA grants NSG 7323, NSG 7312 and NGL 12-001-057.

REFERENCES

Adams, J. B.: 1974, J. Geophys. Res., 79, pp. 4829-4836.
Adams, J. B., and McCord, T. B.: 1970, Proceedings of the Apollo 11 Lunar Science Conference, 3, pp. 1937-1945.
Adams, J. B., and McCord, T. B.: 1972a, Proceedings of the Third Lunar Science Conference, 3, pp. 3021-3034.
Adams, J. B., and McCord, T. B.: 1972b, In Lunar Science III, 1-3, Lunar Science Institute, Houston.
Adams, J. B., and McCord, T. B.: 1973, Proceedings of the Fourth Lunar Science Conference, 1, pp. 163-177.
Becklin, E. E., and Wynn-Williams, C. G.: 1979, Nature, 279, pp. 400-401.
Bell, J. F., Clark, R. N., McCord, T. B., and Cruikshank, D. P.: 1979, Bull. Am. Astron. Soc., 11, p. 570.
Benner, D. C., Fink, U., and Cromwell, R. H.: 1978, Icarus, 36, pp. 82-91.
Bowell, E., and Lumme, K.: 1979, in T. Gehrels (ed.), "Asteroids", Univ. of Arizona Press, Tucson, Arizona, pp. 132-169.
Caldwell, J.: 1978, in D. M. Hunten and D. Morrison (eds.), "The Saturn system," NASA CP 2068, pp. 113-126.
Chapman, C. R., and Gaffey, M. J.: 1979, in T. Gehrels (ed.), "Asteroids", Univ. of Arizona Press, Tucson, Arizona, pp. 655-687.
Charette, M. P., McCord, T. B., Pieters, C., and Adams, J. B.: 1974, J. Geophys. Res., 79, pp. 1605-1613.
Clark, R. N.: 1980, Icarus (in press).
Clark, R. N., and McCord, T.B.: 1980, Icarus, 41, pp. 323-339.
Cruikshank, D. P.: 1979a, Rev. Geophys. Space Phys. 17, pp. 165-176.
Cruikshank, D. P.: 1979b, Icarus 37, pp. 307-309.
Cruikshank, D. P.: 1980a, Icarus 41, pp. 240-245.
Cruikshank, D. P.: 1980b, Icarus 41, pp. 246-258.
Cruikshank, D. P., and Brown, R. H.: 1980, Bull. Am. Astron. Soc. (in press).
Cruikshank, D. P., Degewij, J., and Zellner, B.: 1981, in D. Morrison (ed.), "The Satellites of Jupiter", Univ. of Arizona Press, Tucson, Arizona (in press).
Cruikshank, D. P., Jones, T. J., and Pilcher, C. B.: 1978, Astrophys. J., 225, pp. L89-L92.
Cruikshank, D. P., Pilcher, C. B., and Morrison, D.: 1976, Science, 194, pp. 835-837.
Cruikshank, D. P., and Silvaggio, P. M.: 1979, Astrophys. J., 233, pp. 1016-1020.
Cruikshank, D. P., and Silvaggio, P. M.: 1980, Icarus, 41, pp. 96-102.
Cruikshank, D. P., Stockton, A., Dyck, H. M., Becklin, E. E., and Macy, W., Jr.: 1979, Icarus, 40, pp. 104-114.
Cuzzi, J.: 1978, in D. M. Hunten and D. Morrison (eds.), "The Saturn System", NASA CP 2068, pp. 73-104.
Degewij, J., Andersson, L., and Zellner, B.: 1980a, Icarus (in press).
Degewij, J., Cruikshank, D. P. and Hartmann, W. K.: 1980b, Icarus, (in press).

Degewij, J., and van Houten, C. J.: 1979, In T. Gehrels (ed.), "Asteroids", Univ. of Arizona Press, Tucson, Arizona, pp. 417-435.
Evans, D. L., and Adams, J. B.: 1979, Proc. Lunar Planet. Sci. Conf. 10th.
Fanale, F. F., Brown, R. H., Cruikshank, D. P., and Clark R. N.: 1979, Nature, 280, pp. 761-763.
Fink, U., Dekkers, N. H., and Larson, H. P.: 1973; Astrophys. J., 179, pp. L155-L159.
Fink, U., Larson, H. P., Gautier, N., and Treffers, R.: 1976, Astrophys. J., 207, pp. L63-L67.
Fink, U., Larson, H. P., Lebofsky, L. A., Feierberg, M., and Smith, H.: 1978, Bull. Am. Astron. Soc., 10, 580.
Franz, Otto: 1979, private communication.
Gaffey, M. J., and McCord, T. B.: 1977a, Tech. Rev., pp. 50-59.
Gaffey, M. J., and McCord, T. B.: 1977b, Mercury, 6, pp. 1-9.
Gaffey, M. J., and McCord, T. B.: 1978, Space Sci. Rev., 21, pp. 555-628.
Gaffey, M. J., and McCord, T. B.: 1979, in T. Gehrels, (ed.), "Asteroids", pp. 668-723.
Gehrels, T., ed.: 1971, "Physical Studies of Minor Planets", NASA SP-267, Washington, DC, US Government Printing Office.
Gehrels, T. (ed.): 1979, "Asteroids", Univ. of Arizona Press, Tucson, Arizona.
Golitsyn, G. S.: 1979, private communication.
Greenberg, R.: 1975, Icarus, 24, pp. 325-332.
Greenberg, R.: 1976, Icarus, 29, pp. 427-433.
Greenberg, R.: 1978, Bull. Am. Astron. Soc., 10, p. 585.
Guinness, E. A., Arvidson, R. E., Gehret, D. C., and Bolef, L. K.: 1979, J. Geophys. Res., 84, pp. 8355-8364.
Hapke, B., Danielson, G. E., Jr., Klaasen, K., and Wilson, L.: 1975, J. Geophys. Res., 80, pp. 2431-2443.
Head, J. W., Adams, J. B., McCord, T. B., Pieters, C., and Zisk, S.: 1978, Proc. of the Conference on Luna 24 entitled "Mare Crisium: The View from Luna 24:, in R. B. Merrill and J. J. Papike (eds.), Pergamon Press, New York, pp. 43-74.
Huck, J. O., Jobson, D. J., Park, S. K., Wall, S. D., Arvidson, R. E., Patterson, W. R., and Benton, W. D.: 1977, J. Geophys. Res., 82, pp. 4401-4411.
Hunten, D. M.: 1976, in J. A. Burns (ed.), "Planetary Satellites", Univ. of Arizona Press, Tucson, Arizona, pp. 420-437.
Hunten, D. M.: 1978, in D. M. Hunten and D. Morrison (eds.), "The Saturn System", NASA CP 2069, pp. 127-140.
Johnson, T. V., Mosher, J. A., and Matson, D. L.: 1977a, Proc. Lunar Sci. Conf. 8th, pp. 1013-1028.
Johnson, T. V., Saunders, R. S., Matson, D. L., and Mosher, J. A.: 1977b, Proc. Lunar Sci. Conf. 8th, pp. 1029-1036.
Johnson, T. V., Veeder, G., and Matson, D. L.: 1975, Icarus, 24, pp. 428-432.
Kuiper, G. P.: 1957, Astron. J., 62, p. 245.
Kuiper, G. P., Cruikshank, D. P., and Fink, U.: 1970, Sky Telesc., 39, pp. 14 and 80.

Larson, H. P., and Veeder, G. J.: 1979, in T. Gehrels, (ed.), "Asteroids", Univ. of Arizona Press, Tucson, Arizona, pp. 724-744.
Lebofsky, L. A.: 1978, Mon. Not. Roy. Astron. Soc., 182, pp. 17P-21P.
Lebofsky, L. A., Rieke, G. H. and Lebofsky, M. J.: 1979, Icarus, 37, pp. 554-558.
Matthews, K., Neugebauer, G., and Nicholson, P.: 1978, Bull. Am. Astron. Soc., 10, p. 580.
McCord, T. B.: 1968, California Institute of Technology, Pasadena, California (Ph.D. Thesis).
McCord, T. B., and Adams, J. B.: 1972a, Science, 178, pp. 746-747.
McCord, T. B., and Adams, J. B.: 1972b, Icarus, 17, pp. 585-588.
McCord, T. B., Adams, J. B., and Johnson, T. V.: 1970, Science, pp.1445-1447.
McCord, T. B., Charette, M. P., Johnson, T. V., Lebofsky, L. A., and Pieters, C.: 1972, J. Geophys. Res., 77, pp. 1349-1359.
McCord, T. B., and Clark, R. N.: 1979, J. Geophys. Res., 84, pp. 7664-7668.
McCord, T. B., and Clark, R. N., Hawke, B. R., McFadden, L. A., Owensby, P. D., Pieters, C. M., Adams, J. B.: 1980, J. Geophys. Res. (submitted).
McCord, T. B., Clark, R. N., and Huguenin, R. L.: 1978, J. Geophys. Res., 83, pp. 5433-5441.
McCord, T. B., Grabow, M., Feierberg, M. A., MacLaskey, D., and Pieters, C.: 1979, Icarus, 37, pp. 1-28.
McCord, T. B., Pieters, C., and Feierberg, M. A.: 1976, Icarus, 29, pp. 1-34.
McCord, T. B., Singer, R. B., Clark, R. N.: 1980, J. Geophys. Res., submitted.
Moroz, V. I.: 1965, Astron. Zh. (in Russian), 42, pp. 1287-1295.
Morrison, D., Cruikshank, D. P., Pilcher, C. B., and Rieke, G. H.: 1976, Astrophys. J., 207, pp. L213-L216.
Morrison, D., Jones, T. J., Cruikshank, D. P., and Murphy, R. E.: 1975, Icarus, 24, pp. 157-171.
Nicholson, P. D., and Jones, T. J.: 1980, Icarus, 42, pp. 54-67.
Pearl, J., Hanel, R., Kunde, V., Maguire, W., Fox, K., Gupta, S., Ponnamperuma, C., and Raulin, F.: 1979, Nature, 280, pp. 755-758.
Pilcher, C. B., Ridgway, S. T., and McCord, T. B.: 1972, Science, 178, pp. 1087-1089.
Pilcher, C. B., Chapman, C. R., Lebofsky, L. A., and Kieffer, H. H.: 1970, Science, 167, pp. 1372-1373.
Pieters, C. M.: 1978, Proc. Lunar and Plan. Sci. Conf. 9th, pp. 2825-2849.
Pieters, C. and McCord, T. B.: 1976, Proc. Lunar Sci. Conf. 7th, pp. 2677-2690.
Pieters, C. M., Head, J. W., Adams, J. B., McCord, T. B., Zisk, S. H., and Whitford-Stark, J.: 1980, J. Geophys. Res., 85, pp. 3913-3938.
Pollack, J. B.: 1975, Space Sci. Rev., 18, pp. 3-94.
Rieke, G. H.: 1975, Icarus, 25, pp. 333-334.
Rose, H. J., Jr., Baedecker, P. A., Berman, S., Christian, R. P., Dwornik, E. J., Finkelman, R. B., and Schnepfe, M. M.: 1975, Proc. Lunar Sci. Conf, 6th, 2, pp. 1362-1373.

Sill, G. T., and Clark, R. N.: 1981, in D. Morrison (ed.), "The Satellites of Jupiter", Univ. of Arizona Press, Tucson (in press).
Silvaggio, P. M.: 1977, Cornell Univ. Ithaca, New York (Ph.D Thesis).
Singer, R. B., McCord, T. B., Clark, R. N., Adams, J. B., and Huguenin, R. L.: 1979, J. Geophys. Res., 84, pp. 8415-8426.
Sinton, W. M.: 1977, Science, 198, pp. 503-504.
Smith, B. A.: 1977, Nature, 268, p. 32.
Smith, D. W., Johnson, P. E., and Shorthill, R. W.: 1981, Icarus, (submitted).
Soderblom, L. A., Edwards, K., Eliason, E. M., Sanchez, E. M., and Charette, M. P.: 1978, Icarus, 34, pp. 446-464.
Soifer, B. T., Neugebauer, G., and Matthews, K.: 1980, Astron. J., 85, pp. 166-167.
Soter, S.: 1974, paper presented at IAU Coloquium 28, Ithaca, New York.
Strickland, E. L., III.: 1979, in Lunar and Plan. Sci. X, Lunar and Planetary Institute, Houston, Texas, pp. 1192-1194.
Taylor, S. R.: 1975, "Lunar Science: A Post-Apollo View", Pergamon, New York.
Tepper, L., and Hapke, B.: 1977, Bull. Am. Astron. Soc., 9, p. 532.
Thomas, P., and Veverka, J.: 1981, in D. Morrison (ed.), "The Satellites of Jupiter", University of Arizona Press, Tucson, Arizona, (in press).
Vilas, F., and McCord, T. B.: 1976, Icarus, 28, pp. 593-599.

DISCUSSION FOLLOWING PAPER DELIVERED BY T. B. McCORD

ALLEN: The only asteroid you mentioned was Vesta which, I believe, has an unusual surface. Can you mention what you find on other asteroids, if they've been observed?

McCORD: We find a wide variety of surface compositions among the asteroids. A few percent of the total sample have surfaces composed of metals, but many are dominated by olivine and pyroxene. A large fraction of the objects appear similar to carbonaceous chondritic material. There are many subtle subdivisions in the compositional classification of asteroids which have been extensively reviewed in the literature, for example by Gaffey and McCord in the book Asteroids edited by T. Gehrels, 1979, University of Arizona Press.

THERMAL STUDIES OF PLANETARY SURFACES

David Morrison
University of Hawaii
Honolulu, Hawaii 96822

ABSTRACT

Ground-based and spacecraft observations of planets, satellites, and asteroids in the thermal infrared have provided a wealth of information on planetary temperatures, dimensions, and surface properties. Internal heat sources have been revealed for Jupiter, Saturn, and Neptune, and active volcanism on Io has been discovered and monitored. The thermophysical properties of Mars have been mapped for nearly all the surface by spacecraft, and ground-based observations have given similar information for the Galilean satellites of Jupiter. Infrared radiometry thus sheds important light on significant problems of dynamics, interiors, and surfaces of solar system bodies.

I. INTRODUCTION

The brightest infrared sources in the sky are in the solar system, and studies of this radiation have played an important role in infrared astronomy since the pioneering thermocouple measurements of the 1920s by Coblentz and Lampland at Lowell and Pettit and Nicholson at Mt. Wilson. Indeed, the history can be traced back even further, to the detection of lunar thermal radiation in 1869 by the Earl of Rosse with his 3-foot reflector. The temperature measurements from the premodern era are reviewed by Pettit (1961) and Sinton (1961); the primary results of interest, beyond the simple measurement of planetary temperatures, were the discovery that there was no diurnal infrared temperature variation on Venus, the determination from eclipse cooling and heating curves that the lunar surface has an exceedingly low thermal inertia, and the measurements of the diurnal temperature variations on Mars.

The rapid improvements of the 1960s in infrared instrumentation led to the birth of non-solar-system infrared astronomy. Initially, the bright planetary sources served as calibration standards for other objects, and they are still used for this purpose at the longer

C. G. Wynn-Williams and D. P. Cruikshank (eds.), Infrared Astronomy, 89–106.

wavelengths. However, no truly satisfactory standard has been found, since the planets are generally variable, often have pronounced spectral structure, and have such large angular sizes that they overfill the beam of most photometers. Mars has been used frequently, but it is perhaps the least appropriate calibrator, since its thermal emission can vary substantially with weather as well as in ways more readily modeled. Potentially the most useful solar system standards are the larger asteroids, such as Ceres, which should behave in a predictable way and are easily measured by existing systems at wavelengths from 8 μm to beyond 100 μm. However, today the calibration situation is usually reversed, with hot stars providing the primary reference, and the planets being studied for their own sakes.

Ground-based and spacecraft thermal observations, primarily broad-band photometry, continue to occupy a central place in planetary investigations. Among the highlights of the past decade have been the measurements of the internal energy sources of Jupiter, Io, Saturn, and Neptune, the discovery of temperature inversions in the upper atmospheres of the Jovian planets and the analysis of hydrocarbon trace chemistry; the use of thermal spectra to sound the temperature-pressure structure of planetary atmospheres; measurement of the composition of planetary atmospheres, including the fundamental datum of the hydrogen-to-helium abundance ratio; the identification of the composition of the Martian polar caps; the measurement of thermal inertia for numerous planets and satellites and the systematic mapping of inertia over the surfaces of the Moon and Mars; and the measurement of the sizes and albedos of more than a hundred asteroids and nearly a dozen satellites too small to be resolved. In this chapter I discuss four topics: the energy balance of the Jovian planets; the thermal inertias of surfaces; asteroid diameters and albedos, and the volcanic activity of Io. The composition and structure of planetary atmospheres are discussed elsewhere in this volume by Orton and by Encrenaz and Combes.

II. INTERNAL HEAT SOURCES OF THE JOVIAN PLANETS

One of the most exciting and unexpected early discoveries of the modern era of infrared astronomy was the existence of large internal heat sources in Jupiter and Saturn. A Jovian excess was first suggested by Low (1966), based on the high brightness temperatures measured at 10 and 20 μm, and this excess was confirmed for Jupiter and extended to Saturn through Lear Jet observations extending out to nearly 100 μm (Aumann et al. 1969). These measurements demonstrated that Jupiter radiated more energy than it received from the Sun, but they could not establish the magnitude of the internal heat source since it was not possible to measure the radiation from the night side or from the polar regions. Spacecraft observations from a variety of directions are required to establish a model-independent value for the heat source. In the case of Saturn, the situation has been further complicated by the presence of thermal emission from the rings.

The first spacecraft measurements of thermal radiation from Jupiter were made from Pioneer 10 in 1973 and Pioneer 11 in 1974. Pioneer 10 made the first observations of the night side of the planet, and in addition Pioneer 11 was able to contribute data on the polar temperatures. The descriptions of the individual encounters are given by Chase et al. (1974) and Ingersoll et al. (1975a), with a summary of the results of both missions by Ingersoll et al. (1975b). Although temperature differences were measured between bands and zones, there was no significant cooling toward the poles or at high phase angles, leading to a derived global effective temperature of 125 $\pm$ 3 K, marginally lower than the Earth-based values. Independently calibrated infrared observations from the Voyager Jupiter encounters in 1979 are consistent with the Pioneer results. The source of the internal energy of Jupiter is apparently primarily primordial, representing the slow leakage of heat from a still hot interior. Models of the early evolution of the planet by Bodenheimer (1974), Graboske et al. (1975), and others indicate that the present heat source of about 4×10^{17} watts is easily fitted by standard evolutionary models for the planet.

In the case of Saturn, all of the early infrared observations were made at a time when the rings were widely open and emitting with a brightness temperature of 90-95 K (Morrison 1974a). In large aperture photometry, the rings contributed as much radiation as the planet itself. During the late 1970s the rings closed, lowering both their temperature and their solid angle, while infrared models of the rings (e.g., Cuzzi 1978) permitted corrections to be made for their contribution. Thus Stier et al. (1978) derived an effective temperature for the planet of about 90 K, Erickson et al. (1978) observed 97 K, and Courtin et al. (1979) obtained 95 K, all suggestive of a heat source somewhat smaller than that of Jupiter. In 1979, the Pioneer 11 flyby of Saturn provided an opportunity for more comprehensive measurements, made with the same radiometer used five years earlier to observe Jupiter. The effective temperature found by the Pioneer investigators was 97 $\pm$ 3 K (Ingersoll et al. 1980, as corrected by G. Orton, private communication). The implied internal power of 2×10^{17} watts is substantially larger than expected from evolutionary modeling of the sort that worked for Jupiter (Pollack et al. 1977; Pollack 1978). Apparently an energy source is needed in addition to primordial heating, with the most frequently suggested candidate gravitational separation of hydrogen and helium in the core.

Jupiter and Saturn each radiate most of their thermal energy in spectral regions accessible from the ground, but for Uranus and Neptune the peak of the spectrum moves beyond 30 μm, so that ground-based photometry is less readily interpreted. The first evidence for an internal heat source was presented by Morrison and Cruikshank (1973a) and Rieke and Low (1974), who found Neptune to be brighter than Uranus in the 20-30 μm band in spite of its greater distance from the Sun. This result was much strengthened by observations extending to longer wavelengths obtained from the Kuiper Airborne Observatory

(KAO) by Loewenstein et al. (1977a,b) and from balloon altitudes by Stier et al. (1978). The work clearly showed that Uranus has very nearly the temperature expected for a rapidly rotating planet in equilibrium with the insolation (58 K), while Neptune has essentially the same effective temperature (56 K), and is apparently emitting about twice as much energy as it absorbs from the Sun. The reason for the difference between the two planets is unknown, although it has been suggested by Trafton (1974) that tidal effects from Triton may heat the interior of Neptune.

The current information on heat sources in the giant planets is summarized in Table 1. Although Jupiter has the highest total luminosity, it is interesting that the luminosity-to-mass ratio is greater for Saturn, presumably as a result of an additional energy-producing mechanism in its core. The heat source in Neptune is large only in comparison to its feeble illumination by the Sun; in absolute units it is two orders of magnitude below Jupiter and Saturn, and in specific units one order of magnitude lower. The observational upper limit for an intrinsic luminosity for Uranus is a factor of 2 or 3 below the value derived for Neptune.

TABLE 1

INTERNAL HEAT SOURCES OF THE JOVIAN PLANETS

JUPITER	T_B	$= 127 \pm 3$ K
	Total power	$= (2.0 \pm 0.2)$ solar input
	Internal power	$= 7$ W m^{-2} $= 4 \times 10^{17}$ W
		$= 0.2$ μW/ton
SATURN	T_B	$= 97 \pm 3$ K
	Total power	$= (2.8 \pm 0.4)$ solar input
	Internal power	$= 3.5$ W m^{-2} $= 2 \times 10^{17}$ W
		$= 0.3$ μW/ton
URANUS	T_B	$= 58 \pm 3$ K
	Total power	$\simeq$ solar input
	Internal power	< 0.1 W m^{-2} $= 10^{15}$ W
		< 0.01 μW/ton
NEPTUNE	T_B	$= 56 \pm 3$ K
	Total power	$= (2.5 \pm 0.5)$ solar input
	Internal power	$= 0.4$ W m^{-2} $= 3 \times 10^{15}$ W
		$= 0.03$ μW/ton

All of these measurements of the infrared luminosity of the Jovian planets provide important constraints on their internal structure and thermal evolution. Only for Jupiter does the mechanism that produces the heat appear to be well understood, and even there the processes of energy transfer that produce polar regions as warm as the solar-heated equator are obscure. A great deal of theory remains to be worked out before we can claim to understand these observations.

III. THERMOPHYSICS OF PLANETARY SURFACES

One of the earliest results of infrared astronomy was the determination of the low thermal conductivity of the upper few centimeters of the lunar crust. As is well known, the response of the surface temperature of a semi-infinite, homogeneous slab of material with temperature-independent thermal properties to changing radiative boundary conditions at its upper surface is a function only of the composite parameter $(K\rho c)^{1/2}$, where K is the thermal conductivity, ρ is the density, and c is the heat capacity. This parameter is called the "thermal inertia", by analogy with mechanical inertia. The greater the thermal conductivity, the more slowly the surface temperature can respond to a rapidly changing insolation, and thus the more "inertia" the surface has. Small inertia, then, implies low conductivity and rapidly fluctuating temperatures.

This paper follows the general practice of using I to represent thermal inertia and expressing it in units of 10^{-3} cal cm^{-2} $s^{-1/2}$ K^{-1}, which is equal to 41.84 J m^{-2} $s^{-1/2}$ K^{-1}. In these units, the typical thermal inertia of the Moon is $I \simeq 1$ and of a solid terrestrial rock is $I \simeq 40$.

To determine thermal inertia, one must measure the response of surface temperature to changing insolation. Nature frequently provides two time scales for these changes, one corresponding to the diurnal cycle and one to brief interruptions of sunlight during eclipses. With each time scale is associated a characteristic skin depth related to the thickness of the surface layer that experiences these transient temperature changes. Typically for low-conductivity surfaces, these depths are from a few millimeters to several centimeters. Measurements of thermal inertia thus characterize the physical properties of the uppermost layer of the regolith and are sensitive to conditions on a scale substantially smaller than can be imaged remotely.

The first application of thermal inertia measurements outside the Earth-Moon system was to Mars, using diurnal temperature variations observed by Sinton and Strong (1960). Morrison (1968) used these data to estimate I = 4-6, with suggestion of a larger inertia for dark areas. Most determinations of the thermal inertia, however, have been made from eclipse cooling and heating observations, generally of unresolved objects.

The Galilean satellites of Jupiter undergo eclipses often and are fairly easily observed with Earth-based telescopes. At the very start of the modern era of infrared astronomy, Murray et al. (1965) observed part of the eclipse cooling curve of Ganymede and concluded from the rapid changes in temperature that its thermal inertia must be lunar-like. Morrison et al. (1971) followed the temperature change for Ganymede over its full range and concluded that I was definitely lower than for the Moon. A series of eclipse measurements made at the Hale 5-meter and at Mauna Kea were summarized by Hansen (1973) and Morrison and Cruikshank (1973b), who found that for all the Galilean satellites $I \simeq 0.3$, indicating a regolith of extremely low conductivity. Radiometric eclipse measurements of Mars' satellite Phobos obtained from Mariner 9 by Gatley et al. (1974) yielded a comparably low thermal inertia.

The advent of spacecraft radiometry in the late 1960s allowed the thermal inertia to be mapped over the surface of a planet with good spatial resolution. The Mariner 6 and 7 flybys of Mars established that the seasonal polar caps were composed of CO_2 (Neugebauer et al. 1971). The Mariner 9 orbiter provided coverage of more than 35% of the surface with a resolution of 100 km and established $I = 7$ as an appropriate mean value to characterize the planet, with a range in thermal inertia from 4 to 17 (Kieffer et al. 1973). The much larger value of inertia for Mars as compared with the Moon or the Galilean satellites is a consequence of the atmosphere, which provides additional energy transfer between grains as well as compaction of the surface material by wind.

In 1974, Mariner 10 flew past the night side of Mercury and spatially resolved temperatures were measured. The thermal inertia derived by Chase et al. (1976) from these data ranged between 1.5 and 3.1; the fact that these are higher than the lunar values probably represents in part a radiative contribution to the thermal conductivity at the much higher average subsurface temperatures on Mercury.

The most important application of thermal inertia measurements to date has resulted from several years of systematic orbital observations of Mars carried out with the Viking IRTM (Infrared Thermal Mapper). This multi-channel radiometer was able to map surface areas at a variety of spatial scales and at many points in the diurnal cycle. As the seasons passed on Mars and dust storms developed and subsided, it was also possible to learn how the atmosphere influenced surface temperature. The wealth of information in these observations has emerged in a series of papers beginning with Kieffer et al. (1977), and the analysis is still underway.

In most of the earlier IRTM papers (e.g., Zimbelman and Kieffer 1979), the thermal inertia of surface elements was derived from the predawn temperature; the more the surface cools during the Martian night, the lower the conductivity and the smaller the inertia. (This provides a convenient way to remember the relationship: small

T => small K => small I). The analysis based on predawn temperature alone neglects the effects of albedo on temperature. Later papers (e.g., Palluconi and Kieffer 1981) used the full data set to derive the thermal inertia directly for each spot observed. The results are thereby improved, but not changed in any major way.

Figure 1, taken from Palluconi and Kieffer (1981), is a thermal inertia map of Mars between 60°N and 60°S with a spatial resolution of 2° in latitude and 2° in longitude. The data were sufficient to derive the inertia for 10,171 of the possible 10,800 areas so defined. All of these inertias correspond to clear conditions well separated in time from the major planet-wide dust storms. The total range in derived inertia is $1 \leq I \leq 15$. The distribution of inertias is bimodal, with all values less than 4 associated with northern-hemisphere bright regions (Tharsis-Amazonis, Elysium, and near 330°W, 15°N) comprising 20% of the surface. These regions of low conductivity are probably blanketed by fine deposits of wind-blown dust. The highest inertias generally correspond to dark regions; apparently insulating dust blankets are not composed of dark materials. Probably there are exposures of bare rock in the high-inertia areas, although the observed maximum value of I at this resolution is not high enough to indicate any 2° x 2° area with predominantly bare rock.

One interesting conclusion from the global thermal inertia mapping of Mars relates to the nature of the two Viking lander sites. These sites were selected for smoothness, but both turned out to be rocky: 8% of the surface is rock-covered at VL-1, and 14% at VL-2. The thermal inertias in these areas as shown in Figure 1 are 9 and 8, respectively. This places the landing sites in the higher 20th percentile of the observed inertias. To the extent that thermal inertia measures the extent of bare rocks, it would appear that both sites are atypically rocky. Palluconi and Kieffer (1981) further argue that the two sites have unusually high albedo for their thermal inertias, suggesting they might not be at all like most of the surface. This is a point worth remembering, since it seems likely that the perception of the Martian surface by scientists and lay persons alike will be based for the next generation on pictures taken by these two Viking landers.

The thermal inertia is a quantity that stands midway between a set of temperature measurements and a true characterization of the physical properties of a surface. Even when the observed temperature variations are matched well with a homogeneous heat conduction model, the actual conductivity and porosity of the surface may not be well established. When the data are sparse, we are content to characterize an entire planet by the simple parameter I, but in the case of Mars, it is clear that the surface and the processes modifying it are extremely complex. Thermal inertia mapping is just one of many inputs to developing an understanding of the regolith of a planet.

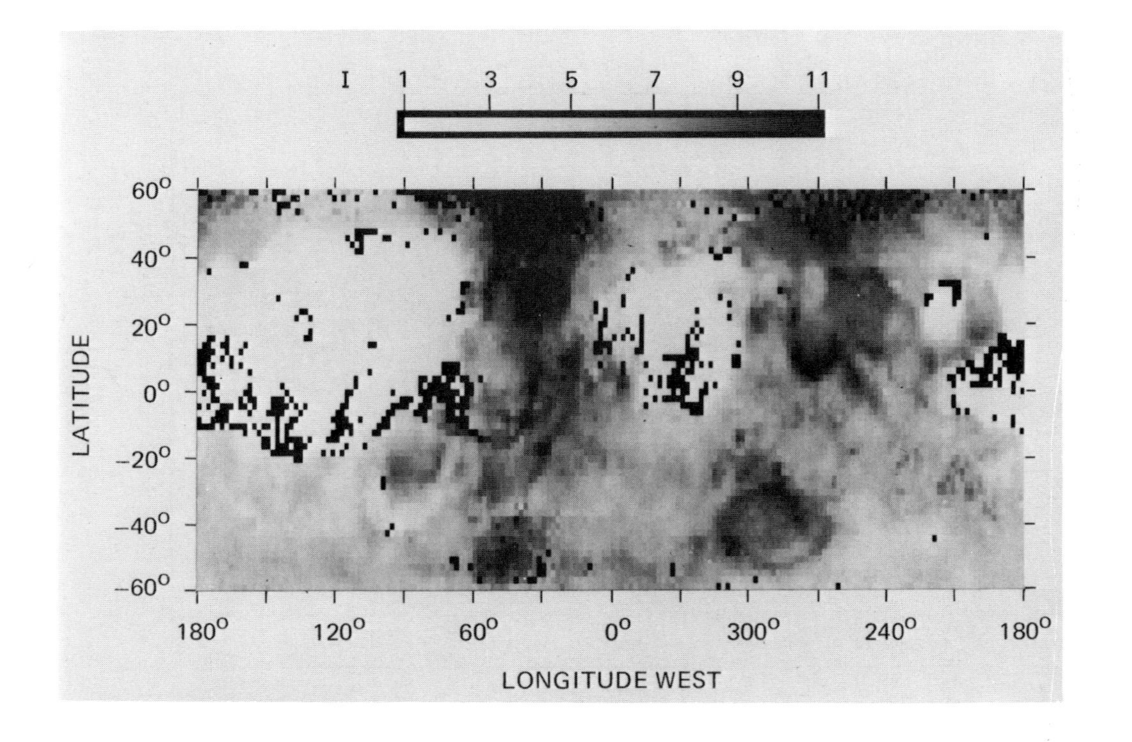

Figure 1: Thermal inertia map of Mars derived from Viking Orbiter 20-μm infrared mapping by Palluconi and Kieffer (1981). The data, all taken under clear atmospheric conditions (L_s 344° to 125°), are analyzed in bins 2° x 2° in size; there are 10,800 such bins in the figure. The bins (5.8% of the total) with no determination of thermal inertia are black. For the other regions, white represents inertia of 1 and dark grey inertia greater than 10 (in units of 10^{-3} cal cm^{-2} $s^{-1/2}$ K^{-1}), as indicated by the grey scale at the top of the figure. The three large northern-hemisphere areas of low inertia are, from the left, the high albedo regions Tharsis-Amazonis, Elysium, and an unnamed region near longitude 330°. Prominent areas of high inertia include the Viking landing sites in Chryse Planetia (22°.5 N, 47°.8 W) and in Utopia Planetia (48°.0 N, 225°.6 W).

IV. DIAMETERS AND ALBEDOS OF MINOR PLANETS AND SATELLITES

In solar system applications as elsewhere, infrared photometry is not generally obtained in order to derive a simple temperature. One of the most productive infrared programs of the past few years has involved the measurement of thermal emission from bodies so small that no diameter is known, and thus no brightness temperature can be defined. Instead, the infrared photometry is combined with visible photometry to derive the albedo and diameter of the object.

This radiometric/photometric technique for determining sizes and albedos of small airless objects was first applied by Allen (1970) to measure the diameter of asteroid 4 Vesta, and most of the subsequent applications of this method have been in asteroid studies. The basic principle is simple. For an object of a given angular size in equilibrium with the insolation, the brightness in reflected light will scale with the albedo, while the thermal reradiation of absorbed sunlight scales in the opposite manner, being roughly proportional to (1 - albedo). Measurements of the reflected and the reradiated energies determine the albedo and the size. In practice, the interpretation is model dependent and requires some knowledge of the photometric properties and the infrared emissivity of the object. In addition, there should be a calibration in terms of objects of known size, since the basic uncertainties in the absolute calibration of the magnitude scales also introduce errors. These calibrations, and the details of the thermal models used to interpret the data, are described in detail by Matson (1971), Morrison (1973), Jones and Morrison (1974), Matson et al. (1978), and Morrison and Lebofsky (1979).

Although it might seem in general that the albedos and diameters derived from radiometry and photometry would be strongly model dependent, there are some interesting special causes where this problem largely disappears. One simplification occurs if the thermal inertia of the surface is low; that is, if the illuminated face is at a temperature nearly in equilibrium with the insolation, while the dark hemisphere is much colder. In this case, nearly all of the absorbed sunlight is radiated from the sun-facing (and therefore Earth-facing, for a superior planet) hemisphere, which is the hemisphere we observe. Low thermal inertias appear to be ubiquitous among airless solar system bodies, as was noted in the previous section, presumably as the result of vacuum welding of fine regolith material produced from meteoric impacts. For the asteroids, theoretical models suggest that an insulating regolith should be maintained for objects down to a few kilometers in diameter; the only major exception might be for a largely metallic asteroid (Housen et al. 1979). A second simplification occurs for dark objects. In general, to interpret the radiometry one needs to know the relationship between Bond albedo and geometric albedo, but for a very dark object this information is not required. In effect, if an object is nearly black we can assume nearly all the incident sunlight goes to heating it, and the details of its photometric behavior are not important. Since most asteroids and many satellites are exceedingly dark, this effect considerably enhances the accuracy of the radiometric/photometric method.

In the satellite systems of the outer planets, infrared radiometry has been used to determine the diameters and albedos of Iapetus and several of the inner satellites of Saturn (Murphy et al. 1972; Morrison 1974b), but these results suffer from the fact that the albedos are not low, and in any case they will be superseded by Voyager results before this book is published. Of more significance is the measurement by Cruikshank (1979) of Hyperion, for which he obtained an albedo of 0.47 ± 0.11 and a diameter of 224 ± 30 km. These values are less likely to be improved upon by Voyager. In the Jovian system, Cruikshank (1977) has used this method to make the only measurements of Himalia (J6) and Elara (J7), for both of which he finds albedos of 0.03. His diameters are 170 ± 20 km for Himalia and 80 ± 20 km for Elara. Cruikshank et al. (1979) used an upper limit of the 20-μm flux of Triton to set an upper limit of 5200 km to the diameter and a lower limit to the geometric albedo of 0.19. This is an important result, inasmuch as these are the only data that tell us that Triton is not as large as Titan. The other outer planet satellites are undetectable with present instruments.

Asteroids have provided the most productive application of the radiometric/photometric technique. There are hundreds of asteroids large enough to have angular diameters of 0.1 arcsec as seen from Earth--too small to be resolved optically, of course, but large enough to produce a strong infrared signal. At the distance from the sun of the main asteroid belt, the temperatures result in peak thermal radiation in the 10 and 20-μm bands, where it is easily observed. In fact, if one were to look at the sky with 10-μm eyes, asteroids would provide a high percentage of the sources--as will become perhaps too apparent when IRAS begins its all-sky survey.

Only two alternative methods have been used to measure asteroid sizes and albedos. Stellar occultations observed photoelectrically from a network of sites provide by far the best values, but predictable events are rare and the area of the Earth's surface from which the observations can be made is small: comparable to that from which a total solar eclipse is visible. Half a dozen occultations have been observed, but the only two that were really successful were of 2 Pallas and 4 Juno. The results for these objects provide the best calibration for the radiometric/photometric technique. The other method is based on optical polarimetry at a variety of wavelengths. It has been applied to more than 50 asteroids, but it requires observations at several phase angles and it becomes unreliable for very dark objects. Most of the known asteroid diameters today were obtained from infrared observations; Morrison (1977) lists values for 187 objects. Most of these are estimated to be accurate to $\pm 10\%$ in the diameter and $\pm 20\%$ in the albedo. In the best observed cases, the accuracy may reach $\pm 5\%$ in diameter, but disparities in the calibration based on the Juno and Pallas occultation results suggest an inherent uncertainty of at least ±5%.

All known asteroid diameters and albedos are listed along with other physical observations in the TRIAD (Tucson Revised Index of Asteroid Data) file, which was published in 1979 as a series of appendices to the University of Arizona volume *Asteroids*. The coordinated collection and interpretation of asteroid observations as represented in TRIAD has played a central role during the past decade in expanding our understanding of these objects. The first such effort to analyze spectral, polarimetric, and radiometric data for a statistically significant group of asteroids was made by Chapman et al. (1975) at a time when radiometric diameters were available for only 47 objects. Even then, it was apparent that most asteroids had extremely low albedos (<6%) and were chemically primitive. Chapman et al. designated these "C-class" asteroids by analogy with the carbonaceous chondrites; the next most populous class, which they called S objects, have higher albedos and spectra that frequently show evidence of iron-magnesium-silicate mineral assemblages. The classification of asteroids and its interpretation are discussed in detail by Bowell et al. (1978), while the actual mineralogy of these objects as derived from infrared reflectance spectroscopy is discussed by McCord and Cruikshank in this book.

The TRIAD file, which included 195 radiometric diameters when published in 1979, has been used for several statistical studies that provide the first reliable picture of the physical nature and distribution of the minor planets (Zellner and Bowell 1977; Zellner 1979; Chapman 1979). We now know what the numbers and distributions over size and distance from the Sun are for asteroids in each of several compositional classes. It is also possible to determine the compositional classes of objects related dynamically and presumed to have a common history. There is no room here to discuss these results, but it is enough to note that thermal radiometry is playing a central role in this fast-moving field as the primary means available today to measure the size of an asteroid and, perhaps even more important, its compositionally sensitive surface albedo. As a single specific example of these results, Table 2 lists the sizes and classifications of all asteroids 200 km or larger in diameter; before the widespread application of the radiometric/photometric technique to the asteroids, such a compilation would have been unthinkable.

V. VOLCANIC ACTIVITY ON IO

In 1979, Voyager discovered that Io is a planetary object of incredible geologic activity, with eruptive plumes rising hundreds of kilometers above the surface, changes in albedo and color within a few weeks that affect areas of thousands of square kilometers, and localized hot spots with temperatures as much as 500 K above ambient

(Morabito et al. 1979; Smith et al. 1979a,b; Hanel et al. 1979). This unique level of volcanism is the outward manifestation of a molten interior, maintained by a major internal heat source. The most probable cause of the internal heating is tidal stressing of Io that results from its non-circular orbit, which is in turn dynamically coupled to the other Galilean satellites. Initial calculations of the magnitude of the internal heat source suggested values for the power dissipated as high as 10^{13} watts (Peale et al. 1979). Luminosity of this scale can be measured with Earth-based infrared techniques.

TABLE 2

THE LARGEST ASTEROIDS

Asteroid	Type	Diam. (km)	Asteroid	Type	Diam. (km)
1 Ceres	C	1025	24 Themis	C	249
2 Pallas	U	583	3 Juno	S	249
4 Vesta	U	555	16 Psyche	M	249
10 Hygeia	C	443	13 Egeria	C	245
704 Interamnia	U	338	216 Kleopatra	CMEU	236?
511 Davida	C	335	165 Loreley	C	228
65 Cybele	C	311	19 Fortuna	C	226
52 Europa	C	291	7 Iris	S	222
451 Patientia	C	281	532 Herculina	S	219
31 Euphrosyne	C	270	250 Bettina	CMEU	211?
15 Eunomia	S	261	702 Alauda	CU	217
324 Bamberga	C	256	747 Winchester	C	208
107 Camilla	C	252	432 Diotima	C	209
87 Sylvia	CMEU	251?	386 Siegena	C	203
45 Eugenia	U	250	375 Ursula	C	200

With the clear perspective of hindsight, it is now clear that infrared observers had detected the effects of Ionian volcanism long before Voyager. In 1973, 10-μm photometry of Io obtained with the Hale 5-meter telescope during eclipses showed a remarkably high flux density, quite different from the cooling curves of the other Galilean satellites (Hansen 1973). At the same time, 20-μm eclipse observations with the Mauna Kea 2.2-meter telescope displayed more normal behavior (Morrison and Cruikshank, 1973b). It is now clear that almost all of the 10-μm radiation seen by Hansen originated in spots at temperatures of 200 K or higher covering a small fraction of the surface. A more direct measurement of a thermal outburst on Io was made in 1978 by Witteborn et al. (1979), who observed a short-lived but dramatic enhancement at 5 μm; unfortunately, they considered but then rejected the volcanic hypothesis. Between the two Voyager flybys Sinton (1980) observed another similar 5-μm event, but by then the correct explanation had become dramatically apparent.

Ground-based observations are capable of detecting some form of volcanic outbursts on Io, but these events are rare and it is not at all clear just what physical mechanisms are involved. More important is the ability of infrared astronomy to measure the total emitted power from Io. If a source of magnitude 10^{13} watts resulted in a uniform heat flow through the crust, such a measurement would not be possible, since the total rise in surface temperature would be only about 1 K. However, we are fortunate that the escape of energy from Io takes a more readily measurable route. Voyager 1 (Hanel et al. 1979) measured discrete hot spots, with temperatures hundreds of degrees above the background, and it is similar hot spots that contribute to the anomalous thermal behavior of Io. The elevation of the 10-μm temperature during eclipse has already been noted, and even outside of eclipses it is clear that the brightness temperature of Io increases markedly toward shorter wavelengths. Matson et al. (1980) have analyzed these spectral data in terms of a hot-spot model and calculated that 2 ± 1 watts m^{-2} was being released through hot spots, corresponding to an internal power of about 10^{14} watts.

The effects of the hot spots are even more dramatic at shorter wavelengths when Io is in eclipse, as discussed by Sinton et al. (1980): even at wavelengths as short as 2 μm the satellite remains detectable, indicating that some areas on the surface have temperatures as high as 600 K. Eclipse photometry covering the entire range from 3 to 30 μm was obtained in April 1980 by Morrison and Telesco (1981). These measurements of the hot spots yield an average heat flow of 1.5 ± 0.5 watts m^{-2} or an internal power of $(6 \pm 2) \times 10^{13}$ watts. The models fit to the spectrum suggest that there is a broad range of temperatures, from about 500 K (covering a few millionths of the surface) down to about 200 K (covering nearly 1% of the surface).

The observational determinations of the Ionian heat flow as represented by the hot spots are summarized in Table 3, together with the results of dynamical calculations. It is important to remember that the observational values refer only to areas with temperatures substantially above the mean; the temperature increase due to a more uniform heat flow could not be detected. Thus the numbers given represent a lower limit to the internal energy source, if we adopt the assumptions that the recent observations of the Jupiter-facing hemisphere of Io are representative of the entire surface over long periods of time. This lower limit is in excess of 10^{13} watts, and may be as high as 10^{14} watts. Yoder (1979,1980) has argued that the dynamics of the Jovian satellite system do not permit the deposition of more than a few times 10^{13} watts of tidal energy in Io on an equilibrium basis. Clearly, the infrared measurements not only are fascinating for the information they provide us on volcanic processes; in addition, they appear to challenge the basic mechanics of origin of the internal heat of Io. These topics are pursued in detail in chapters by Greenberg (1981), Cassen et al. (1981), and Pearl and Sinton (1981) in the forthcoming book *The Satellites of Jupiter*.

TABLE 3

VALUES FOR THE INTERNAL HEAT SOURCE OF IO

Method	Luminosity (watts)	Heat Flow (W m^{-2})	Reference
Theory:	$\lesssim 4 \times 10^{12}$	$\lesssim 0.1$	Peale et al. (1979)
Theory:	$10^{12} - 10^{13}$	0.02 - 0.2	Yoder (1979)
Theory:	$2 \times 10^{12} - 4 \times 10^{13}$	0.05 - 1.0	Yoder (1980)
Spectrum:	$(8 \pm 4) \times 10^{13}$	2 ± 1	Matson et al. (1980)
Voyager:	$\sim 10^{14}$	~2	Pearl (1980)
Eclipse:	$(7 \pm 3) \times 10^{13}$	1.8 ± 0.8	Sinton (1981)
Eclipse:	$(6 \pm 2) \times 10^{13}$	1.5 ± 0.5	Morrison & Telesco (1981)

VI. CONCLUSIONS

As the foregoing examples indicate, thermal infrared studies of planets and smaller solar system bodies are used to investigate a variety of scientific problems. In a few cases, such as determining the nature of the phase changes on the sublimating Martian polar cap, the actual temperatures are of direct interest. However, most of these investigations have as their goal the study of the more fundamental physical nature of planetary surfaces and interiors, and of the processes influencing them. It seems certain that such studies, carried out from the ground, from high in the terrestrial atmosphere, from Earth orbit, and from planetary flybys and orbiters, will continue to play an important part in efforts to explore and understand the planets and their origins.

For their advice and assistance I thank H. H. Kieffer, T. Z. Martin, F. D. Palluconi, and W. M. Sinton, and I especially am grateful to D. P. Cruikshank for his encouragement and aid in the preparation of this chapter. This research was supported in part by NASA Grants NGL 12-001-057 and NSG 7633.

REFERENCES

Allen, D.A.: 1970, Nature 227, pp. 158-159.
Aumann, H.H., Gillespie, C.M., and Low, F.J.: 1969, Astrophys. J. 157, pp. L69-L72.
Bowell, E., Chapman, C.R., Gradie, J.C., Morrison, D., and Zellner, B.: 1978, Icarus 35, pp. 313-335.
Bodenheimer, P.: 1974, Icarus 23, pp. 319-325.
Cassen, P.M., Reynolds, F.T., and Peale, S.J.: 1981, in D. Morrison (ed.), "The Satellites of Jupiter," Univ. of Arizona Press, Tucson (in press).
Chapman, C.R.: 1979, in T. Gehrels (ed.), "Asteroids," Univ. of Arizona Press, Tucson, pp. 25-60.
Chapman, C.R., Morrison, D., and Zellner, B.: 1975, Icarus 25, pp. 104-130.
Chase, S.C., Miner, E.D., Morrison, D., Münch, G., and Neugebauer, G.: 1976, Icarus 28, pp. 565-578.
Chase, S.C., Ruiz, R.D., Münch, G., Neugebauer, G., Schroeder, M., and Trafton, L.M.: 1974, Science 183, pp. 315-317.
Courtin, R., Lena, P., de Muizon, M., Rouan, D., Nicollier, C., and Wijnbergen, J.: 1979, Icarus 38, pp. 411-419.
Cruikshank, D. P.: 1977, Icarus 30, pp. 224-230.
Cruikshank, D.P.: 1979, Icarus 37, pp. 307-309.
Cruikshank, D.P., Stockton, A., Dyck, H.M., Becklin, E.E., and Macy, W.: 1979, Icarus 40, pp. 104-114.
Cuzzi, J.: 1978, in D.M. Hunten and D. Morrison (eds.), "The Saturn System," NASA CP-2068, pp. 73-104.
Erickson, E.F., Goorvitch, D., Simpson, J.P., and Strecker, D.W.: 1978, Icarus 35, pp. 61-73.
Gatley, I., Kieffer, H., Miner, E., and Neugebauer, G.: 1974, Astrophys. J. 190, pp. 497-503.
Graboske, H.C., Pollack, J.B., Grossman, A.S., and Olness, R.J.: 1975, Astrophys. J. 199, pp. 265-281.
Greenberg. R.: 1981, in D. Morrison (ed.), "The Satellites of Jupiter," Univ. of Arizona Press, Tucson (in press).
Hanel, R., and the Voyager IRIS Team: 1979, Science 204, pp. 972-976.
Hansen, O.L.: 1973, Icarus 18, pp. 237-246.
Housen, K.R., Wilkening, L.L., Chapman, C.R., and Greenberg, R.J.: 1979, in T. Gehrels (ed.), "Asteroids," Univ. of Arizona Press, Tucson, pp. 601-627.
Ingersoll, A.P., Münch, G., Neugebauer, G., Diner, D.J., Orton, G.S., Schupler, B., Schroeder, M., Chase, S.C., Ruiz, R.D., and Trafton, L.M.: 1975a, Science 188, pp. 472-473.
Ingersoll, A.P., Münch, G., Neugebauer, G., and Orton, G.S.: 1975b, in T. Gehrels (ed.), "Jupiter," Univ. of Arizona Press, Tucson, pp. 197-205.
Ingersoll, A.P., Orton, G.S., Münch, G., Neugebauer, G., and Chase, S.C.: 1980, Science 207, pp. 439-443.
Jones, T.J., and Morrison, D.: 1974, Astron. J. 79, pp. 892-895.
Kieffer, H.H., Chase, S.C., Miner, E., Münch, G., and Neugebauer, G.: 1973, J. Geophys. Res. 78, pp. 4291-4312.

Kieffer, H.H., Martin, T.Z., Peterfreund, A.R., Jakosky, B.M., Miner, E.D., and Palluconi, F.D.: 1977, J. Geophys. Res. 82, pp. 4249-4291.
Loewenstein, R.F., Harper, D.A., Moseley, S.H., Telesco, C.M., Thronson, H.A., Hildebrand, R.H., Whitcomb, S.E., Winston, R., and Stiening, R.F.: 1977a. Icarus 31, pp. 315-324.
Loewenstein, R.F., Harper, D.A., and Moseley, S. H.: 1977b, Astrophys. J. 218, pp. L145-L146.
Low, F.J.: 1966, Astron. J. 71, p. 391 (abstract).
Matson, D.L.: 1971, in T. Gehrels (ed.), "Physical Studies of Minor Planets," NASA SP-267, pp. 45-50.
Matson, D.L., Veeder, G.J., and Lebofsky, L.A.: 1978, in D. Morrison and W.C. Wells (eds.), "Asteroids: An Exploration Assessment," NASA CP-2053, pp. 127-144.
Matson, D.L., Ransford, G.A., and Johnson, T.V.: 1980, J. Geophys. Res. (in press).
Morabito, L.A., Synnott, S.P., Kupferman, P.N., and Collins, S.A.: 1979, Science 204, p. 972.
Morrison, D.: 1968, Smithsonian Astrophys. Obs. Special Report #284.
Morrison, D.: 1973, Icarus 19, pp. 1-14.
Morrison, D.: 1974a, Icarus 22, pp. 57-64.
Morrison, D.: 1974b, Icarus 22, pp. 51-56.
Morrison, D.: 1977, Icarus 31, pp. 185-220.
Morrison, D., and Cruikshank, D.P.: 1973a, Astrophys. J. 179, pp. 329-331.
Morrison, D., and Cruikshank, D.P.: 1973b, Icarus 18, pp. 224-236.
Morrison, D., and Lebofsky, L.A.: 1979, in T. Gehrels (ed.), "Asteroids," Univ. of Arizona Press, Tucson, pp. 184-205.
Morrison, D., and Telesco, C.: 1981, Icarus, submitted.
Murphy, R.E., Cruikshank, D.P., and Morrison, D.: 1972, Astrophys. J. 177, pp. L93-L96.
Murray, B.C., Westphal, J.A., and Wildey, R.L.: 1965, Astrophys. J. 141, pp. 1590-1592.
Neugebauer, G., Münch, G., Kieffer, H.H., Chase, S.C., and Miner, E.: 1971, Astron. J. 76, pp. 719-728.
Palluconi, F.D., and Kieffer, H.H.: 1981, Icarus (in press).
Peale, S.J., Cassen, P.M., and Reynolds, R.T.: 1979, Science 203, pp. 892-894.
Pearl, J.C.: 1980, paper given at IAU Colloquium No. 57.
Pearl, J.C., and Sinton, W.M.: 1981, in D. Morrison (ed.), "The Satellites of Jupiter," Univ. of Arizona Press, Tucson (in press).
Pettit, E.: 1961 in G.P. Kuiper and B.M. Middlehurst (eds.), "Planets and Satellites," Univ. of Chicago Press, Chicago, pp. 400-428.
Pollack, J.B.: 1978, in D. M. Hunten and D. Morrison (eds.), "The Saturn System," NASA CP-2068, pp. 9-28.
Pollack, J.B., Grossman, A.S., Moore, R., and Graboske, H.C.: 1977, Icarus 30, pp. 111-128.
Rieke, G.H., and Low, F.J.: 1974, Astrophys. J. 193, pp. L147-L148.
Stier, M.T., Traub, W.A., Fazio, G.G., Wright, E.L., and Low, F.J.: 1978, Astrophys. J. 226, pp. 347-349.

Sinton, W.M.: 1961, in G.P. Kuiper and B.M. Middlehurst (eds.,) "Planets and Satellites," Univ. of Chicago Press, Chicago, pp. 429-441.
Sinton, W.M.: 1980, Astrophys. J. 235, pp. L49-L51.
Sinton, W.M.: 1981, J. Geophys. Res. (in press).
Sinton, W.M., and Strong, J.: 1960, Astrophys. J. 131, pp. 459-469.
Sinton, W.M., Tokunaga, A., Becklin, E.E., Gatley, I., Lee, T.J., and Lonsdale, C.: 1980, Science (in press).
Smith, B.A. and the Voyager Imaging Team: 1979a, Science 204, pp. 951-972.
Smith, B.A. and the Voyager Imaging Team: 1979b, Science 206, pp. 927-950.
Trafton, L.: 1974, Astrophys. J. 193, pp. 477-480.
Witteborn, F.C., Bregman, J.D., and Pollack. J.B.: 1979, Science 203, pp. 643-646.
Yoder, C.F.: 1979, Nature 279, pp. 767-770.
Yoder, C.F.: 1980, paper given at IAU Colloquium No. 57.
Zellner, B.: 1979, in T. Gehrels (ed.), "Asteroids," Univ. of Arizona Press, Tucson, pp. 783-808.
Zellner, B., and Bowell, E.: 1977, in A. H. Delsemme (ed.), "Comets, Asteroids, and Meteorites," Univ. of Toledo, Toledo, pp. 185-197.
Zimbelman, J.R., and Kieffer, H.H.: 1979, J. Geophys. Res. 84, pp. 8239-8251.

DISCUSSION FOLLOWING PAPER DELIVERED BY D. MORRISON

BEER: Do the inferences about the compactness of the surface of Callisto support the old ideas about the structure of regoliths *in vacuo*? That is, can we expect a "fairy castle" structure?

MORRISON: The inference about the compactness comes from the modeled thermal conductivity of the surfaces of the satellites. The thermal conductivity of Callisto appears to be within a factor of two or three of that of the Moon, and I believe we can assume that the surface structure is therefore similar to that of the Moon. You will recall that the Moon has very low thermal conductivity and a rather compact surface, and that the surface has substantial bearing strength.

WERNER: How do your inferences from the ground-based measurements of the surface properties of the Jovian satellites compare with the interpretation of the Voyager photographs?

MORRISON: At present these two data sets appear to be orthogonal. The disparity between our thermal measurements of surfaces on a length scale of a few centimeters and the maximum Voyager resolution of about 1 kilometer is very great. Perhaps we can come closer by looking at the results of the Viking measurements of Mars where we have a lander that can help connect the two data sets. In any case, I think there is room for a great deal of interpretation in the analysis of these data.

ALLEN: A decade ago when infrared measurements of asteroid diameters and albedos were first made I felt that one would never achieve better than 10% accuracy on diameters--hence 20% on albedos--due to the effects of the irregular surface of the asteroids and uncertainties in the amount of energy radiated from the dark side. Thus it doesn't surprise me that the accuracy you are attaining compared to occultation measures is 10%. Now, I suspect that these effects (especially shape) will increase with smaller asteroids. Now I see that you are pushing down below 50 km diameters, and I'd like to ask how small you think you can go before the errors become too large to be useful.

MORRISON: This is an excellent question. Of course, 10% accuracy in diameter and 20% in albedo, or even worse, is quite enough for the broad compositional classification, so our interest in pushing for higher accuracy is either for aesthetic reasons or interest in determinations of the densities of asteroids. If, however, the regolith of an asteroid is completely removed, a much higher fraction of the incident solar radiation is emitted from the anti-solar side, so that the computational scheme used so far in this work is no longer completely valid. Calculations indicate that the non-regolith limit might be on the order of 10 km diameter, so we might be able to see the effects on the small Earth-approaching asteroids. There is some evidence that a few asteroids of small size, particularly Betulia, emit substantially less infrared thermal radiation than expected. That is, their diameters determined by the radiometric technique are substantially smaller than those found by other means. These may represent cases of small asteroids without regoliths. For some small asteroids the radiometric technique seems to work, however.

JOSEPH: Why is gravitational contraction adequate to account for the excess power radiated by Jupiter (over that absorbed from the Sun), but not for Saturn?

MORRISON: I have not personally made the calculations, but Pollack, Bodenheimer, and others predict a weaker internal heat source for Saturn than for Jupiter. My intuitive feeling is that because Saturn is a smaller object we should expect less initial heating per unit mass as it collapses gravitationally, whereas the observations indicate a greater power per unit mass for Saturn. Thus an additional energy source is indicated.

INFRARED SOURCES IN DENSE MOLECULAR CLOUDS

Neal J. Evans II
Electrical Engineering Research Laboratory
and Department of Astronomy
The University of Texas at Austin
Austin, Texas 78712

I. A Short History of Molecular Clouds

The study of infrared sources in molecular clouds necessarily places the practitioner at the interface between two rapidly evolving fields of study. In such a situation, yesterday's heresy often becomes today's dogma. Since it is hard enough to keep up with even one field of study, I thought it might be helpful to recount a bit of history regarding molecular clouds. The objective is to put various notions about molecular clouds into proper context.

In the beginning (around 1969), only a few interstellar molecules were known to exist and those only in a few sources. Interstellar molecules were still a mere curiosity and could be safely ignored by theoreticians of the interstellar medium, among whom the two-phase model still held sway. Beginning in the early 70's, radio telescopes began to operate at wavelengths of 2-4 mm, where many new molecules were discovered, most importantly carbon monoxide (CO). Maps of CO emission showed the existence of large clouds of primarily molecular material. These large clouds (envelopes) often contained cores which were hot and dense, and which often coincided with infrared sources and H II regions. Others, mostly nearby dark clouds, showed no such hot core.

The situation around 1974 was summarized by Zuckerman and Palmer (1974). The typical molecular cloud had an envelope which was 10 pc in size, and a smaller core 0.5 pc in size. The total mass of the cloud was 10^5 $M_{\odot}$. The prevailing theoretical view was that clouds were collapsing freely (Goldreich and Kwan 1974; Scoville and Solomon 1974), although a minority viewpoint (Zuckerman and Evans 1974) asserted that most clouds could not be in free fall collapse. Zuckerman and Palmer (1974) pointed out that if all such clouds were collapsing freely to make stars, the inferred star formation rate would be more than 10 times the observed one. Zuckerman and Evans (1974) suggested that the suprathermal line widths in molecular clouds were

C. G. Wynn-Williams and D. P. Cruikshank (eds.), Infrared Astronomy, 107–123.

more likely to arise from turbulence, that the clouds were quite irregular, and that star formation was likely to occur near a surface of the cloud.

The next major synthesis occurred around 1978, largely based on the results of CO surveys of the galactic plane (Burton and Gordon 1978; Solomon, Sanders, and Scoville 1979). The discovery of extensive CO emission was interpreted by Solomon, Sanders, and Scoville (1979) to imply that at a galactic radius of 5 kpc, 90% of the mass of the interstellar medium resided in Giant Molecular Clouds or GMC's. These GMC's were also located near essentially all OB associations (Blitz 1980; Sargent 1977, 1979). A typical GMC is elongated, with a greatest linear extent of 90 pc and a mass of 10^5 $M_\odot$.

Because of the enormous amount of mass residing in the GMC component, even the former proponents of cloud collapse were now converted and some (Solomon, Sanders, and Scoville 1979, Scoville and Hersh 1979, Kwan 1979) derived cloud lifetimes, $\tau_{c\ell} > 2 \times 10^9$ years, from considerations of GMC formation. Such a long life-time implies that a GMC would survive several passages through the spiral arms and support for the long lifetimes was found in the apparent failure of the CO to manifest spiral structure. A minority viewpoint was expressed by Bash and Peters (1976) and by Bash, Green, and Peters (1977), based on their model of molecular cloud formation in the spiral density wave shock. They found a best match to the CO surveys if $\tau_{c\ell} = 3 \times 10^7$ years (later revised to 4×10^7 years, Bash 1979), in agreement with the age at which CO emission disappears from young clusters.

In the last year, Cohen et al. (1980) completed their fully sampled CO maps of the first and third galactic quadrants, in which they claim to see clear evidence of spiral structure, implying lifetimes in accord with those suggested by Bash, Green, and Peters (1977). The mass fraction of the interstellar medium in molecular form would then be a more seemly 50% or less. Stark (1979) finds that CO emission in M31 is strongly concentrated in the spiral arms, but this does not seem to be the case in M51 (Rickard, private communication). Detailed studies of individual clouds have indicated a wide range in molecular cloud properties such as size, mass, peak temperature, peak density, and chemical composition (see Evans 1980 for a review of cloud properties). Large amounts of atomic hydrogen have been found in and around molecular clouds (cf. Sato and Fukui 1978). In addition, considerable structure is seen within molecular clouds, on various size scales. Turbulence is now the favored dynamical description, as self-absorbed and complex profiles have been seen in many clouds (cf. Langer et al. 1978). The observed linewidths are correlated with size and mass in a way expected for turbulence (Larson 1980).

One should note that the currently favored lifetime for clouds still exceeds the free-fall time significantly. Rotation (Field 1978), magnetic fields (Mouschovias 1978), and turbulence (Zuckerman and

Evans 1974) have been suggested to provide support against gravity. In the case of turbulence, it has been argued that supersonic turbulence should decay too rapidly to significantly increase $\tau_{c\ell}$ Goldreich and Kwan 1974). This argument is now being questioned on general grounds (Scalo, private communication); also, in the presence of a magnetic field, dissipation may also be decreased via Alfven waves (Zuckerman and Evans 1974; Bash, Hausman, and Papaloizou 1980). Finally, the turbulence may be regenerated by collisions with smaller clouds (Bash, Hausman, and Papaloizou 1980), by stellar winds from T Tauri stars (Norman and Silk 1979), or by nearby H II regions and supernova remnants (Wheeler, Mazurek, and Sivaramakrishnan 1980).

II. Classification of Molecular Clouds

One question which is often raised is which kinds of molecular clouds form stars? The question presupposes the existence of a classification system for molecular clouds; in fact, no satisfactory classification system has ever been developed. Most of the terms used to define cloud types are left-overs from the pre-molecular era. Thus, people refer to dark clouds, dust clouds, diffuse clouds, etc. I have discussed elsewhere why I think this terminology is unnecessarily confusing (Evans 1978). Fortunately, the term "molecular cloud" has come into increasing use as a generic identifier for objects in which molecules are common. The most common sub-division of the genus molecular cloud, is into giant molecular clouds (GMC's) and "dark clouds". Since the latter term merely represents a selection effect (i.e., the cloud is close enough to show up as visual obscuration), nearby GMC's are also dark clouds (e.g., the Orion molecular cloud complex). Therefore, a more useful subdivision may be into "big clouds" and "little clouds", using some arbitrary length scale as the discriminant. Even here, one encounters difficulties because big clouds usually have complicated sub-structure and may equally well be referred to as "cloud complexes". Also the size must be defined consistently; maps of different molecular lines will have different sizes.

Despite this unfortunate confusion, it does appear that big clouds form massive (O and B) stars while little clouds may or may not. It may be that an isolated globule can collapse to form a single massive star, but no example of this presently occurring is known. Rather one finds young clusters of O and B stars near big molecular clouds, with an average size of 90 pc (Blitz 1980). Lower mass stars seem less discriminating in their choice of parent cloud, being found near molecular clouds of all sizes. Even these statements are not entirely secure because selection effects are potentially severe.

Another classification which is sometimes useful is based on the peak gas temperature (T_K) as measured by CO (Evans 1978; Rowan-Robinson 1979). Group A clouds (T_K < 20K throughout) may be explained without invoking local heat sources. Group B clouds (T_K > 20K) require additional heat sources; most Group B clouds have evidence of recent star formation as manifested by compact H II regions, infrared sources, or masers. Thus the question asked to begin this section can be answered in a modified form: Group B clouds are likely to be currently engaged in the formation of massive stars; Group A clouds are not. This answer is not very useful, unless one is looking for star formation regions, because the elevated temperature is a symptom, not a cause of star formation.

Deciding <u>in advance</u> which molecular clouds will form massive stars is at present beyond our capabilities. Much more detailed study of an unbiased sample of molecular clouds is a prerequisite in dealing with this question.

III. Location of Star Formation in Molecular Clouds

Can we specify where, in a given molecular cloud, star formation is occurring? A naive analysis would suggest that stars would form near the center of mass of the molecular cloud, as gravity pulled the outer parts inward in a collapse. This analysis may be correct for a small, isolated, spherical cloud. Indeed, some analysis of globules suggests just such a centrally condensed structure (see Villere and Black 1980). The naive analysis is probably incorrect for larger clouds for two reasons: big clouds are generally very non-uniform and they contain many Jeans masses. In this case, gravitational collapse of randomly located dense regions would tend to produce a more random pattern of star formation, with several different centers of activity, perhaps in different stages of evolution. Such a picture is very commonly seen. In this case, many centers of star formation will lie near the outside of the molecular cloud (Zuckerman and Evans 1974). Even for a uniform density sphere 50 percent of the mass lies in the outer 20 percent of the radius. Elongated and irregular molecular clouds have much larger surface to volume ratios, so that star formation is very likely to occur near <u>some</u> surface region of the cloud. Once formed, the stars are often able to dissipate the remaining layer of molecular material, breaking out to a much lower density medium (cf. Whitworth 1979, Mazurek 1980).

In view of these facts, it is hardly surprising that many young stars are seen near the surface of clouds. Many H II regions can be interpreted in terms of a blister model (Zuckerman 1973, Israel 1978) and such phenomena as Herbig-Haro objects suggest a similar situation for less massive stars. These phenomena in themselves cannot be used to argue that external triggers are needed to induce stellar birth.

On somewhat larger scales, regularities are seen which do suggest the presence of external forces. OB associations are commonly, though not always, arranged in a linear sequence of sub-groups which increase in age with increasing distance from an elongated molecular cloud (cf. Blaauw 1964, Blitz 1980). Elmegreen and Lada (1977) have argued that the expanding H II region from the initial sub-group drives a shock into the molecular cloud, triggering the formation of the next sub-group by compressing a layer of the molecular cloud adjacent to the H II region. This "sequential star formation" finds considerable support in the morphology of many OB associations and the molecular clouds that have produced them. Other external triggers which have been proposed for particular situations are supernovae (Herbst and Assousa 1977), cloud-cloud collisions (Loren 1976), and the spiral density wave (Woodward 1976).

It is difficult to study the location of star formation within a molecular cloud by studying the location of well-developed H II regions and optically visible objects like OB associations and Herbig-Haro objects. If a star has produced a well-developed H II region, then it has probably disrupted a sizeable portion of the molecular cloud. The location of optically visible objects will be selectively near the front edge of the cloud. Thus infrared observations offer two advantages: objects which are more deeply embedded can be detected; and the location of massive stars can be studied before they have altered too drastically the surrounding terrain. Unfortunately, complete surveys of molecular clouds are extremely tedious and hence rare, especially at the longer wavelengths where identification of young objects is more clear-cut. Nadeau (unpublished) has covered large areas in several molecular clouds at 10μm; generally speaking, no sources were found outside already known centers of activity, as marked by CO hot spots or compact H II regions. Unfortunately, a significant number of very cool sources have now been found (Beichman, Becklin, and Wynn-Williams 1979) which would not be detected at 10μm. Complete surveys at longer wavelengths will probably await the success of IRAS.

Meanwhile, the most useful work has been done by Beichman (1979) who surveyed 19 hot spots in Group B molecular clouds at 20μm. About half the clouds were found to contain embedded infrared objects. If the hot spot was also known to be dense, the detection probability was near unity. While hardly free of selection effects, this study does support the view that stars are forming in the hottest, densest parts of molecular clouds. Even within the region searched at 20μm, Beichman found that the sources were tightly concentrated in a single center of activity. The projected separation of the infrared sources from the edge of the cloud is larger than the thickness of the shocked layer, where one would expect to see the sources if their formation were induced by the nearby H II region. Thus, these observations do not support the detailed predictions of the sequential star formation scenario, and Beichman (1979) argues that the apparent tendency for stars to form near the surface of a cloud is based on selection effects caused by the short wavelengths previously used to find the young stars.

Beichman (1979) also found that multiple infrared sources were often found within a single hot spot (see also Wynn-Williams et al. 1980). The mean separation between sources is 0.17 pc, very similar to that found in compact clusters of O and B stars, like the Trapezium, and not much greater than the mean separation of wide visual binaries. He argues that fragmentation during gravitational collapse provides the best explanation, both of wide binaries and of compact clusters, while bifurcation of a rotating fragment leads to close visual binaries, as also suggested by the work of Abt and Levy (1978).

On a larger scale, Elmegreen (1980) has used I band photography to study a $2^\circ \times 3^\circ$ field including W3 and most of W4. About 135 stars were found which were brighter at I than at R; Elmegreen suggests that most are young stars embedded in the cloud. They tend to cluster near regions of peak CO, optical, or far-infrared activity. This technique may prove useful for studying the fainter, but less obscured members of a nascent open cluster.

IV. Properties of Infrared Sources

A. Size

Determining the size of infrared sources, one encounters problems similar to those which make it hard to determine the size of molecular clouds. The size clearly depends on the wavelength. At millimeter and sub-millimeter wavelengths the emission may be coextensive with the molecular cloud. At such long wavelengths the emission is close to Rayleigh-Jeans and even the low temperature envelopes of clouds may emit. Such extensive emission may be very hard to detect unless large beams and very large chopper throws are employed. In this regime, the dust emission is optically thin and studies of dust-to-gas ratios, and molecular depletion may be pursued by comparing the optical depth, τ_λ, to N_{13}, the ^{13}CO column density. Moving to shorter wavelengths, the source shrinks, as only the relatively warm dust grains emit substantially. Somewhere around 10μm the typical infrared source becomes smaller than can be resolved normally (ignoring extended H II regions). Here occultations and Michelson or speckle interferometry must be used to supply a size. The data are difficult to obtain and only exist for a few sources.

In some cases, the infrared source is resolved into a double source (W3IRS5 Low 1980; Mon R2IRS3 Beckwith et al. 1976). In other cases, (GL2591, the Becklin-Neugebauer object, S255, and S140) the source remains single and is resolved (Low 1980; Foy et al. 1979; Neugebauer, private communication). Generally speaking, the available sizes are consistent with observed luminosities and inferred dust temperatures. The size is a crucial input into the question of evolutionary state since massive stars will eventually produce an H II region which expands into an extended infrared source, even in the near-infrared. Thus the size and structure of extremely compact sources should be related roughly to an evolutionary sequence.

Studies of the shape of compact infrared sources are also of interest, since deviations from spherical symmetry are suggested by the presence of infrared reflection nebulae (Werner et al. 1979). Other studies have also invoked aspherical geometries (Harvey, Campbell, and Hoffman 1977; Beichman, Becklin, and Wynn-Williams 1979) and these are also expected on theoretical grounds if rotation plays a role in the collapse.

B. Energy Distribution and Luminosity

The difficulties in establishing sizes spill over into the effort to determine the energy distribution, especially in the presence of multiple sources. Typically several near infrared sources are embedded within a region of extended far-infrared and even more extended sub-millimeter emission. With no a priori way to apportion the far-infrared energy to the various near-infrared sources, one can only define the energy distribution and luminosity of the source complex. Thus the search for the elusive simple, solitary source is a worthwhile, if frustrating, one.

In this situation, only a few generalities are possible. The most well established is that the bulk of the energy from molecular clouds does appear in the far-infrared. Typical energy distributions are shown by Werner, Becklin, and Neugebauer (1977). The shape of the energy distribution shortward of about 50μm shows considerable variation from source to source. The spectra of some sources drop very steeply with decreasing wavelength. Others show a nearly power law energy distribution through the near-infrared (cf. Beckwith et al. 1976). As shown by Kwan and Scoville (1976) the latter behavior can be understood if the emission arises from dust at a wide range of temperatures. This situation is often present in well-evolved compact H II regions embedded in molecular clouds. Here the dust in the H II region reaches temperatures of several hundred kelvins and is extended sufficiently to contribute substantially to the energy distribution.

In contrast, the sources with steep energy distributions in the near-infrared are often radio-quiet or have only a very compact H II region. It is tempting to assign these sources to an earlier evolutionary state, but caution must be exercised. Heavy overlying extinction can redden the energy distribution substantially. In general, it is very hard to distinguish an energy distribution which is intrinsically steep from one which has been selectively extinguished. This can be illustrated with the spectra of Type II OH masers which can have extremely steep energy distributions and deep silicate features even though they are not embedded in molecular clouds (Evans and Beckwith 1977). By the same token, some compact H II regions have quite steep spectra, suggesting heavy extinction. In this case, specific spectral lines such as Brα and Brγ can be used to estimate the extinction and hence allow de-reddening of the energy distribution. For sources without such lines, the intrinsic spectrum cannot be disentangled from the overlying extinction. Indeed, there may be no real distinction between them in the very early evolutionary stages, before dust shells and H II regions form.

Finally, another type of energy distribution can be identified: the far-infrared emission dominates as usual, but the source is reasonably strong in the near-infrared. The energy distribution may be fairly flat in the near-infrared, and in some cases a star is even visible optically. The striking feature of these sources is that emission in the middle-infrared (5-30μm) is weak. Thus, these sources must have little dust at temperatures from 100-500 K. Since this is hard to explain with a continuous distribution of dust, it is likely that inhomogeneities play a role. The extreme of this class occurs when a normal main-sequence star exists near a molecular cloud, but does not have significant dust in a shell around it. These sources are often marked by reflection nebulae, and a nice example is NGC 2023 (Harvey, Thronson, and Gatley 1980). Because powerful optical spectroscopy techniques can be brought to bear on these objects, they promise to be very useful sources for separating processes which are entangled in embedded sources.

C. Spectral Features

The difficulty of separating the intrinsic spectrum from effects of overlying extinction increases the importance of specific spectral features. The most directly useful are recombination lines and molecular bands. If the source has produced a significant volume of ionized gas, recombination lines such as Brα and Brγ may be detectable, even when the H II region is too compact to show radio continuum emission. If more than one line can be detected, recombination theory can be invoked to predict the relative strength of the lines, and the observed lines can then be used to determine both the extinction and the flux of ionizing photons (cf. Thompson and Tokunaga 1979). Measurement of the total luminosity could then allow us to begin to place young stars on a modified HR diagram (see Thompson review). Molecular bands have been less widely used, owing to the weakness of most sources at relevant wavelengths. The 2.3μm CO overtone band has been used to distinguish post-main-sequence objects from young ones, while studies of the 4.7μm CO fundamental band with sufficient resolution and sensitivity can indicate the amount of molecular gas in front of the object. Such studies have also revealed large mass motions around the Becklin-Neugebauer object (Hall _et al._ 1978) and can help to distinguish collapse from expansion. The remarkable phenomena of the H_2 emission in vibration-rotation lines (see Beckwith review) and emission from CO rotational lines as high as $J=22\rightarrow21$ (Watson _et al._ 1980) are precursors of the exciting discoveries awaiting extension of spectroscopy throughout the infrared. Much of this work will use the infrared source as a convenient background lamp for doing absorption spectroscopy of the intervening molecular cloud, but studies have also shown that phenomena such as expanding shells and shocks associated with activity of the source itself are able to heat some of the gas enough to produce emission in the infrared.

Another category of spectral features are the broad band features, presumably associated with broad-band resonances in the dust grains. While these are primarily diagnostic of the dust material (see review by Aitken), they have also been used to obtain information on the source itself. For example, the 9.7μm absorption feature, generally attributed to silicates, has been used to estimate extinction, as calibrated by studies of stars whose intrinsic spectrum is known or can be deduced. Values of $A_V/\tau_{9.7}$ range from 8 (Becklin et al. 1978) to 13 (Gillett et al. 1975) and 23 (Rieke 1974). Direct application to sources deeply embedded in molecular clouds is suspect because the dust immediately surrounding the ultimate energy source may produce the feature either in emission or in absorption as illustrated by the radiative transport of Kwan and Scoville (1976). Nonetheless, the depth of the 9.7μm feature is useful as a qualitative indicator, at least, of how deeply buried an object must be. Other spectral features (e.g., the 3.1μm "ice" feature and the 6.0 and 6.8μm un-identified features) seem to be particularly diagnostic of extinction by cold molecular cloud material and may be useful measures of extinction, if they can be calibrated.

D. Evolutionary State

It is tempting to arrange the various types of energy distributions discussed above into an evolutionary sequence, but differences in stellar mass, geometry and initial conditions may complicate matters greatly (cf. Habing and Israel 1979). To begin with the most general view, one would want to consider pre-main-sequence, main-sequence, and post-main-sequence stars as possible infrared sources in molecular clouds. Post-main-sequence stars have generally been discounted, although the presence of an SiO maser in the Orion infrared cluster has been used to raise the possibility that a post-main-sequence object is present. The energy distribution alone cannot always be used to rule out post-main-sequence objects. Type II OH masers are post-main-sequence objects with energy distributions which can be disturbingly similar to those seen in molecular clouds (Evans and Beckwith 1977). As isolated objects, they are relatively weak in the far-infrared (Werner et al. 1980) but this would not necessarily be true if one were embedded in a molecular cloud.

Given the density of giants of all types of 6×10^{-4} pc^{-3}, the average GMC could contain over 20 giant stars purely by accident. Thus, the possibility that a given infrared source in a GMC is a post-main-sequence object cannot be discounted. The chance that a giant star accidently lies in a molecular cloud core or in a nearby small cloud (both with typical diameters of 1 pc) is much less: $\sim 3 \times 10^{-4}$.

Another possibility is that a star might exhaust its main sequence life without leaving its parent cloud. If the age of a molecular cloud is 4×10^7 years, the mass of such a star would have to be greater than $\sim 10\ M_{\odot}$. If the star had a random velocity component of 1 km s^{-1}, it would travel 40 pc in its lifetime, probably escaping the cloud. To be confined within the 1 pc diameter of the dense core the star would have to live $< 10^6$ years, implying a mass $> 40\ M_{\odot}$. Such a massive star would surely disrupt the cloud core during its lifetime. Thus it appears safe to conclude that infrared sources in dense cloud cores are not post-main-sequence objects, but that sources found more generally in molecular cloud envelopes could be.

The more difficult question is then to separate main-sequence stars from pre-main-sequence stars or protostars. Since the terminology can be confusing, let us concentrate on the question of whether ionizing photons are being produced yet. If the star in question is massive enough, this question can be answered in principle by observations in the radio continuum or by observation of infrared recombination lines. If no evidence of ionizing photons can be found, at a level below that expected for a main-sequence star of the same luminosity, the term "protostar" has often been applied. (Recently more modest descriptions, such as "extreme youth" have been favored.) Note that application of this criterion depends on having sufficient sensitivity to detect radio or recombination line fluxes expected from main-sequence stars. Further, the luminosity, usually concentrated in the far-infrared, must be assumed to arise from a single object (or somehow be apportioned among a number) in order to predict the flux of ionizing photons from a main-sequence star. Finally, note that stars which have only recently begun to produce ionizing photons may have H II regions so compact as to be optically thick at radio frequencies and hence weaker than expected, as well as heavy dust shells which attenuate substantially even the infrared recombination lines. The upshot is that conclusive proof that an object is not yet producing ionizing photons is very difficult to obtain. A candidate "protostar" is always susceptible to the ravages of time and improved sensitivity. Thus the Becklin-Neugebauer object, long considered the prototype of this group, fell from grace with the discovery of Brα emission (Grasdalen 1976).

The other question involved in separating protostars from main-sequence stars is whether material is still accreting. High spectral resolution studies have been able to identify outflow around some objects, and continued work in this area may locate sources with infalling material.

V. Energetics

It was suggested earlier that one useful classification system for molecular clouds was to ask whether they contain a hot spot ($T_K > 20$ K). Those with hot spots usually are engaged in active star formation. It seems natural then to suggest that young or forming stars supply the energy for the hot spot. Goldreich and Kwan (1974) suggested a specific mechanism: the dust in the molecular cloud absorbs photons emitted by the star or by the H II region, if present, and is warmed to $T_D \sim 20 - 100$ K; collisions of gas molecules with these warm grains leads to an equipartition of energy and the gas molecules leave with $3/2\ k\ T_D$ translational energy. Thus, as long as $T_D > T_K$, the gas will be heated by these collisions. A consequence of this picture is that copious far-infrared emission should be produced by the warm dust grains.

In order to formulate a more quantitative test of this picture, Evans, Blair, and Beckwith (1977) used CO and ^{13}CO observations of the S255 molecular cloud to predict far-infrared luminosities. They assumed that $T_D = T_K$, a minimum condition, and that the shells of various T_K defined by the CO observations emitted like gray-bodies at T_D with an optical depth τ_{FIR}. From data on a variety of sources, they found that $\tau_{FIR} = 10^{-18}\ N_{13}$ was roughly satisfied. The far-infrared emission from the S255 molecular cloud has been observed (Werner, private communication; Sargent et al. 1980). Blair et al. (1978) applied the same technique to the S140 molecular cloud and Harvey, Campbell, and Hoffman (1978), Rouan et al. (1977), and de Muizon et al. (1980) observed far-infrared emission at about the expected level. Similar results were found in several other clouds [S235 (Evans and Blair 1980; Evans et al. 1980); W3-W4 (Thronson, Campbell, and Hoffman 1980)]. The dust-gas heating scheme appears to be plausible at least in dense, hot cores of molecular clouds.

On the other hand, some discrepancies have also been found. Crutcher, Hartkopf, and Giguere (1978) applied the techniques of Evans, Blair, and Beckwith (1977) to the NGC 2264 molecular cloud and predicted about 10 times the observed (Harvey, Campbell, and Hoffman 1977) far-infrared luminosity. To check this discrepancy, the region should be observed with a large beam in the far-infrared, because Sargent and Van Duinen (paper presented at this conference) have found far-infrared luminosities to be as much as 10 times larger when measured with large beams. Even in the case of S140, Blair et al. noted that the far-infrared emission was predicted correctly only from the regions very close to the middle-infrared source, S140IR. The extended plateau of modest T_K was not seen by Harvey et al. (1978). Subsequent observations by de Muizon et al. (1980) indicate far more extensive far-infrared emission and they conclude that $T_D > T_K$ as far as 20' from S140IR. Still, the densities thought to exist in those regions are insufficient to effectively couple T_D and T_K. The energetics of these extended plateaus of warm gas are still not adequately explained. More striking discrepancies have been found by Lada and Wilking (1980a). The ρ Ophiuchi molecular cloud has regions of quite high T_K (cf. Loren, Evans, and Knapp 1979) which do not show substantial far-infrared emission (Fazio et al. 1976).

A particularly hard limit is available at a position of ^{13}CO self-absorption (Lada and Wilking 1980a) where $N_{13} = 2.5 \times 10^{17}$ cm^{-2} and $T_K \geq 35$ K. Predictions of far-infrared fluxes from dust at $T_D = 35$ K and with τ_{FIR} inferred from molecular lines far exceed the observed upper limit of 10 Jy. In other cases, no stellar or proto-stellar heat sources have been found which are luminous enough to balance dust cooling rates inferred from the molecular data. An example is B35 (Lada and Wilking 1980b). Since some infrared sources have not appeared until wavelengths as long as 20μm were used to search for them, these results should be viewed with caution until corroborating far-infrared observations are made. Nonetheless, it is certainly plausible that other energy sources exist and may produce CO hot spots (cf. Elmegreen, Dickinson, and Lada 1978). Shock waves, magnetic ion-slip, and dissipation of turbulence are all processes which can heat the gas directly, thus avoiding the much greater energy requirements involved in heating the dust.

VI. Future Prospects

I will conclude by reviewing some of the areas in which future work may be of the greatest value.

Our understanding of infrared sources in molecular clouds is impeded by the absence of systematic surveys. These are now becoming available in CO from the Columbia telescope, but surveys in other lines would also be of value, as would surveys from the sourthern hemisphere. A number of surveys exist in the infrared, but none are completely satisfactory. The IRAS survey promises to revolutionize this field. With an unbiased sample in hand, we can hope to develop useful classification schemes, determine the location of star formation in molecular clouds, and determine a luminosity function for infrared sources in molecular clouds.

The size and shape of the dust distributions around newly formed stars should be determined and compared to theoretical models. In the near-infrared and middle-infrared, interferometric techniques will be needed to achieve adequate resolution. Larger telescopes, or separate telescopes, will be needed to provide appropriate baselines. The techniques have been pioneered and suitable telescopes are being planned. In the far-infrared, a large airborne or space telescope is essential for obtaining adequate resolution. The observations must be interpreted with radiative transport models to determine the distribution of dust temperature and optical depth. Comparably high resolution in the radio molecular lines, again with appropriate radiative transfer models, is needed to determine gas temperature and density distributions; with these data in hand, a full understanding of the energetics of both gas and dust will become possible. A more immediately achievable goal is to measure the total luminosity of molecular clouds with large beams and large chopper throws.

A third area where important advances are likely is in spectroscopy. With improved resolution and sensitivity, the velocity, temperature, and density structure of molecular clouds can be probed on a very fine spatial scale, using the infrared source as a background lamp. High excitation lines may also be used to study the role of shocks in molecular clouds. Finally, the long-standing question of the nature of the dust grains may be answered by higher resolution studies of some of the new, unidentified features.

This research has been supported in part by NSF Grant AST 79-20966, by NASA Grants NSG-7381 and NSG-2345, and by the Research Corporation. Colleagues too numerous to mention individually have provided preprints and helpful discussion.

REFERENCES

Abt, H. A. and Levy, S. G. 1978: Astrophy. J. (Suppl.), 36, 241.
Bash, F. N. 1979: Astrophys. J., 233, 524.
Bash, F. N., Green, E., and Peters, W. L., III, 1977: Astrophys. J., 217, 464.
Bash, F., Hausman, M., and Papaloizou, J. 1980: Astrophys. J., in press.
Bash, F. N., and Peters, W. L., III 1976: Astrophys. J., 205, 786.
Becklin, E. E., Matthews, K., Neugebauer, G., and Willner, S. P. 1978: Astrophys. J., 220, 831.
Beckwith, S., Evans, N. J., II, Becklin, E. E., and Neugebauer, G. 1976: Astrophys. J., 208, 390.
Beichman, C. 1979: Unpublished Ph.D. thesis, University of Hawaii.
Beichman, C. A., Becklin, E. E., and Wynn-Williams, C. G. 1979: Astrophys. J. (Letters), 232, L47.
Blaauw, A. 1964: Ann. Rev. Astron. Astrophys., 2, 213.
Blair, G. N., Evans, N. J., II, Vanden Bout, P., and Peters, W. L., III 1978: Astrophys. J., 219, 896.
Blitz, L. 1980: Giant Molecular Clouds in the Galaxy, P. M. Solomon, and M. G. Edmonds (Oxford: Pergamon).
Burton, W. B., and Gordon, M. A. 1978: Astr. and Ap. 63, 7.
Cohen, R. S., Cong, H., Dame, T. M., and Thaddeus, P. 1980: preprint.
Crutcher, R. M., Hartkopf, W. I., and Giguere, P. T. 1978: Astrophys. J., 226, 839.
de Muizon, M., Rouan, D., Lena, P., Nicollier, C., and Wijnbergen, J. 1980: Astr. and Ap., 83, 140.
Elmegreen, B. G., Dickinson, D. F., and Lada, C. J. 1978: Astrophys. J., 220, 853.
Elmegreen, B. G., and Lada, C. J. 1977: Astrophys. J., 214, 725.
Elmegreen, D. M. 1980: Astrophys. J., in press.
Evans, N. J., II, 1978: Protostars and Planets, ed. T. Gehrels (Tucson: University of Arizona Press) pg. 153.
Evans, N. J., II, 1980: IAU Symposium 87, Interstellar Molecules, ed. B. Andrew (Dordrecht: Reidel) pg. 1.

Evans, N. J., II, and Beckwith, S. 1977: Astrophys. J., 217, 729.
Evans, N. J., II, Beichman, C., Gatley, I., Harvey, P., Nadeau, D., and Sellgren, K. 1980: in preparation.
Evans, N.J., II, and Blair, G. N. 1980: in preparation.
Evans, N. J., II, Blair, G. N., and Beckwith, S. 1977: Astrophys. J., 217, 448.
Fazio, G. G., Wright, E. L., Zeilik, M., II, and Low, F. J. 1976: Astrophys. J. (Letters), 206, L165.
Field, G. 1978: Protostars and Planets, ed. T. Gehrels (Tucson, Univ. of Arizona Press), pg. 243.
Foy, R., Chelli, A., Sibille, F., and Lena, P. 1979: Astr. and Ap., 79, L5.
Gillett, F. C., Jones, T. W., Merrill, K. M., and Stein, W. A. 1975: Astr. and Ap., 45, 77.
Goldreich, P., and Kwan, J. 1974: Astrophys. J., 189, 441.
Grasdalen, G. L. 1976: Astrophys. J. (Letters), 205, L83.
Habing, H. J., and Israel, F. P. 1979: Ann Rev. Astron. and Astro-Phys., 17, 345.
Hall, D. N. B., Kleinman, S. G., Ridgway, S. T., and Gillett, F. C. 1978: Astrophys. J., 223, L47.
Harvey, P. M., Thronson, H. A. Jr., and Gatley, I. 1980: Astrophys. J., 235, 894.
Harvey, P. M., Campbell, M. F., and Hoffman, W. F. 1978: Astrophys. J., 219, 891.
Harvey, P. M., Campbell, M. F., and Hoffman, W. F. 1977: Astrophys. J., 215, 151.
Herbst, W., and Assousa, G. E. 1977: Astrophys. J., 217, 473.
Israel, F. P. 1978: Astr. and Ap., 70, 769.
Kwan, J. 1979: Astrophys. J., 229, 567.
Kwan, J., and Scoville, N. 1976: Astrophys. J., 209, 102.
Lada, C. J., and Wilking, B. A. 1980a: Astrophys. J., in press.
Lada, C. J., and Wilking, B. A. 1980b: Astrophys. J., in press.
Langer, W. D., Wilson, R. W., Henry, P. S., and Guelin, M. 1978: Astrophys. J. (Letters), 225, L139.
Larson, R. B. 1980: preprint.
Loren, R. B. 1976: Astrophys. J., 209, 466.
Loren, R. B., Evans, N. J., II, and Knapp, G. R. 1979: Astrophys. J., 234, 932.
Low, F. 1980: "IR Interferometry", in Optical and Infrared Telescopes for the 1990's, (Tucson: Kitt Peak National Observatory) pg. 825.
Mazurek, T. J. 1980: Astr. and Ap., in press.
Mouschovias, T. Ch. 1978: Protostars and Planets, ed. T. Gehrels, (Tucson: University of Arizona Press), pg. 209.
Norman, C., and Silk, J. 1979: Astrophys. J., in press.
Rieke, G. H. 1974: Astrophys. J. (Letters), 193, L81.
Rouan, D., Lena, P. J., Puget, J. L., de Boer, K. S., and Wijnbergen, J. J. 1977: Astrophys. J. (Letters), 213, L35.
Rowan-Robinson, M. 1979: Astrophys. J., 234, 111.

Sargent, A. I. 1977: Astrophys. J., 218, 736.
Sargent, A. I. 1979: Astrophys. J., 233, 163.
Sargent, A. I., Nordh, H. L., Van Duinen, R. J., and Aalders, J. W. G. 1980: Astr. and Ap., in press.
Sato, F., and Fukui, Y. 1978: Astron. J., 83, 1607.
Scoville, N. Z., and Hersh, K. 1979: Astrophys. J., 229, 578.
Scoville, N. Z., and Solomon, P. M. 1974: Astrophys. J. (Letters), 187, L67.
Solomon, P. M., Sanders, D. B., and Scoville, N. Z., 1979: in IAU Symposium 84, The Large Scale Characteristics of the Galaxy, ed. W. B. Burton (Dordrecht: Reidel) pg. 3.
Stark, A. 1979: unpublished dissertation, Princeton.
Thompson, R. I., and Tokunaga, A. T. 1979: Astrophys. J., 231, 736.
Thronson, H. A., Jr., Campbell, M.F., and Hoffman, W. F. 1980: Astrophys. J., in press.
Villere, K. R., and Black, D. C. 1980: Astrophys. J., 236, 192.
Watson, D. M., Storey, J. W. V., Townes, C. H., Haller, E. E., and Hansen, W. L., 1980: Astrophys. J. (Letters), in press.
Werner, M. W., Becklin, E. E., Gatley, I., Matthews, K., Neugebauer, G., and Wynn-Williams, C. G. 1979: M. N. R. A. S., 188, 463.
Werner, M. W. , Becklin, E. E., and Neugebauer, G. 1977: Science, 197, 723.
Werner, M. W., Beckwith, S., Gatley, I., Sellgren, K., Berriman, G., and Whiting, D. L. 1980: Astrophys. J., in press.
Wheeler, J. C., Mazurek, T. J., and Sivaramakrishnan 1980: Astrophys. J., in press.
Whitworth, A. 1979: M.N.R.A.S., 186, 59.
Woodward, P. R. 1976: Astrophys. J., 207, 484.
Wynn-Williams, C. G., Bechlin, E. E., Beichman, C. A., Capps, R., and Shakeshaft, J. R. 1980: preprint.
Zuckerman, B. 1973: Astrophys. J., 183, 863.
Zuckerman, B., and Evans, N. J., II 1974: Astrophys. J. (Letters), 192, L149.
Zuckerman, B., and Palmer, P. 1974: Ann. Rev. Astron. and Astrophys. 12, 279.

DISCUSSION FOLLOWING PAPER DELIVERED BY N. J. EVANS

SARGENT: You have referred to Beichman's assertion that he finds protostars to be too deeply embedded in molecular clouds for the mechanism formalized by Elmegreen and Lada to be valid. Beichman used the present positions of visible H II regions to derive his result. Certainly in the case of the Cepheus OB3 association, if the positions of the stars are projected backwards in time to their birthsites the scales determined by Elmegreen and Lada hold well.

RICKARD: Two comments about the interpretation of galactic plane CO surveys: First, Liszt and Burton (1980, preprint) have shown that, because of blending, the GMC interpretation of the first quadrant data (i.e., domination by >50 pc clouds) is not self-consistent. Intercomparisons of models and data suggest that most of the molecular gas is in smaller (~20 pc) clouds. Second, it may be fair to say that the CO surveys show as much spiral structure as the H I surveys. But, as Burton has shown, the apparent patterns are purely kinematical phenomena. There is little evidence for real enhancements in the density of CO clouds within spiral arms. M31 may be different; but even there, the arm-interarm contrast is not large (less than a factor of three).

EVANS: The question of the sizes of molecular clouds is a real problem because their structure is so complicated. You can equally well describe something as a complex of clouds or as a group of unrelated, overlapping clouds.

T. L. WILSON: The CO cloud sizes are biased, perhaps, to lower density gas. In cloud mass and lifetime calculations the gas densities in the cores of these clouds should be considered.

EVANS: It is true that the CO can mislead us because it is so extended compared to everything else, but the dense cores are much more confined, typically 1 pc in size.

SCOVILLE: In my opinion much of the confusion on the question of molecular clouds in spiral arms or out of them has resulted merely from imprecise definition of the question and what constitutes a regular spiral pattern. Thus, although the CO data in both our survey (Sanders, Solomon, and Scoville 1980) and that done at Columbia (Cohen, Dane, and Thaddeus 1980) show regular arms in the outer galaxy (e.g., the Perseus arm), the two groups do disagree on the interpretation in the interior of the galaxy. If one is willing to take liberty with the definition and the required regularity for the "spiral" arms, then it is of course possible to connect up most of the CO features in the inner galaxy into a pattern. In our opinion such a procedure does not fairly test the coherence of the pattern since the observations themselves have been used in a circular fashion to define the pattern. For example, the "arm" traced out at lowest longitude ($\ell = 30°$) in the ℓ/V picture of Cohen *et al*. does not really have a reasonable spatial/kinematic behavior to be a density wave spiral arm. The most clearcut test of whether the clouds exist in both arms and interarms occurs for the emission at the terminal velocity of each longitude. The observed continuity near the terminal velocity out to $\ell = 50°$ is inconsistent with what one would expect from clouds confined to just one or two arms here. This is not to say that there are not large structures (~300 pc long) in the cloud distribution. It is simply that the pattern is irregular, perhaps like that in an Sc galaxy.

EVANS: I think that the question is still open, and depends on whether you emphasize the fact that there is some contrast which indicates spiral arms, or whether you emphasize the fact that you sometimes see CO in places there are not supposed to be any spiral arms.

RICKARD: Scoville's comment that CO spiral arms may only be well-defined outside the solar circle has a parallel in the Westerbork H I map of M51, in which the arms are not seen in strong contrast until outside the bright optical disk of the galaxy.

HABING: You commented that IRAS was needed to make an unbiased survey for regions of star formation. I suggest that masers are also ideal for this purpose since, as Downes and Genzel have shown, they are easy to map out throughout the Galaxy.

EVANS: This is clearly a useful technique, but I have questions as to whether the lifetime of the maser phenomenon is long enough to ensure that you find all the star formation regions.

GLOBULES, DARK CLOUDS, AND LOW MASS PRE-MAIN SEQUENCE STARS

A. R. Hyland
Mount Stromlo and Siding Spring Observatories
The Australian National University
Canberra, Australia

ABSTRACT

The current observational and theoretical literature on Bok globules and their relationship to star formation is reviewed. Recent observations of globules at optical, infrared, and far infrared wavelengths are shown to provide important constraints on their structure and evolutionary status, and the suggestion that many globules are gravitationally unstable is seriously questioned.

Dark clouds associated with T associations are well-known sites of recent and continuing star formation. In recent years molecular observations and far infrared surveys have provided maps of such regions from which possible sites of star formation may be identified. Optical (Hα) and near infrared surveys have enabled a clear identification of pre-main sequence (PMS) objects within the clouds. Methods of distinguishing these from background objects and the nature of their infrared excesses are examined in the light of recent observations in the near and far infrared. The perennial question as to the existence of anomalous reddening within dark clouds is also investigated.

1. GLOBULES

The large globules (Bok and Reilly 1947) are fascinating to observer and theoretician alike, for, although their status is not crystal clear, recent evidence from optical and radio studies (e.g., Dickman 1977) suggests that the majority of globules are in a state of gravitational collapse, which may eventually lead to star formation (Bok 1977).

There is also direct observational evidence on this question to be drawn from the discovery by Herbst and Turner (1976) of possible embedded sources in Lynds 810, and the association of two Herbig-Haro objects with the southern globule 210-6A (Schwartz 1977; Bok 1978). On the theoretical side, there are certain difficulties in collapse

C. G. Wynn-Williams and D. P. Cruikshank (eds.), Infrared Astronomy, 125–151.

calculations, which have been succinctly spelled out by Buff et al. (1979), who conclude that "[given the practical impossibility of knowing the initial spectrum of perturbations] there will be no unique solution for a collapsing cloud."

With these problems in mind it would appear that increased knowledge of the evolutionary status of globules depends to a large extent on continuing observational studies. It is important, therefore, that these observational studies be able to determine the correct physical parameters for the globules, which is not as simple as might be supposed.

1.1 Description and Methods of Study

The large globules are generally round in appearance, dense (i.e., opaque), and come in a variety of apparent sizes from a few minutes to half a degree in diameter. Photographs of a selection of globules are given by Bok (1977) and Bok et al. (1971), and in Figure 1, Coalsack globules 1, 2, and 3 (Bok et al. 1977) are shown.

Typical methods of study in the optical are from the reddening of background objects, and a variety of methods of star counts. These have been beautifully summarised by Bok and Cordwell (1973), and have been the basis for most optically determined parameters for globules, e.g., Bok and McCarthy 1974; Tomita et al. 1979; Schmidt 1975. Unfortunately, the single direct method of determining the density

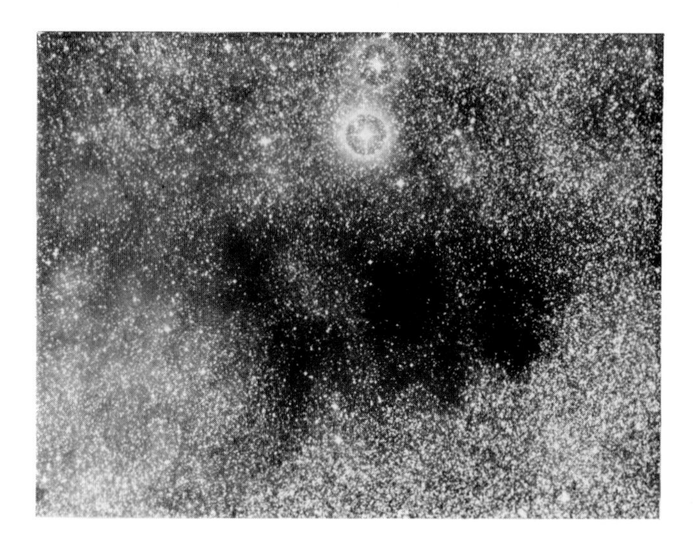

Figure 1. A positive print of the Coalsack globules 1, 2, and 3. Globule 2 is the very dense symmetrical central cloud. North is up and east to the left. The print covers an area approximately 56' x 42'. Taken from a IV N Schmidt plate courtesy of the UK Schmidt Telescope Unit.

distribution within a globule by optical means (i.e., by reddening determinations) can only be applied to the less opaque members of the class. Even the indirect star count methods run into difficulty towards the cores of the globules, as the number of observable objects tends to zero.

The advent of sensitive radio observations (particularly of the ^{13}CO molecule) has revolutionised the study of globules (Martin and Barrett 1978; Dickman 1978a, 1978b). From these observations it has been possible to determine not only the mean density of the globules, but also their gas temperatures (on the order of 10 K), and an indication of their dynamical properties. In particular, the existence of suprathermal broadening of the CO lines has been interpreted in terms of velocity gradients symptomatic of collapse. Dickman (1978a) obtained similar masses for individual globules from both optical and radio techniques with the assumptions that

$$CO/N_{H_2} = 2 \times 10^{-6} \text{ and } N_{H_2}/A_V = 1.25 \times 10^{21} \text{ cm}^{-2} \text{ mag}^{-1} \quad (1)$$

(Jenkins and Savage 1974; Dickman 1978b). The latter value has been shown to be consistent for A_V up to 5 and probably as high as 10. Beyond this nothing is known. For $^{13}CO/N_{H_2}$ however, values as low as 100 times smaller than Dickman's value have been determined in dense clouds by Wootten et al. (1978), by combining CO and H_2CO measurements with chemical equilibrium models. The dynamical collapse models of Villere and Black (1980) also suggest lower values of $^{13}CO/N_{H_2}$ in regions of high density as is shown in Figure 2.

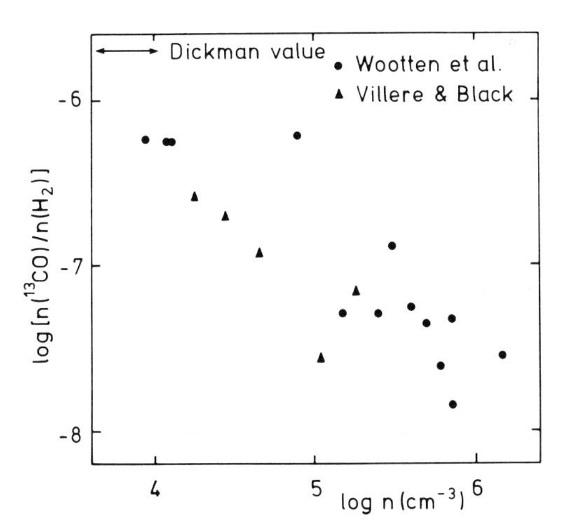

Figure 2. Relationship between the number density of ^{13}CO molecules and that of molecular hydrogen ($^{13}CO/N_{H_2}$), as a function of the total particle density. Filled circles are observationally deduced points from Wootten et al. (1978); triangles are from the best fit dynamical models of Villere and Black (1980) for well-observed globules. Dickman's (1978b) mean value for low densities is also given.

TABLE 1

OBSERVED PARAMETERS OF SELECTED GLOBULES

Globule	Adopted Distance* (pc)	Temperature (K)	Radius (pc)	Mass ($M_\odot$)			
				Tomita *et al.*	Dickman	Bok *et al.*	Others
B34	200[a]		0.6	3.7 (6)†		8[d]	
L134	200[a]	13	0.7	41 (65)	66		280[f]
B134	400[a]	9	0.5	4.3 (7)	19	1[e]	
B335	400[b]	9	0.3	3.6 (5.8)	≥23	12[e]	22[f], 170[g]
B361	350[a]		1.0	36 (58)		17[d]	86[h]
Coalsack 2	175[c]		0.3			25[c]	11[i]

NOTES: *The mass and radii have been corrected for the distance adopted.
†Numbers in parentheses are Tomita *et al.*'s values adjusted upward by the factor 1.61 (see text).

REFERENCES:
[a]Tomita *et al.* 1979
[b]Dickman 1978a
[c]Bok 1977
[d]Bok *et al.* 1971
[e]Bok and McCarthy 1974
[f]Martin and Barrett 1978
[g]Villere and Black 1980
[h]Schmidt 1975
[i]Jones *et al.* 1980

1.2 Derived Globule Parameters

The crucial parameters sought in all globule studies are their total masses, radii, and temperatures. Once these are known it is possible to compare the observed position in a mass/radius diagram with predictions of the virial theorem at a given temperature. It might be expected that the observational situation in globules is clear-cut, but unfortunately, nothing could be further from the truth. It is exceedingly disconcerting to see the disparity in values of the mass for a given globule, in the published literature.

In Table 1 are collected together a sample of the derived physical characteristics of some of the best-studied globules. As one can readily see the masses obtained by different authors from similar techniques differ by factors up to 50. A detailed examination of these papers reveals that masses are obtained from visual absorption data alone in slightly different ways: see e.g. Dickman's (1978a) equation 12, Tomita et al.'s (1979) equation 2, and Schmidt's (1975) equation 3. These all give consistent values when applied to the same data, if Tomita et al.'s value of $\alpha\delta/Q$ is adjusted upwards by a factor 1.61. Their values corrected for this factor are also given in Table 1 in brackets. While this generally improves the agreement, some cases remain widely disparate. Figure 3 presents the mass and radius values for the globules tabulated in Table 1.

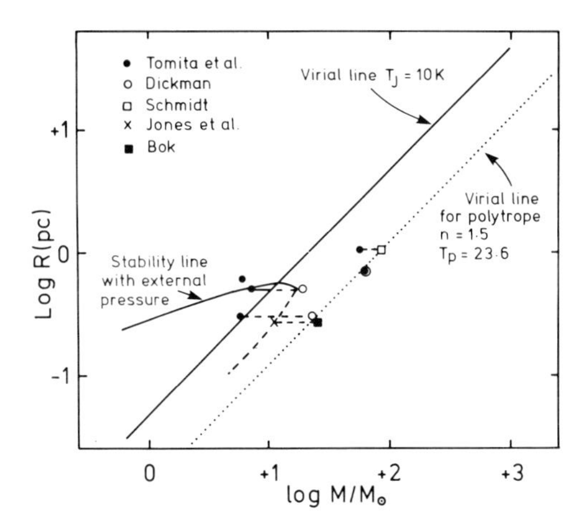

Figure 3. The mass/radius relationship for the globules listed in Table 1. Symbols are as shown in the figure. Horizontal dashed lines join globules for which two independent masses have been derived. Also shown are two virial lines (see text), and the stability line calculated by Tomita et al. (1979) for a 10 K sphere of gas with external pressure (see text). The solid part of the curve refers to stable equilibrium states, while the dashed portion refers to unstable equilibrium and is similar to normal virial lines.

It would appear that _reliable_ derived masses for globules are almost nonexistent! Data in the opaque cores of the globules are of undoubted importance in resolving this issue. While several optical and near IR (Schmidt 1975; Tomita _et al_. 1979; Jones _et al_. 1980) studies suggest a plateau to the absorption in the inner regions of many globules, the data are inconclusive, and radio data suffer from a lack of spatial resolution. The dynamical collapse models on the other hand predict exceedingly dense cores and absorptions so large as to be unobservable (Villere and Black 1980), and consequently predict much higher masses than empirically determined. It would be of great interest to attempt to fit the density distribution from such models with the observations of less opaque globules.

It has been the simplistic comparison with predictions of the virial theorem which led to the conclusions that the large globules are gravitationally unstable. Figure 3 shows virial lines for T = 10 K derived from

$$M/M_{\odot} = (T/0.24\mu)\ (R/R_{pc}) \qquad (2)$$

and for a polytrope with internal pressure with $n = 1.5$, $T = 23.6$, $\mu = 1$ (Kenyon and Starrfield 1979). In the standard view globules lying above the lines will continue to expand while those below are gravitationally unstable and will contract. Tomita _et al_. (1979) undertook a different approach in considering a mass of gas with external pressure, and have examined the conditions where $d^2I/dt^2 < 0$, $= 0$, or > 0 for T = 10 K, where they have taken these to be the criteria that the object is in a state of expansion, equilibrium, or contraction respectively. Their curve for the equilibrium condition is shown in Figure 3, where the sign of d^2I/dt^2 is positive and negative respectively inside and outside of the curve. According to their analysis, objects more massive than a critical value (where lines defining the stable and unstable states meet) are gravitationally unstable regardless of their radii, under the influence of external pressure.

While from their data alone it is concluded that many globules are stable configurations, the lack of consistent masses poses a great problem in the interpretation. The further derivation of stability lines for more realistic globule models will be of great value in the interpretation of new data as they become available. Certainly the belief that a large majority of these objects are unstable to gravitational contraction should perhaps be suspended until more reliable data are available.

1.3 An Infrared Role in the Study of Globules

1.3.1 _Near infrared studies_. It has been shown that optical and radio techniques for estimating the physical characteristics of globules have their limitations (e.g., opacity, resolution). The near infrared method adopted by Jones _et al_. (1980) essentially suffers from neither of these problems, and offers a true direct method of deriving the density distribution within a globule. They undertook a 2-μm survey of Globule No. 2 in the Coalsack (Tapia 1975) down to a limit of K ~ 9.5 in the outer regions and K ~ 11.3 in the inner 4' × 4' block. JHK photometry of each of the 75 sources discovered was then obtained and is shown in Figure 2 of their paper. It was an advantage for the object under consideration to be close to the galactic plane, since it allowed a large enough sample of background sources to be measured through the globule for determining the density distribution of the globule. From the data it is possible to derive the reddening E(J-K) on the assumption that each of the stars was spectral type M3III, a plausible assumption in terms of luminosity function calculations for the region. Figure 4 shows the derived density distribution as a function of radius from the optical center of the globule. It is clear that there is a constant background value of extinction over the whole globule which may be due to a general Coalsack sheet of obscuration, or to extinction just beyond the globule. For comparison are shown the density profiles of the modified polytropes of Kenyon and Starrfield (1979) with indices of 1.0 and 2.5. These have been claimed by Kenyon and Starrfield to fit (respectively) Schmidt's (1975) data for B361, and Bok's data for Coalsack globule 2. It is clear that the optical and near infrared data give conflicting results close to the center

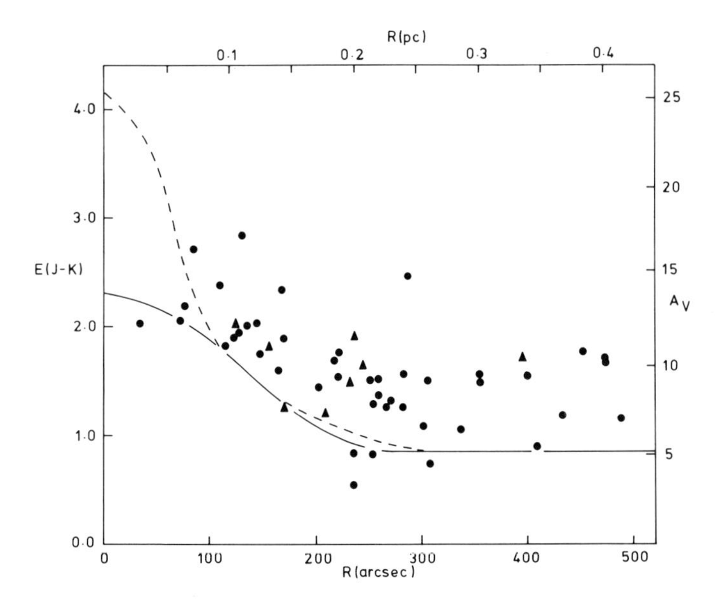

Figure 4. The color excess E(J-K) as a function of distance from the center of Coalsack globule 2, as described in the text. Filled triangles refer to stars for which spectral types had been obtained from CVF spectroscopy. The solid and dashed lines are the density profile of a polytropic model with indices n = 1.0 and 2.5 respectively (Kenyon and Starrfield 1979) (from Jones _et al_. 1980).

of the globule. It is, however, interesting to note that in the outer parts of the globule, these data give $\rho \propto r^{-3}$, as did both Schmidt and Tomita et al., different from the expected $\rho \propto r^{-2}$ calculations of dynamical collapse models.

Integrating to obtain the total mass, and using the usual value $N_{H_2}/A_V = 1.25 \times 10^{21}$, $M \sim 11\ M_\odot$ is derived, as shown on Figure 3. Globule 2 thus lies right on the dividing line between expansion to a stable state and gravitational contraction. Temperature measurements would be invaluable in evaluating the exact situation.

1.3.2 Far infrared observations. One of the most important parameters for determining the stability of globules is their temperature. Essentially all the information available has been obtained from molecular line emission (e.g., Martin and Barrett, and Dickman). An alternative approach is to search for thermal radiation from the globules, but this is difficult because of the very cool temperatures expected ($T \sim 10$ K), and the low luminosities involved ($L \sim 2\ L_\odot$) if globules are externally heated by the interstellar radiation field (Werner and Salpeter 1969).

The fundamental observation of this nature is that by Keene et al. (1980), who obtained the far infrared and submillimeter spectrum of B335 from Mauna Kea and the Kuiper Airborne Observatory. The measurements they obtained between 150 and 450 μm are shown in Figure 5 where they are compared with curves of the form $B(\nu)$ (for T = 16 K) and $B(\nu)$ (for T = 13 K), both of which give an acceptable fit to the data. The

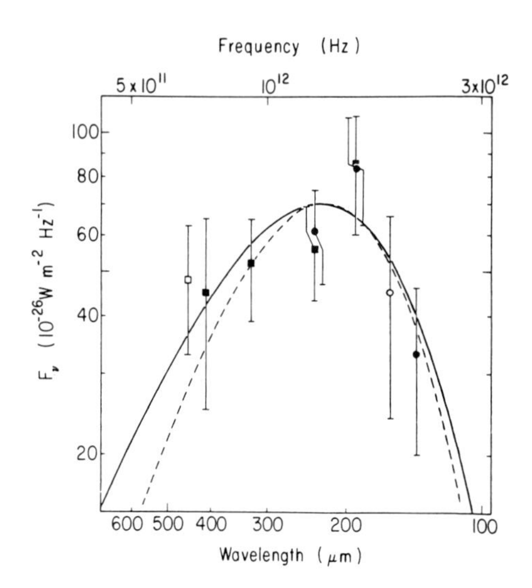

Figure 5. The observed energy distribution of B335. The solid and dashed curves have the forms $\nu B(\nu,\ 16\ K)$ and $\nu^2 B(\nu,\ 13\ K)$ respectively (from Keene et al. 1980).

data are also reminiscent of the silicate ice model energy distributions given by Spencer and Leung (1978).

The main conclusions of Keene _et al_. (1980) may be summarized as follows:

a) radiation from the globule with a luminosity of ~5 $L_\odot$ (at 400 pc) is optically thin thermal emission, $\tau_{225} \sim 0.010$ with T lying in the range 13-16 K.

b) the dust mass lies between 0.07 $M_\odot$ and 0.17 $M_\odot$ in agreement with values given in Table 1 for a gas/dust ratio of ~100.

Keene _et al_. show that it is unnecessary to invoke the presence of an internal luminosity source within the globule as heating by the interstellar radiation field (particularly if modified in the manner of Jura [1979]) is by far the most important source of heating. It would appear from this first far infrared approach that the physical parameters of B335 are firmly established, and that they all consistently suggest that B335 is gravitationally bound.

The aim in this discussion has been to show that infrared observations at both short and long wavelengths have a vital role to play, in producing a clear picture of the structure and physical processes involved in the formation and evolution of globules. Probably the most puzzling aspect of the study of Bok globules, and the question with which I would like to complete the discussion, is what is the next phase of evolution, and do we see it? Why is it, for instance, that we do not appear to see globules with a wide variety of radii and similar masses? After all, they are expected to evolve downwards in the mass-radius diagram (Figure 3) yet no high-mass small-radii globules appear to have been observed. This perplexing question suggests two perhaps heretical alternatives: (a) our techniques are not sufficiently refined to obtain the parameters of such objects should they exist, or (b) globules are intrinsically stable, and have nothing to do with star formation.

2. DARK CLOUD REGIONS OF LOW MASS STAR FORMATION

2.1 Introduction

In this section of the paper I shall give an overview of infrared work on nearby dark cloud complexes and, in section 3, will discuss the IR characteristics of PMS objects.

It is my object to concentrate on those dark cloud regions where low mass (1-2 $M_\odot$) star formation is occurring and where violent processes do not appear to be dominant. Associated with these are the well-known emission line T Tauri stars, initially recognized as a class by Joy (1945). It is now universally accepted that T Tauri stars, and associated objects embedded within the dark cloud complexes, are low mass pre-main sequence objects.

Detailed optical studies of dark cloud regions have been hampered by obscuration within the clouds. In the last decade, however, infrared observations have made tremendous strides in penetrating the veil of these clouds, which has in turn led to a much better understanding of the population of newly formed stars enshrouded within them.

2.2 Near Infrared Surveys of Dark Cloud Regions

2.2.1 Historical background. The fundamental and pioneering set of near infrared observations was a 2 μm survey of part of the ρ Ophiuchus dark cloud by Grasdalen *et al*. (1973) which was extended by Vrba *et al*. (1975). Their work revealed for the first time a group of obscured embedded sources near the center of the cloud, which they interpreted as members of an embedded young cluster within the cloud, the possible existence of which had been suggested more than a decade previously (Bok 1956). Spurred on by this success, this group of workers undertook a series of comprehensive 2 μm surveys in a variety of dark cloud regions, and combined these with optical spectroscopy and photometry (K. Strom *et al*. 1975, 1976; S. Strom *et al*. 1976; Vrba *et al*. 1976). In parallel with this undertaking, Rydgren *et al*. (1976) published a major study of the properties of optical T Tauri stars, and S. Strom *et al*. (1974) investigated in detail the properties of the related Herbig-Haro objects (Herbig 1969). Many of the conclusions of these studies have been summarized in two review papers by S. Strom *et al*. (1975) and S. Strom (1977).

Following the success of the initial study of ρ Oph, it was disappointing to find that in many regions (e.g., L1630, L1517, R CrA) the obscured cluster hypothesis was not viable. Although a number of 2 μm sources were found in each survey (some undoubtedly associated with the dark cloud complex) no further *major* population of PMS objects was found. In particular, study of the nearby R CrA region provided no evidence for an embedded cluster (Vrba *et al*. 1976) despite the overwhelming evidence of recent star formation in the region (Knacke *et al*. 1973; Glass and Penston 1975).

The one major criticism which can be levelled at the interpretation of the early surveys was the failure to take careful account of confusion by background sources. Although several of the regions investigated lie well above the plane, the general galactic background of old disk stars is still significant. For example at the position of ρ Oph, in a region covering 0.2 square degrees, and down to a K magnitude of 10, one expects (from a simple luminosity function and exponential disk model of the galaxy) to find 41 background sources! This is well over half the number of sources found in the survey of Vrba *et al*. (1975) and has profound conseqences on any conclusions from that survey.

2.2.2 Taurus, Ophiuchus, and the background sources at 2 μm. The first person to realize the importance of the background contribution to 2 μm survey material was Elias (1978a,b,c) in his studies of IC 5146, and the Ophiuchus and Taurus dark clouds. In the latter two cases,

Elias undertook massive surveys, covering respectively 18 and 20 square degrees, and detecting some 400 and 200 stars respectively down to a K magnitude of +7.5. Examination of his Figure 1b (Elias 1978b,c) shows clearly that the sources were essentially evenly distributed over both clouds, and it was therefore essential to distinguish between objects within the cloud and background field stars. This was undertaken first by restricting the sample to objects with H-K $\geq$ 0.70 and K < 7.0. Elias then undertook the daunting task of obtaining CVF spectroscopy and broad band colors for the selected sources, and used the spectroscopic data to identify the sources associated with the cloud. Figure 6 shows the 2-2.5 μm spectra obtained by Elias (1978c) in Taurus. The basis for the identification is that according to the model of the galactic stellar distribution (Elias 1978a), almost all the field stars should be G8 giants and later, and identifiable by their CO absorption beyond 2.3 μm. The association objects, however, are likely to have smooth energy distributions either as a result of their earlier spectral type, or as a result of excess thermal or gaseous emission in the 2 μm region.

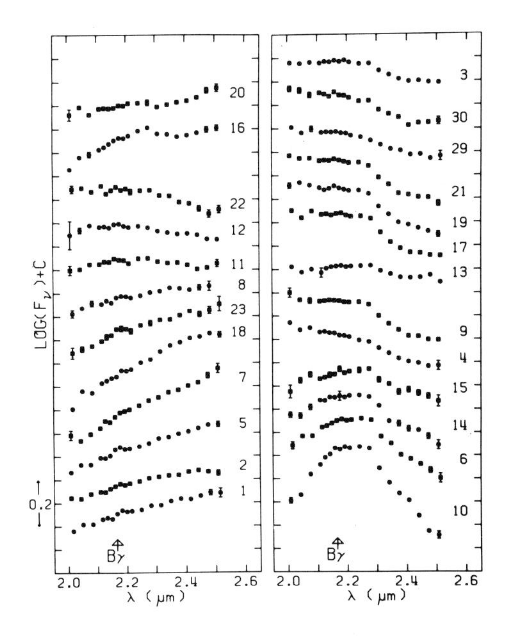

Figure 6. 2 μm spectra of selected objects in the Taurus dark cloud complex. All the objects in the right half of the figure plus Nos. 16 and 20 have been classified as background field stars (from Elias 1978c).

As an example of the photometric properties of the embedded sources, Figure 7 shows photometry of the Oph sources by Elias (1978b). It can be seen that these objects exhibit a wide variety of characteristics, the common feature being that most exhibit some kind of infrared excess. By examining background objects, Elias concluded that the infrared reddening law in Taurus and Ophiuchus was identical and the same as the normal interstellar law. Of particular interest are the 3 objects shown at the right of Figure 7. All 3 of these are extremely red objects, appear extended at 1.6 μm, and are interpreted by Elias as obscured reflection nebulae formed within regions where the minimum molecular hydrogen density is 10^5 molecules cm^{-3}. Both polarization measurements and high resolution maps at 1.6 and 2.2 μm are needed to elucidate the properties of these interesting objects further.

There is an alternative method of determining the membership of PMS sources within a cloud which is almost 100% effective, i.e., their position in a JHK diagram. This technique, being significantly faster than CVF spectroscopy, would have enabled Elias to enlarge his sample of sources dramatically. Figure 8 shows a plot of his JHK photometry in a J-H vs H-K diagram for both Ophiuchus (right) and Taurus (left). The association and field stars as identified by Elias are plotted as different symbols. Almost all the association objects can be easily identified by their (H-K) excesses, and certainly <u>none</u> of the Taurus field stars (apart from the peculiar variable RV Tauri) fall in that part of the diagram. They follow a remarkably tight locus in the diagram for which as yet no satisfactory explanation has been advanced, although it is probably tied up wtih the nature and evolution of the circumstellar shells, which are one explanation for the infrared colors of PMS objects (see 3.1).

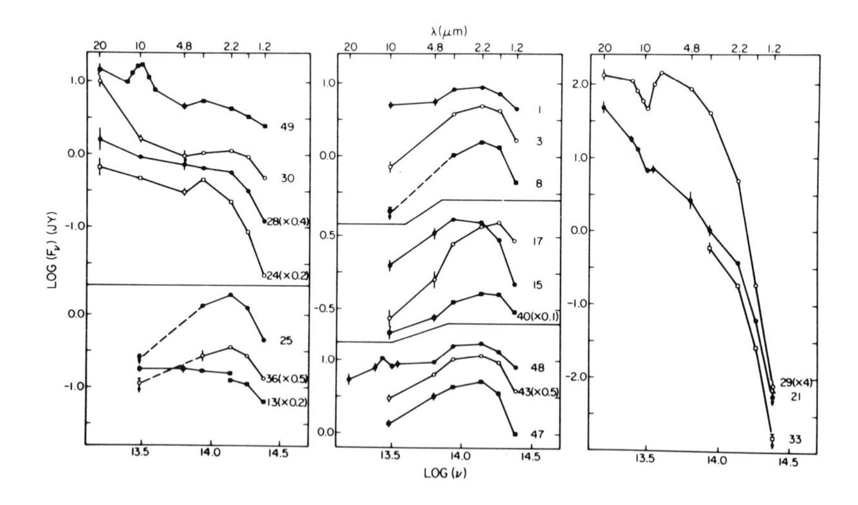

Figure 7. Energy distributions of selected objects in the ρ Oph dark cloud between 1 and 20 μm (from Elias 1978b).

The work of Elias showed that there are clear differences between Ophiuchus and Taurus, which are apparent in the embedded population as well as in the optical stars. The exact reasons for these differences obviously lie in the obscure details of the star formation triggering process (see 2.3). It is unfortunate that these surveys did not go to fainter magnitudes, especially in Taurus, as the presence in large numbers of low luminosity embedded sources and their spatial distribution might substantially alter current thinking on star formation within that cloud.

2.2.3 Studies of the Chamaeleon dark cloud complex. The Chamaeleon T association was first recognized by Henize (1963) and Hoffmeister (1962) and consists of an elongated dust cloud containing a large sample of emission line stars and three conspicuous reflection nebulae. Henize and Mendoza (1973) discussed the spectra of some 32 emission line stars within the association, while subsequently Feast and Glass (1973) showed one to be an R Mon type object, Schwartz (1977) discovered several Herbig-Haro objects in the cloud, and Appenzeller (1977, 1979) reported a high percentage of YY Ori type spectra among the emission line stars. In all ways, Chamaeleon is an ideal region for the detailed study of low mass PMS objects and their interaction with their placental dark cloud: it is close ($\lesssim$ 200 pc), is substantially out of the galactic plane (b ~ -15°) and contains a significant population of young objects.

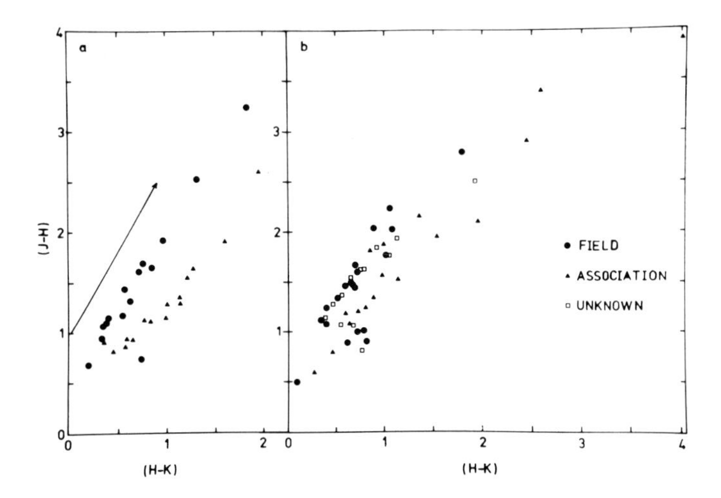

Figure 8. (J-H) vs (H-K) diagrams for the objects measured by Elias (1978b,c) in ρ Oph (right) and Taurus (left). Filled circles and filled triangles refer to field and association stars defined by Elias on the basis of spectroscopy. Open squares are objects for which there is no spectroscopic data. The association objects are separated very clearly by an excess E(H-K) in this diagram.

Grasdalen *et al*. (1975) and more recently Rydgren (1980) obtained photometry and spectra of the emission line objects, and derive different distances to the association on the basis of an anomalous reddening law within the cloud (R = 5.5 and 5.0 respectively). Glass (1979) obtained JHKL photometry of almost all the HM emission line stars, as well as a selection of apparently background objects. The latter appeared to follow a normal interstellar reddening curve well (in the near infrared), while the emission line objects occupied a similar locus in (J-H) vs (H-K) diagram as the T Tauri stars in Taurus (Figure 8).

Hyland and Jones (1980) have undertaken a systematic 2 μm survey and follow-up observations to search for previously unrecognized embedded PMS objects, and to map the interstellar reddening through the cloud by observations of the background stellar population. The area covered in the initial phase of this survey was approximately 0.32 square degrees; the (J-H) vs (H-K) diagram for the sources discovered is shown in Figure 9.

Several interesting features may be noted in Figure 9: (a) the "background" sources follow the normal interstellar reddening curve in the near infrared (Jones and Hyland 1980); (b) the embedded sources follow the locus of similar objects in Taurus, and include several objects which were too faint to be recognized in optical surveys; (c) there is a large group of stars clustered in the region (J-H) ~ 0.6 (H-K) ~ 0.3 which are absent from the similar figure for background objects in the Coalsack region. Spectroscopic observations of several of these (Schwarz 1977; Hyland and Jones 1980) reveal weak Balmer emission, suggesting that these may be a population of less extreme PMS objects lying closer to the main sequence. If so, the population of PMS objects within Chamaeleon, including the newly found obscured sources, may need to be increased by as much as a factor of two over the number of optically discovered emission line stars.

It is evident that the intrinsic (i.e., reddening-free) colors of the embedded sources are greatly influenced by extinction within the cloud and it is necessary to devise a means of taking this into account. Figure 10 is a map of the positions of background stars and embedded (IR excess) sources found in the survey. With the assumption that the mean spectral type of the background sources is Kl III (as predicted by a simple exponential disk model of the galaxy in the manner of Elias 1978a), values of E(J-K) and A_V = 5.6 E(J-K) were obtained. These were then combined to form the approximate contour map shown in Figure 11, which can be used to obtain upper limits to the effect of cloud extinction on the colors of the embedded sources.

Figure 11 also provides an interesting and unexpected insight into the spatial distribution of the young embedded sources within the cloud. There is clearly a small grouping of PMS objects at the position $11^h08^m30^s$, -76°19', right at a steep density gradient. The triangle indicates the position of the Be star HD 97300 which illuminates the bright reflection nebula. This clustering is highly suggestive that the more massive early-type star has caused compression of the gas in its neighborhood, creating the conditions for and triggering further star formation. While it is less apparent in the data presently available, there is a similar group of embedded sources at 11^h07^m, -77°29', which also lie on the boundary of the possible interaction of HD 97048 with the dense cloud material.

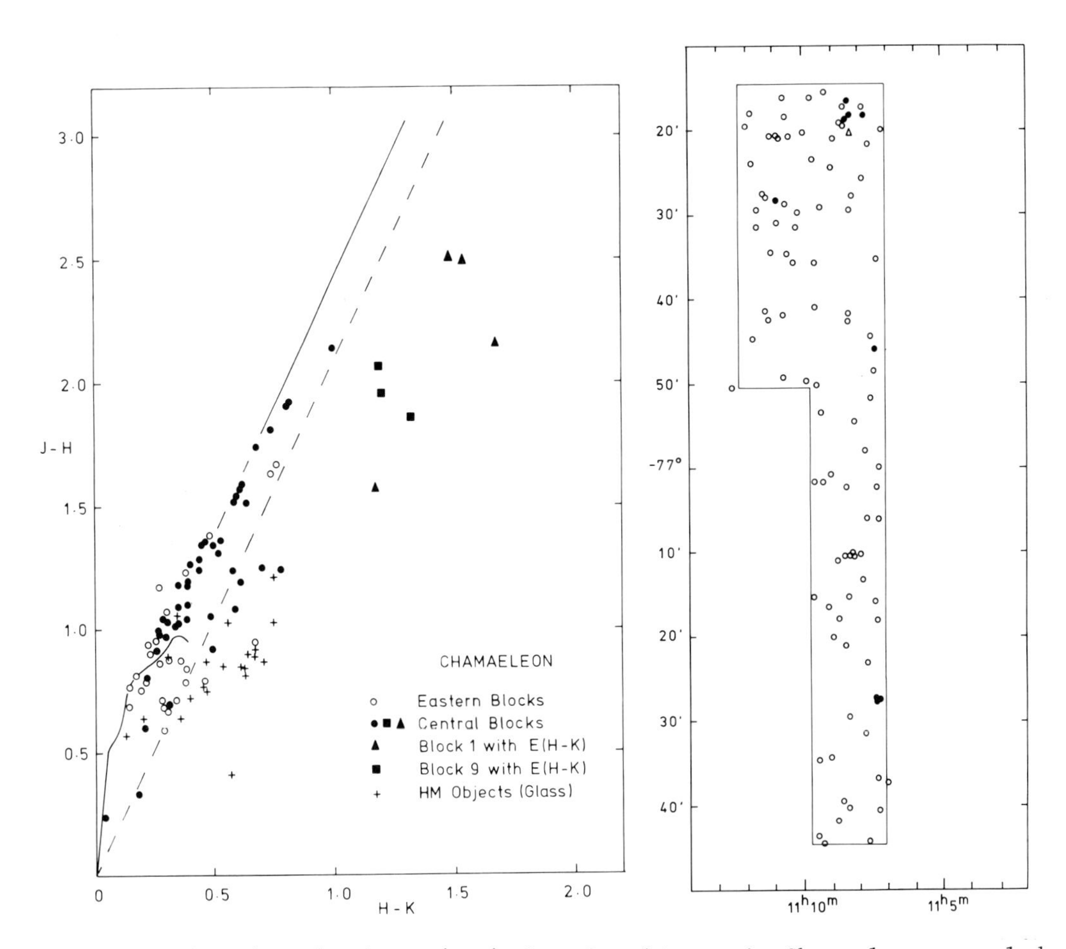

Figure 9 (left). (J-H) vs (H-K) for the objects in Chamaeleon revealed by the 2 μm survey (Hyland and Jones 1980), plus the Henize and Mendoza objects measured by Glass (1979) which lie outside the survey region. Objects with extreme excesses E(H-K) are shown as filled triangles and filled squares.

Figure 10 (right). The distribution of objects found in the 2 μm survey in Chamaeleon. Filled symbols refer to objects with extreme excesses of E(H-K).

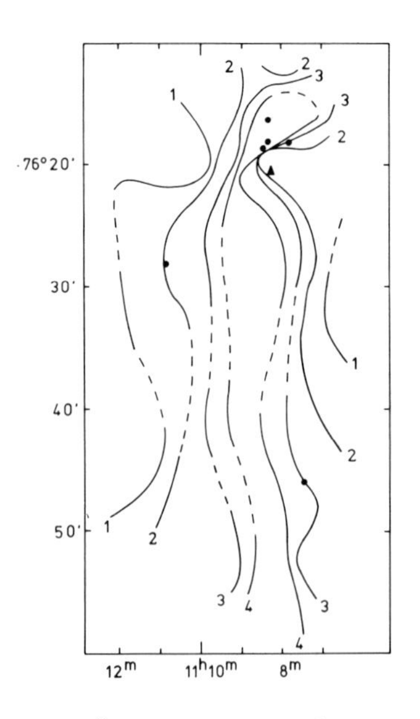

Figure 11. Map of the distribution of reddening E(J-K) in the northern region of the Chamaeleon 2 μm survey, derived from background field stars. The filled circles refer to objects with extreme E(H-K) excesses. The contours refer to the following values of E(J-K): #1 = ≤ 0.40, #2 = 0.65, #3 = 0.90, #4 = ≥ 1.15.

Deep 2 μm surveys when combined with photometry and optical data appear to provide an extremely powerful probe of dark cloud regions, which, despite all the work which has so far been undertaken, has yet to realize its full potential.

2.3 Spatial Clustering and Triggering Mechanisms in Low Mass Star Formation Regions

The idea of sequential star formation has been advanced to explain the nature of OB associations (Elmegreen and Lada 1977). In their model, ionization-driven shock fronts from an initial group of OB stars propagate into a dense molecular cloud; eventually the gas which has accumulated between the shock and ionization fronts becomes gravitationally unstable, initiating the formation of OB stars. While this picture has been eminently successful in delineating high mass star formation processes in giant molecular clouds, the picture in dark clouds associated with low mass ($M \sim 1\ M_\odot$) star formation is unclear. In these there is no obvious population of first generation objects with high enough stellar winds to compress the interstellar gas, and the spatial distribution of optical T Tauris is generally fairly even over the cloud's extent (although embedded sources may have been missed). There is, however, some information on clustering of PMS objects and triggering mechanisms within such clouds, which may throw some light on the processes of star formation in them.

1. The infrared observations of Elias (1978b) where background objects have been eliminated provides evidence on the presence of a considerable embedded cluster of stars at the center of the ρ Oph cloud. This is the only cloud for which an extensive cluster has yet been identified.

2. IC 5146 (Elias 1978a) is similar in many respects to ρ Oph, although many of the cluster stars have already reached the main sequence. The spectral type of its earliest star is BOV. From CO observations which show several intense bright spots, Lada and Elmegreen (1979) have found evidence for clustering of star formation within the cloud complex. They suggest that star formation of stars with spectral type later than B is occurring and "has been most intense in a small (~ 5 pc), dense, possibly compressed molecular core located at one end of the more extended (~ 20 pc) cloud complex."

3. Elias (1979c) finds no evidence for widespread clustering within the Taurus cloud (though it should be remembered that his survey was not very deep), but there are small clusterings of young objects. L 1551 is a case in point where two visible T Tauri stars (HL and XZ Tau), 3 Herbig-Haro objects, and an embedded infrared source are found in close proximity. Fridlund et al. (1980) have recently made balloon-borne far infrared observations of the region, and identify a 65 $L_{\odot}$ point source with the embedded source (IRS 1). They conclude that it too is T Tauri in nature.

4. Evidence for small scale clustering of PMS objects in Chamaeleon (Hyland and Jones 1980) has already been mentioned (2.2.3). Further observations to complete the whole cloud are required to examine the overall picture.

5. CO observations of the R CrA region (Loren 1979) show the greatest heating of the CO occurs in the regions close to the B star TY CrA and the PMS object R CrA. The population of embedded objects so far discovered is extremely small, and the CO observations further imply a nonhomologous collapse of the core of the molecular cloud, while the cloud is also elongated at all densities up to $4 \times 10^5 cm^{-3}$.

While these data provide some evidence on star formation mechanisms in dark cloud regions, there is no coherent picture regarding triggering mechanisms, and it may be that there are as many mechanisms as there are clouds to contemplate. For example, in ρ Oph, Vrba (1977) has argued that the magnetic field geometry and spatial distribution of some of the youngest stars is indicative of external shock compression. Elias (1978c) points out that some of the embedded objects may postdate the age given to the initial shock compression (~ 6×10^6 yrs) and speculates that an event such as a supernova (Herbst and Assousa 1977) which produced the runaway ζ Oph may also have triggered the most recent star formation.

In Taurus, Elias (1978c) suggests that star formation has been spontaneous, notwithstanding the small scale clustering mentioned earlier. In R CrA, Loren (1979) interprets the lack of CO line splitting in the denser star formation regions of the cloud as conclusive evidence that dynamical effects can be ruled out. He suggests that Vrba's (1976) hypothesis that the evolution of the cloud is controlled by flow along magnetic field lines is correct, and indicates that star formation should and does occur in the region of the cloud closest to the galactic plane. On the other hand, IC 5146 resembles ρ Oph and M17 where regions of active star formation are located in the dense tips of elongated molecular clouds at the end furthest from the galactic plane. Lada and Elmegreen (1979) suggest externally applied pressure as the trigger in all three cases.

The picture of triggering mechanisms for regions of low mass star formation is thus in a state of great flux and is very much a problem for the 1980s.

2.4 The Reddening Law in Dark Cloud Regions

A knowledge of the extinction law in dark clouds is of crucial importance for determining the intrinsic energy distributions and luminosities of the sources within the clouds. Furthermore, the derived distances to many clouds depend critically upon the value of R [A_V/E(B-V)] used to deredden the photometry of the stars. It has long been suggested on the basis of optical photometry (e.g., Walraven and Walraven 1960; Hardie and Crawford 1961) that values of R considerably greater than the normal interstellar value may exist. Carrasco _et al_. (1973) produced evidence from optical and infrared photometry for a wide variety of R values (ranging from 3 to 5) within the Ophiuchus dark cloud complex, and showed that R increased with the total extinction. They concluded from their observations that "the mean particle size increases in the denser parts of the Rho Oph cloud." It is certainly not clear to what extent the large values of R adduced from deeply embedded stars can be applied to other objects within the cloud, yet this procedure has been adopted in certain cases with possible misleading results. Cohen and Kuhi (1979) have argued strongly against this kind of approach, and advance good arguments for rejecting large values of R for the less embedded optical T Tauri stars in Taurus and Ophiuchus.

Grasdalen _et al_. (1975) derived R = 5.5 from optical and near infrared observations of HD 97300, the AOV illuminating star of Cederblad 112 in the Chamaeleon region. Using this value to derive the distance to the cloud forced many of the young emission line T Tauri stars to fall _below_ the zero age main sequence. More recent observations by Rydgren (1980) suggest a marginally smaller value of R = 5.0, but even he cautions against blind adoption of this value, since both stars used in the derivation are probably very young and may have weak circumstellar dust shells.

The problem with all determinations of the reddening law in dark cloud regions discussed earlier is that they have been obtained from observations of young stars intimately associated with or deeply embedded within the clouds. These are just those stars which might be expected to possess extensive circumstellar dust shells as probable remnants of their recent protostellar phase of evolution, and we will see in fact that many do so. However, a number of authors (Elias 1978a,b; Glass and Penston 1975; Glass 1979; Jones and Hyland 1980; and

Jones *et al*. 1980) have observed background field stars seen through the Taurus, Ophiuchus, R CrA, Chamaeleon, and Coalsack clouds, and have shown that the reddening law in the near infrared is normal. They also show that embedded objects generally have infrared excesses consistent with the presence of circumstellar shells (see 2.2.3). Unfortunately, none of the reddening determinations of background stars have been tied to optical photometry, and in the most opaque regions of the clouds it is unlikely that such observations will ever be attempted.

However, although the mean infrared extinction law appears to be unchanged within dark cloud regions, this does not preclude a change in the reddening law at shorter wavelengths. Some exciting new results obtained from near infrared photometry of O stars in the Carina nebula by Smith (1980) are particularly relevant. A combination of the new infrared data with the optical results of Feinstein *et al*. (1973) gives unequivocal evidence for an anomalous reddening law, with a *unique* value of R = 5. Interestingly, the data can be interpreted in terms of a change in the absorption characteristics of the grains in the B filter pass band *alone*. Thus, for objects which appear to follow the normal reddening law in the near infrared, it is probably safer to derive A_V from the normal value of $A_V/E(V-K)$, and to disregard B entirely. Clearly, however, no definitive statement on the interstellar reddening law in dark clouds can be made at this time, and it remains one of the most pressing problems within the larger context of dark clouds and star formation.

3. THE INFRARED CHARACTERISTICS OF PMS SOURCES

3.1 Nature of Near IR Continua

T Tauri stars have been noted for their large infrared excesses since the early work of Mendoza (1966, 1968). As each new group of T Tauris and embedded PMS stars has been observed, more sources with large infrared excesses have been found (e.g., Cohen 1973a,b; Vrba *et al*. 1975; Knacke *et al*. 1973; Grasdalen *et al*. 1975; Glass 1979). While initial interpretation of these was made in terms of thermal emission from circumstellar dust, the striking emission lines in the optical spectra of T Tauri stars led Grasdalen *et al*. (1975) to propose free-free emission from a hot ($T \sim 2 \times 10^4$) gaseous envelope as the major mechanism producing the infrared continuum (see also S. Strom 1977).

The ability to distinguish between the two proposed mechanisms is in no small measure dependent upon obtaining the correct reddening law to use. In particular, at J, H, and K, reddening within the clouds may be substantial, and is of great importance for determining the "intrinsic" (i.e., reddening-free) position of individual stars in the (J-H) vs (H-K) diagram. Any means whereby the total cloud reddening can be mapped, e.g., optical star counts, using background objects as outlined in 2.2.3, or from ^{13}CO measurements, will prove invaluable in setting upper limits to the cloud contribution to the stellar colors.

Also, it is possible with little error to deredden at infrared wavelengths, since the ratios E(J-H)/E(H-K), etc., have been directly determined by reference to background objects.

Rydgren (1976) computed (J-H) vs (H-K) diagrams for models incorporating a stellar photosphere plus free-free and free-bound emission from a hot gaseous envelope to interpret the colors of T Tauri stars and to separate interstellar reddening from the gaseous component. Baschek and Wehrse (1977) pointed out that the inclusion of line emission would reduce the effect of the envelope emission on the colors. Rydgren (1976) and Strom (1977) consider the inphase variation of optical and infrared brightness a strong argument in favor of the gaseous envelope model. However, it now appears that while such a model might be the explanation for the colors of some T Tauri and PMS objects, there are many whose extreme colors require an alternative explanation, presumably in terms of hot circumstellar dust.

Cohen and Kuhi (1979) have undertaken the most recent thorough examination of this question and clearly favor the hypothesis of thermal emission from dust for the large majority of T Tauri stars. Their Taurus-Auriga observations in a (1.6-2.3) vs (2.3-3.5) diagram [(H-K) vs (K-L)], corrected for interstellar reddening, are shown in Figure 12. Thermal excesses are emphasized in such a figure compared with shorter wavelength data. It is clear that although a number of the objects lie close to the photosphere + hot-gas-emission line, the majority have even larger excesses and require thermal emission to be present. Cohen and Kuhi also claim that lack of an observed emission Balmer jump, in contrast to the predictions of the envelope models, strongly suggests that the infrared excess does not arise from gaseous emission.

One important conclusion of the thermal emission hypothesis demanded by JHKL magnitudes is that there is a predominance of hot grains (T ~ 1500 K) in the circumstellar shells, suggesting either that the inner boundary of the shell is very close to the star, or that mechanical heating is important. Furthermore, the grain mix must be such that it contains a sizeable percentage of refractory grains such as graphite. Some objects (e.g., T Tau) show strong evidence for cool (T < 500 K) emission and many have excesses out to 20 μm (Cohen 1973a,b). It is probable that the considerable differences between the characteristic energy distributions of PMS stars and those of late type giants with mass loss nebulae reflects a fundamental difference in the nature and history of their circumstellar shells.

3.2 Far Infrared Observations

Probably the most exciting advance in the last year in the understanding of young emission line PMS stars and their relationship to remnant protostellar material has been the far infrared study of young emission line objects by Harvey *et al.* (1979), in which 50 and 100 μm observations of 8 young emission line stars (including two classical T Tauri stars) were obtained. The spectral energy distributions of these objects from 1 to 100 μm are shown in Figure 13. The remarkable

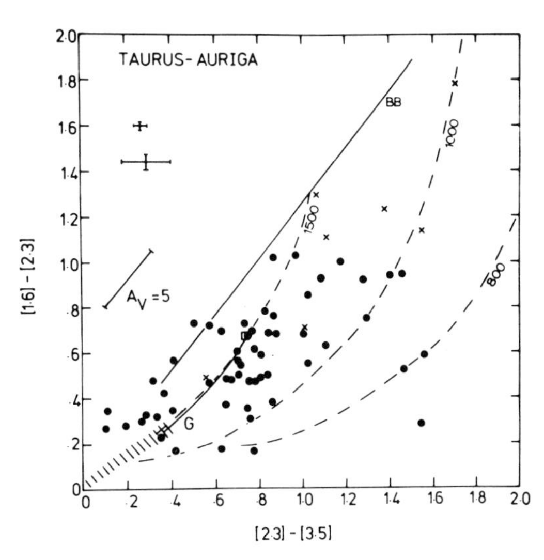

Figure 12. Intrinsic 2 color near IR plot for stars in the Taurus-Auriga clouds. Filled circles refer to stars for which a continuum extinction can be determined; crosses refer to continuum stars for which observed rather than intrinsic colors are plotted. 1 σ error bars are shown for the brighter and more typical stars. The hatched area represents the locus of main-sequence stars. The solid line (BB) represents the locus of blackbodies. The solid curve (G) terminated by an open square indicates the locus of a K7 photosphere with increasing superposed optically thin free-free, free-bound, and bound-bound gas continuum emission. The dashed lines refer to a K7 photosphere overlaid by thermal emission from hot dust grains at 800, 1000, and 1500 K (from Cohen and Kuhi 1979).

feature of these data is the extremely broad energy distribution and unexpected strong far infrared emission. Harvey et al. (1979) interpret these observations in terms of very extensive circumstellar dust shells with optical depths $\tau \sim 0.01$ at 100 μm, implying $\tau \lesssim 1$ at visual wavelengths. It is necessary to invoke a slow radial dependence of dust density ($\rho \propto r^{-1}$), high maximum temperatures for the inner regions of the shell, and very cool minimum temperatures at the outermost regions of the shell ($T \lesssim 20$ K), to model in a simplistic way the flat type of

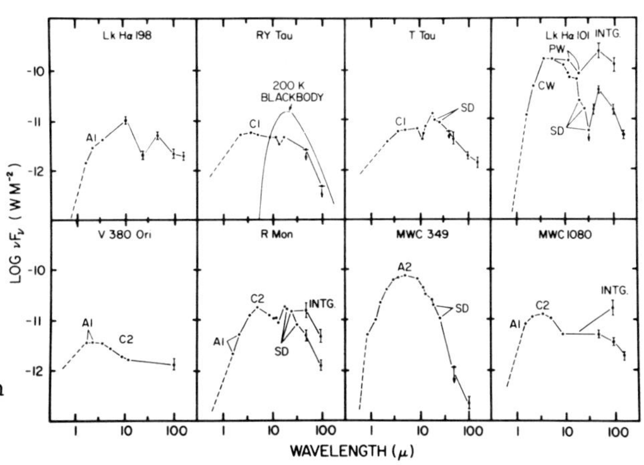

Figure 13. Energy distributions of 8 emission line stars discussed by Harvey et al. (1979). Total integrated far infrared fluxes are given for the extended sources associated with Lk Hα 101, R Mon, and MWC 1080. A 200 K blackbody spectrum is also shown on the RY Tau graph to illustrate the fact that many of the stellar energy distributions are much broader than that of a single-temperature blackbody (from Harvey et al. 1979).

the shell, and very cool minimum temperatures at the outermost regions of the shell T $\lesssim$ 20 K, to model in a simplistic way the flat type of energy distribution. The observations thus set strong constraints on any circumstellar dust shell model for T Tauri energy distributions.

The extensive nature of the shells is further emphasized by the observation that two of the sources are resolved at 100 μm, and have typical sizes ~ 50-60". In physical terms the outer radii of the shells range from ~ 10^{16} to 10^{18} cm which, although large, are clearly small compared with the overall dark cloud dimensions. The r^{-1} dependence of dust density suggests strongly that the shells are remnants of the protostellar clouds from which the objects were formed, and is unlike normal mass loss nebulae (where $\rho \propto r^{-2}$).

These observations suggest that a systematic study of far infrared emission from a wide range of obscured PMS and optical T Tauri stars may be expected to provide important new clues on the development of remnant protostellar clouds with age.

3.3 Infrared Spectroscopy

The CVF ($\lambda/\Delta\lambda$ ~ 100) observations of Cohen (1975) covering 2-4 μm, and similar (2-2.5 μm) observations of the Taurus and Ophiuchus sources by Elias (1978a,b), are the only substantial spectroscopic contributions beyond 1 μm to the present time. Hyland and Jones (1980) have obtained a number of 2-2.5 μm spectra of embedded Chamaeleon and R CrA sources, while Mould and Ridgway (1980) have recently obtained high-resolution (1 cm^{-1}) data for some 7 T Tauri stars.

All these observations confirm the generally smooth infrared continua in the 2-4 μm range (see Figure 6), reminiscent of smooth dust continua. The new high-resolution data reveal for the first time the weak emission features Bγ and 4733 cm^{-1} (a possible line of He I) in several of the sources. It will be possible to use the strength of Bγ in conjunction with optical emission lines to place further constraints on the importance of gaseous emission in the near infrared. Cohen (1975) found a measurable 3.1 μm ice absorption feature in only 1 of 21 objects observed (HL Tau), and a weak 3.3 μm emission feature in Lk Hα — 198. Blades and Whittet (1980) have recently found a striking unidentified emission feature at ~ 3.5 μm in the spectrum of the Chamaeleon PMS star HD 97048, which clearly originates in the star's own circumstellar shell. Beckwith et al. (1978) searched for H_2 emission in T Tauris, the only positive detection being T Tauri itself.

Definitive evidence on the nature of the circumstellar dust material around PMS stars comes only from the narrow band photometry and CVF spectroscopy in the 8-14 μm region (Cohen 1980; Rydgren et al. 1976). Figure 14 shows three examples of silicate emission from PMS stars taken from the work of Cohen. These show similar features which match well with the Trapezium-like grains for temperatures between 200 and 300 K. Of the stars so far observed, only HL Tauri shows evidence for interstellar silicate absorption.

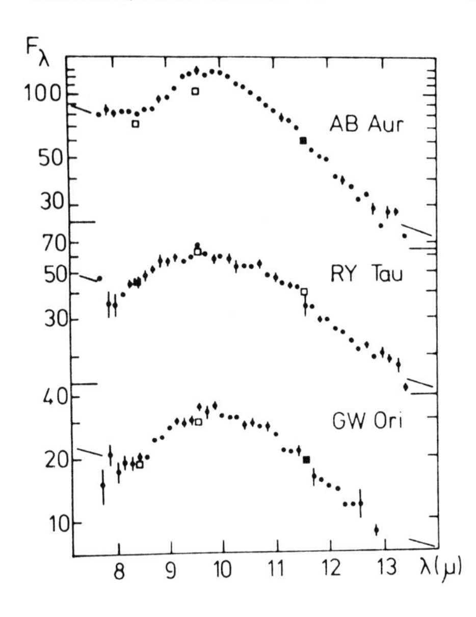

Figure 14. Observed spectrophotometry of 3 T Tauri stars. Open squares indicate independent bound-bound flux calibrations. Tick marks at the edges of the spectra indicate the interpolated linear continua used to define 10 μm features. Where plottable, 1 σ photometric errors are shown at each wavelength. Units of F_λ are 10^{-18} W cm^{-2} μm^{-1} (from Cohen 1980).

3.4 Age and Evolutionary Status of PMS Objects

Among the important parameters which one would like to determine for the PMS stars are their masses, radii, ages, and hence the age spread and mass function within a given association. This can at the present time only be done for optically observable PMS objects, and even then the systematic errors leave one in doubt as to the true value of any such approach. The placement of such objects accurately within an HR diagram depends upon: the accuracy with which the spectrum can be determined (given the problems of emission lines extending into the spectrum), the distance to the cloud (and its dependence on R), and the bolometric correction taking into account all infrared and UV radiation. These then need to be compared with appropriate theoretical evolutionary tracks; Cohen and Kuhi (1979) have made the most systematic attempt so far to use this technique. They used a variety of sources to obtain consistent convective-radiative tracks with which to compare the observations, since the dynamical evolutionary tracks of Larson (1972) predict no visible objects in the region of the HR diagram occupied by PMS objects. Their Figures 2 and 6 show HR diagrams for PMS stars in Taurus and Ophiuchus compared with the theoretical tracks, while their Figure 7 shows the theoretical tracks with isochrones added.

It is immediately apparent that there is a wide age spread in each of the clouds examined, although the mean age for both Taurus and Ophiuchus is around 8×10^5 years. Furthermore, the "growth" parameter of Cohen and Kuhi (1979), which is a timescale for increasing the visible stellar population by a factor of 10, is almost the same for Taurus and Ophiuchus. It would appear that the differences between the two groups mentioned earlier cannot be attributed to age. It is unfortunate that the mass spectrum for Ophiuchus is too sparse for comparison with that of Taurus, although on the surface it would appear that Taurus has a relatively larger population of 0.8-1.0 $M_{\odot}$ objects.

4. CONCLUSION

It is unfortunate that, due to limitations of space, many topics have only been covered in the briefest manner, and some important areas, such as the efficiency of star formation, have been omitted. In particular, the efficiency of star formation, which varies widely from cloud to cloud, is of great interest in delineating differences between regions, and Vrba (1977) has postulated that the efficiency directly relates to the detailed dynamical processes which trigger star formation, and to the strength of magnetic fields which may conspire to prevent or delay collapse of the clouds.

The role of infrared observations in the advancement of our insight into such processes has never been greater, and is in a state of continued expansion. It will be apparent from the foregoing discussion that there are not only significant gaps in our knowledge of these processes but that the basic parameters on which our understanding is based leave much to be desired. One of our major blind spots appears to be a coherent picture of the triggering of low mass star formation and the effects of magnetic fields within the clouds. Clearly infrared polarization measurements will play an increasing role in the understanding of field geometry and its relationship to collapse prevention.

High-resolution spectroscopy, increased molecular mapping, and the mass of mid and far infrared data expected from IRAS will undoubtedly spark major new analyses of the temperature, density, and velocity structure of dark cloud regions. The results of these and follow-up observations from SIRTF should contribute greatly to our understanding of the differences and similarities of these complexes.

In addition to these observational approaches, the time is now ripe for improved models for the photospheres and circumstellar shells of PMS objects, and for a search for dynamical collapse models which satisfy the present observations of pre-main-sequence evolution. With the wealth of opportunities now available to advance the field significantly, it is clear that globules, dark clouds, and PMS objects are going to provide good hunting for observer and theoretician alike in the coming decade.

REFERENCES

Appenzeller, I.: 1977, Astron. Astrophys. 61, pp. 21-26.
Appenzeller, I.: 1979, Astron. Astrophys. 71, pp. 305-309.
Baschek, B., and Wehrse, R.: 1977, Publ. Astron. Soc. Pacific 89, pp. 345-346.

Beckwith, S., Gatley, I., Matthews, K., and Neugebauer, G.: 1978, Astrophys. J. Letters 223, pp. L41-L43.
Blades, C.J., and Whittet, D.C.B.: 1980, Monthly Notices Roy. Astron. Soc. 191, pp. 701-709.
Bok, B.J.: 1956, Astron. J. 61, pp. 309-316.
Bok, B.J.: 1977, Publ. Astron. Soc. Pacific 89, pp. 597-611.
Bok, B.J.: 1978, Publ. Astron. Soc. Pacific 90, pp. 489-490.
Bok, B.J., and Cordwell, C.S.: 1973, in M.B. Gordon and L.E. Snyder (eds.), "Molecules in the Galactic Environment," John Wiley and Sons, New York, pp. 54-90.
Bok, B.J., Cordwell, C.S., and Cromwell, R.H.: 1971, in B.T. Lynds (ed.), "Dark Nebulae, Globules, and Protostars," University of Arizona Press, Tucson, pp. 33-55.
Bok, B.J., and McCarthy, C.C.: 1974, Astron. J. 79, pp. 42-44.
Bok, B.J., and Reilly, E.F.: 1947, Astrophys. J. 105, pp. 255-257.
Bok, B.J., Sim, M.E., and Hawarden, T.G.: 1977, Nature 266, pp. 145-147.
Buff, J., Gerola, H., and Stellingwerf, R.F.: 1979, Astrophys. J. 230, pp. 839-846.
Carrasco, L., Strom, S.E., and Strom, K.M.: 1973, Astrophys. J. 182, pp. 95-109.
Cohen, M.: 1973a, Monthly Notices Roy. Astron. Soc. 161, pp. 97-104.
Cohen, M.: 1973b, Monthly Notices Roy. Astron. Soc. 161, pp. 105-111.
Cohen, M.: 1975, Monthly Notices Roy. Astron. Soc. 173, pp. 279-293.
Cohen, M.: 1980, Monthly Notices Roy. Astron. Soc. 191, pp. 499-509.
Cohen, M., and Kuhi, L.V.: 1979, Astrophys. J. Suppl. 41, pp. 743-843.
Dickman, R.L.: 1977, Sci. Am. 236(6), pp. 66-81.
Dickman, R.L. 1978a, Astron. J. 83, pp. 363-372.
Dickman, R.L. 1978b, Astrophys. J. Suppl. 37, pp. 407-427.
Elias, J.H.: 1978a, Astrophys. J. 223, pp. 859-875.
Elias, J.H.: 1978b, Astrophys. J. 224, pp. 453-472.
Elias, J.H.: 1978c, Astrophys. J. 224, pp. 857-872.
Elmegreen, B.G., and Lada, C.J.: 1977, Astrophys. J. 214, pp. 725-741.
Feast, M.W., and Glass, I.S.: 1973, Monthly Notices Roy. Astron. Soc. 164, pp. 35P-38P.
Feinstein, A., Marrace, H.G., and Muzzio, J.C.: 1973, Astron. Astrophys. Suppl. 12, pp. 331-350.
Fridlund, C.V.M., Van Duinen, R.J., Nordh, H.L., Sargent, A.I., and Aalders, J.W.G.: 1980, Astron. Astrophys. (submitted).
Glass, I.S.: 1979, Monthly Notices Roy. Astron. Soc. 187, pp. 305-310.
Glass, I.S., and Penston, M.V.: 1975, Monthly Notices Roy. Astron. Soc. 172, pp. 227-233.
Grasdalen, G., Joyce, R., Knacke, R.F., Strom, S.E., and Strom, K.M.: 1975, Astron. J. 80, pp. 117-124.
Grasdalen, G.L., Strom, K.M., and Strom, S.E.: 1973, Astrophys. J. Letters 184, pp. L53-L57.
Hardie, R.H., and Crawford, D.L.: 1961, Astrophys J. 133, pp. 843-859.
Harvey, P.M., Thronson, H.D., and Gatley, I.: 1979, Astrophys. J. 231, pp. 115-123.
Henize, K.G.: 1963, Astron. J. 68, p. 280.
Henize, K.G., and Mendoza, V.E.E.: 1973, Astrophys. J. 180, pp. 115-119.

Herbig, G.H.: 1969, in L. Detre (ed.), "Non-Periodic Phenomena in Variable Stars," Academic Press, Budapest, pp. 75-82.
Herbst, W., and Assousa, G.E.: 1977, in T. Gehrels (ed.), "Protostars and Planets," University of Arizona Press, Tucson, pp. 368-383.
Herbst, W., and Turner, D.G.: 1976, Publ. Astron. Soc. Pacific 88, pp. 308-311.
Hoffmeister, C.: 1962, Z. Astrophys. 55, pp. 290-300.
Hyland, A.R., and Jones, T.J.: 1980, Paper presented at IAU Symposium 96.
Jenkins, E.B., and Savage, B.D.: 1974, Astrophys. J. 187, pp. 243-255.
Jones, T.J., and Hyland, A.R.: 1980, Monthly Notices Roy. Astron. Soc. (in press).
Jones, T.J., Hyland, A.R., Robinson, G., Smith, R., and Thomas, J.A.: 1980, Astrophys. J. (in press).
Joy, A.H.: 1945, Astrophys. J. 102, pp. 168-195.
Jura, M.: 1979, Astrophys. Letters 20, pp. 89-91.
Keene, J., Harper, D.A., Hildebrand, R.H., and Whitcomb, S.E.: 1980 (preprint).
Kenyon, S., and Starrfield, S.: 1979, Publ. Astron. Soc. Pacific 91, pp. 271-275.
Knacke, R.F., Strom, K.M., Strom, S.E., Young, E., and Kunkel, W.: 1973, Astrophys. J. 179, pp. 847-854.
Lada, C.J., and Elmegreen, B.G.: 1979, Astron. J. 84, pp. 336-340.
Larson, R.B.: 1972, Monthly Notices Roy. Astron. Soc. 157, pp. 121-145.
Loren, R.B.: 1979, Astrophys. J. 227, pp. 832-852.
Martin, R.N., and Barrett, A.H.: 1978, Astrophys. J. Suppl. 36, pp. 1-51.
Mendoza, V.E.E.: 1966, Astrophys. J. 143, pp. 1010-1014.
Mendoza, V.E.E.: 1968, Astrophys. J. 151, pp. 977-989.
Mould, J.R., and Ridgway, S.: 1980, Private communication.
Rydgren, A.E.: 1976, Publ. Astron. Soc. Pacific 88, pp. 111-115.
Rydgren, A.E.: 1980, Astron. J. 85, pp. 444-450.
Rydgren, A.E., Strom, S.E., and Strom, K.M.: 1976, Astrophys. J. Suppl. 30, pp. 307-336.
Schmidt, E.G.: 1975, Monthly Notices Roy. Astron. Soc. 172, pp. 401-409.
Schwartz, R.D.: 1977, Astrophys. J. Suppl. 35, pp. 161-170.
Smith, R.: 1980, Private communication.
Spencer, R.G., and Leung, C.M.: 1978, Astrophys. J. 222, pp. 140-152.
Strom, K.M., Strom, S.E., Carrasco, L., and Vrba, F.J.: 1975, Astrophys. J. 196, pp. 489-501.
Strom, K.M., Strom, S.E., and Vrba, F.J.: 1976, Astron. J. 81, pp. 308-316.
Strom, S.E.: 1977, "Star Formation," in T. De Jong and A. Maeder (eds.), IAU Symposium 75, D. Reidel Publ. Co., Dordrecht, pp. 179-197.
Strom, S.E., Grasdalen, G.L., and Strom, K.M.: 1974, Astrophys. J. 191, pp. 111-142.
Strom, S.E., Strom, K.M., and Grasdalen, G.L.: 1975, Ann. Rev. Astron. Astrophys. 13, pp. 187-216.
Strom, S.E., Vrba, F.J., and Strom, K.M.: 1976, Astron. J. 81, pp. 314-316.
Tapia, S.: 1975, in Greenberg and Van de Hulst (eds.), IAU Symposium 52, "Interstellar Dust and Related Topics," D. Reidel Publ. Co., Dordrecht, pp. 43-51.

Tomita, Y., Saito, T., and Ohtani, H.: 1979, Publ. Astron. Soc. Japan 31, pp. 407-416.
Villere, K.R., and Black, D.C.: 1980, Astrophys. J. 236, pp. 192-200.
Vrba, F.J.: 1976, Ph.D. Thesis, University of Arizona, Tucson.
Vrba, F.J.: 1977, Astron. J. 82, pp. 198-208.
Vrba, F.J., Strom, K.M., Strom, S.E., and Grasdalen, G.L.: 1975, Astrophys. J. 197, pp. 77-84.
Vrba, F.J., Strom, S.E., and Strom, K.M.: 1976, Astron. J. 81, pp. 317-319.
Walraven, Th., and Walraven, J.H.: 1960, Bull. Astron. Inst. Neth. 15, pp. 67-83.
Werner, M.W., and Salpeter, E.E.: 1969, Monthly Notices Roy. Astron. Soc. 145, pp. 249-269.
Wootten, A., Evans, N.J., II, Snell, R., and Vanden Bout, P.: 1978, Astrophys. J. Letters 225, pp. L143-L148.

DISCUSSION FOLLOWING PAPER PRESENTED BY A. R. HYLAND

NORDH: The dust temperature, 13-16 K, quoted for B335 is higher than previously quoted by Keene _et al._, and also higher than existing theoretical predictions of dust temperatures of dark clouds illuminated by the ISM (see Leung, Ap.J. 199, 340, 1975). Do you or Keene have any comments to this?

KEENE: There are new observations which give a slightly higher temperature than 10 K. I don't believe there is a significant difference between the temperature we observe and the theoretical models.

LADA: The far-IR observations of Keene _et al._ are in very good agreement with theoretical predictions. For example, a few years ago Leung calculated dust temperature distributions for clouds heated by the general background radiation field. He found that the dust temperatures range from ~12 K to 2 K from the surface to core of the clouds. Therefore the observed temperature T_D = 13-16 K for B335 is entirely compatible with expectations. Also for these objects the gas and dust temperatures are not necessarily coupled. Gas is heated by cosmic rays and cooled by line emission with equilibrium gas temperatures of ~8-12 K. Dust is heated by starlight and cooled by thermal re-radiation to temperatures between 7 and 14 K. The fact that the two are roughly equal is mostly fortuitous and not an indication of a common physical basis for heating.

SHERWOOD: We have detected B68 at 1 mm (300-330 GHz). The observations are in agreement with thermal emission from dust at 10 K.

INFRARED SPECTROSCOPY OF PROTOSTELLAR OBJECTS

Rodger I. Thompson
University of Arizona
Tucson, Arizona 85721

A. INTRODUCTION

The process of star formation occurs in regions of space not accessible to the traditional techniques of optical spectroscopy and photometry. Star formation begins with a density enhancement in a cold molecular cloud. Stellar gestation then occurs inside an obscuring cocoon of natal gas and dust, which for high mass stars may still be in place when the star reaches the Zero Age Main Sequence (ZAMS). During this time the new star gathers the material which will make up its total mass, solves its angular momentum problem, achieves hydrostatic equilibrium and is making the conversion from gravitational to nuclear energy. The only direct view of this remarkable process is by radio and infrared techniques. Unlike many areas of infrared astronomy, which are extensions of previous optical studies to other wavelengths, information about star formation is primarily gained through infrared and radio studies.

This review concentrates on the contributions of infrared spectroscopy to the subject of star formation. Several topics such as the extensive work done on the Orion Nebula, molecular hydrogen emission lines, dust emission, and evolved H II regions are covered in detail by other reviews in this book and will not be discussed extensively here. This review will concentrate on results rather than techniques as a review of techniques would need a separate chapter. The results have been obtained with filter wheel, grating, Fourier transform (FTS), and heterodyne spectrometers. When it is important the technique will be mentioned.

The structure of this review is divided according to physical processes rather than by wavelength region. There is, however, a general division of studies according to wavelength which is imposed by the nature of the star formation process. The early stages of star formation are generally accessible only by radio and far infrared spectroscopy due to the low excitation temperature of molecular clouds. Low excitation forbidden lines of neutral atoms and molecules such as

C. G. Wynn-Williams and D. P. Cruikshank (eds.), Infrared Astronomy, 153–165.

CI and CO can be observed in these regions. The continuum shape of the emission can also be studied. More evolved stages can be studied at intermediate (5 - 30μm) infrared wavelengths. Dust and several unknown broad emission and absorption features appear at these wavelengths as well as moderate excitation forbidden lines. Near (1 - 5μm) spectroscopy is generally limited to the later stages of star formation. Recombination lines from surrounding H II regions or absorption lines against hot dust can be observed in this spectral region. All of these topics are discussed below. The limitations on space and time for this review necessitate that it be illustrative rather than comprehensive. No attempt has been made to include all work in the field but rather to give illustrative examples.

B. FORBIDDEN LINE OBSERVATIONS

The observation of forbidden lines is certainly not unique to the infrared spectral region; however, fine structure lines with very low excitation levels which may act as cooling lines in cool gas clouds can only be observed in the infrared. Lines detected to date include [CI] (610μm), [CII] (157μm), [OI] (63μm), [OIII] (51.8μm, 88.3μm), [NIII] (57μm),[SIII](18.71μm) as well as several lines at intermediate wavelengths such as [NeII] (12.8μ), [SIV] (10.51μm) and [ArIII] (8.99μm). The low excitation potential of these lines make them relatively insensitive to temperature in most regions (Simpson 1975) but they are sensitive to density.

The far infrared lines must be observed from airborne or balloon borne platforms in order to reduce the absorption due to telluric water vapor. Most of the current work has centered on the [OIII] 88.35μm line which has been observed in a large number of star formation regions (see references). This line has an excitation potential of 163° K and is therefore easily collisionally excited. Of particular interest for cool clouds are the recent observations of the [CI] 610μm line by Phillips, Huggins, Kuiper and Miller (1980) with heterodyne techniques and the [CII] 157μm line by Russell, Melnick, Gull and Harwit (1980) with a cooled grating spectrometer. The 157μm [CII] line has an excitation potential of 92° K and has been discussed as a prime coolant of gas clouds at low temperature (Dalgarno and McCray 1972, Jura 1978). A high luminosity was detected in this line from M42 (80 $L_\odot$) and NGC 2024 (50 $L_\odot$) although it does not appear to compete with dust in the energetics of either of these clouds. The CI line at 610μm was detected in 8 molecular clouds which are known to have associated H II regions. Phillips _et al_. (1980) found the spatial and velocity distribution of the CI emission to be similar to that of CO. The lower brightness temperature of CI in spite of an emission coefficient almost the same as CO (1-0) indicates that most of the carbon may be tied up in CO or dust.

C. CONTINUUM AND BROAD UNIDENTIFIED FEATURES

The usefulness of infrared spectroscopic studies is not necessarily limited to the detection of atomic or molecular lines. Accurate descriptions of the continuum flux are extremely valuable in constraining models for radiative transfer in the gas and dust around protostellar objects. Moderate resolution continuum spectra have also revealed broad absorption and emission features which are thought to be associated with water ice, silicates and other not yet identified materials.

Far infrared moderate resolution continuum spectra of several star formation regions have been obtained with FTS techniques from the NASA aircraft by Erickson, Pipher and others (see Bibliography). Of particular interest in these spectra are the deviations from single temperature blackbody curves. The emergent far infrared spectrum, which contains most of the power in protostellar objects, is dependent on the central source, the optical properties of the dust and ices in the source, and the geometrical temperature and density distribution of the material. It has not been possible, to date, to solve uniquely for each of these parameters but the available spectra indicate that there is a variation among sources and that optical depth effects play an important role.

Continuum spectra especially in the intermediate infrared and near infrared down to 3μm often show broad absorption or emission bands when studied at resolutions of ≈ 100 with filter wheel spectrometers. Some of these bands have reasonable identifications, such as the 9.7μm silicate feature and the 3.08μm ice band. Other bands at 3.3μm, 6.0μm, 6.8μm etc. are unidentified and may represent important diagnostic tools in determining the compositions and column density of the dust surrounding protostars. It is not yet certain whether these bands are due to solid or gaseous components. Recent work by Tokunaga and Young (1980) has shown that at a resolution of 3000 the 3.3μm emission band does not break up into resolved features. They rule out a gaseous absorption on these grounds and postulate a solid state resonance.

D. MOLECULAR ABSORPTION AND EMISSION LINES

Near infrared spectroscopy has been used to gain information about the surrounding medium as well as about the nature of the central protostellar object. The main source of molecular absorption lines in the near infrared is the system of vibration-rotation transitions of the CO molecule. The fundamental, first overtone and second overtone bands of CO occur at 5, 2.3 and 1.5μm respectively. Each band consists of a very large number of lines with lower levels which vary from the ground state to high excitation temperatures. The oscillator strengths for the individual lines decrease by a factor of 100 between the fundamental and the first and second overtone. The variation of oscillator strengths and excitation temperature make it possible to find unsaturated lines for observation.

Infrared absorption line studies complement the CO emission line studies in that the radio $^{12}C^{16}O$ lines are usually saturated and often observed at low angular resolution. Infrared lines on the other hand can be observed in non-saturated regions to monitor a narrow column determined by the angular extent of the infrared source. The infrared CO lines can provide both temperature and column density information if lines at several excitation temperatures are observed. In some cases it may be possible to gain information on the $^{12}C/^{13}C$ ratio in molecular clouds. This ratio is very important in determining total abundances from radio observations of the emission from the $^{13}C^{16}O$ molecule. The main work on CO absorption and emission lines has been carried out in the Orion region by Hall and his collaborators with a high resolution FTS. This work is described in detail by Scoville in this volume.

At present the most studied infrared molecular emission lines are those of molecular hydrogen. Molecular hydrogen is the most abundant molecule in the galaxy and has no radio transitions. A detailed account of the observations of molecular hydrogen is given by Beckwith in this volume. Very recently, however, the higher rotational transitions of CO have been observed in emission at far infrared wavelengths from the Kleinmann Low nebula by Watson, et al. 1980 and Storey et al. 1980. They have observed four lines of CO with upper J levels in the range between 21 and 30 at wavelengths near 100μm. The lines appear to be optically thin and tentatively seem to imply that most of the carbon is in the form of CO which is consistent with the CI observations discussed in Section B. These same workers also have a possible detection of the $\pi_{3/2}$ 5/2→3/2 doublet of the OH radical at 119μm. If the infrared CO lines are truly optically thin they should provide an important comparison with the radio CO lines. Unfortunately the levels observed may not be in LTE due to their high radiative transition probabilities.

E. RECOMBINATION LINE INFRARED SPECTRA

Interpretation of recombination line spectra is a well understood method of analysis which has been very successfully used in optical spectroscopy. The absolute strength of the hydrogen recombination lines is directly proportional to the value of Ne^2V in an H II region. The value of Ne^2V is in turn a direct measure of the Lyman continuum flux from the central object. Boundary conditions on the temperature of the central object can also be obtained by the ratio of hydrogen to He I and He II recombination lines.

The lack of optical flux from most protostellar objects limits the measurement of line fluxes to the infrared and radio regions. Information about the central objects is given by the total luminosity as well as the radio continuum or infrared line flux which give the proportion of the total luminosity which lies in the Lyman continuum if the lines are formed by recombination. If it is assumed that the central object is a ZAMS star then either the total luminosity or the ratio of Lyman to total luminosity uniquely fixes the spectral type of the central star. The radio and infrared measurements have different advantages and disadvantages. The radio measurements are not affected by extinction but are often subject to optical depth problems. The infrared lines, generally the Brackett series of hydrogen, are almost always optically thin but are subject to extinction. If both the radio and infrared flux is optically thin the ratio of the Brackett γ line at 2.16μm to the 5 GHz radio continuum is given by

$$f_{5\ GHz} = 1.15 \times 10^{14}\ f_{B\gamma}\ Jy \qquad (1)$$

where the $B\gamma$ flux is given in Watts m^{-2} (c.f. Wynn-Williams _et al._ 1978). The current detection limit on $B\gamma$ is about 10^{-17} watts m^{-2} which is equivalent to a radio flux of 1.15 mJy. If a 20 mag visual extinction is assumed then the minimum detectable infrared line is equivalent to a radio flux of about 6 mJy which is near the upper limits quoted for many radio studies. The two techniques are therefore roughly equal in sensitivity with the radio excelling for very high extinction and the infrared excelling for optically thick compact H II regions.

Several studies of the infrared recombination lines of hydrogen and helium have been carried out with filter wheel, grating and FTS techniques (e.g. Simon, Simon and Joyce 1979, Soifer, Russell and Merrill 1976, Thompson and Tokunaga 1979, and other references in the bibliography). Most of these studies have emphasized either the $B\gamma$ and $B\alpha$ lines with some observations of the He I line at 2.0581μm and the 1.7 to 1.2μm region. A summary of these observations is given in Figure 1.

Figure 1 is a plot of the Ne^2V value for an object as derived from the Brackett line fluxes or radio measurements versus the luminosity derived from a combination of near, intermediate and far infrared photometry. The parameter Ne^2V has been used to unify the measurements under the assumption of free-free emission and Menzel's Case B recombination radiation. It is of course possible for other emission mechanisms to contribute to either the radio continuum or the infrared line flux. The plots for ZAMS stars and luminosity Class I stars from Panagia 1973 are also shown on the figure with the spectral classes marked at various points along the lines. The top legend indicates the percentage of the total luminosity that appears in Lyman continuum for ZAMS stars. The directions that errors in extinction, luminosity and distance will move the points is indicated on the figure. Most of the points on the figure were obtained from $B\gamma$ measurements (+). Sources in which only $B\alpha$ was used are indicated with a ($\oplus$). The point marked for BN assumes that the KL nebula is not powered by BN. If the KL luminosity is included the point will move up as indicated by the arrow. S140/IR, MonR2/IRS3 and GL2591 have upper limits on the line flux which are indicated by the arrows moving to the right. Both

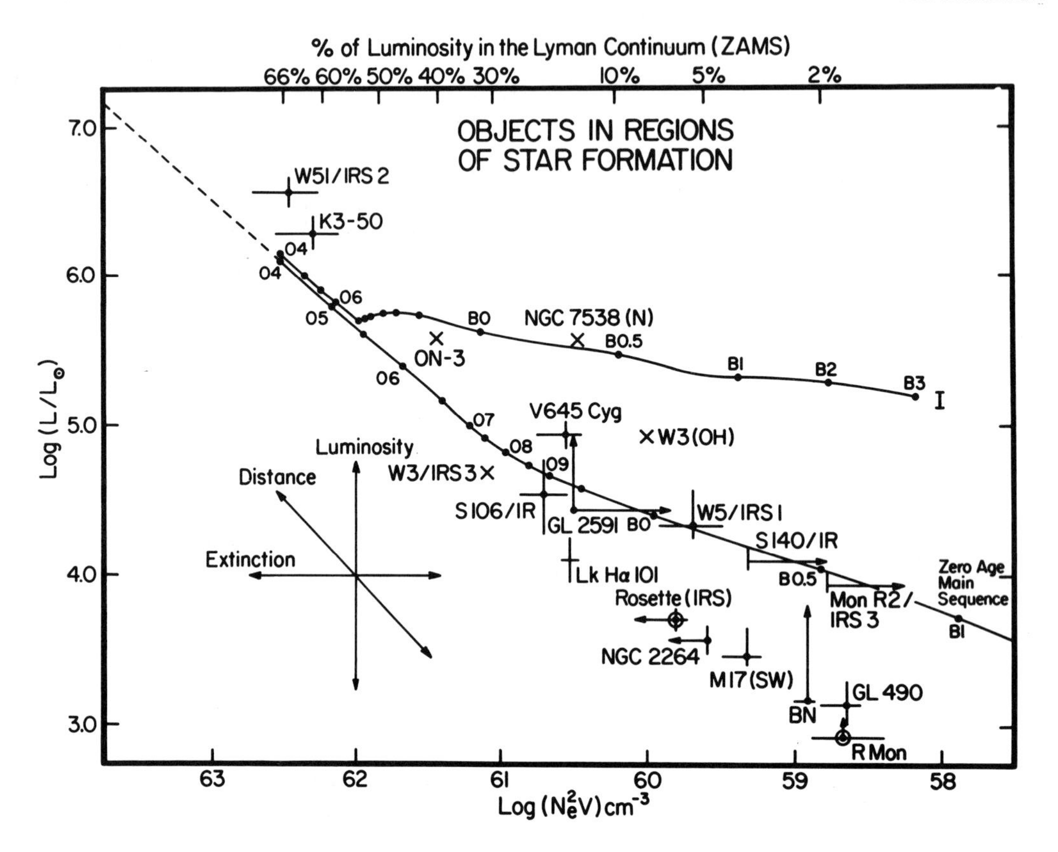

Figure 1. Plot of the Ne^2V value determined observationally versus the measured luminosity for several protostellar objects. See text for a detailed discussion.

S140/IR and MonR2/IRS3 are now thought to be multiple objects; therefore, their lack of Bγ flux is not surprising. A few radio sources from Thronson and Harper (1979) (×) have been plotted for comparison. The Bγ points listed for K3-50 and W51/IRS2 are in agreement with the radio values for those objects.

The basic conclusion to be drawn from Figure 1 is that the simple model of a ZAMS star surrounded by a dust free H II region inside a cocoon of dust does not fit most of the points in the figure. This conclusion is not new for the high luminosity objects but has not been observed before in intermediate luminosity objects. Details of this discrepancy are discussed in Section F.

F. EXCESS LINE FLUXES IN INTERMEDIATE LUMINOSITY OBJECTS

Most of the objects in Figure 1 do not fit the simple model of a ZAMS star surrounded by an H II region inside a dust cocoon. Stars with luminosities higher than 10^5 $L_\odot$ such as W51/IRS2 and K3-50 have line fluxes or Ne^2V values less than those expected for a ZAMS star of the same luminosity. This result has been well established by comparison of the radio emission to the observed far infrared flux (e.g. Jennings 1975). This effect is thought to be due to absorption of ionizing photons by dust inside the H II region. The infrared line studies are consistent with this hypothesis.

Objects with luminosities less than 5×10^4 $L_\odot$ show a trend opposite to the high luminosity objects. Many of the intermediate luminosity objects have line fluxes or Ne^2V values significantly in excess of the value expected from a ZAMS star of the same luminosity. It is interesting to note that most of the objects with excess line fluxes (as well as W5/IRS1 which has a normal line flux) have very compact H II regions with no measurable radio flux. Most of the objects also have no optical flux; therefore the effect is only observable at infrared wavelengths. Since these results are clearly in variance with the simple model either the observations do not imply the luminosity of Ne^2V values ascribed to them or a fundamental property of protostellar objects is being observed. Although the latter is far more exciting it is useful to consider the former first.

Figure 1 indicates the direction that points move in the diagram under the influence of various likely errors. Quantitative measures of line fluxes and hence Ne^2V must include a measurement of the extinction at the line wavelength. Extinctions are calculated from a combination of factors which include the relative strengths of the Brackett lines, the depth of the 9.7μm silicate feature, and photometric data at IR and visual wavelengths. Generally the smallest possible extinction value has been used to calculate the position of the objects in Figure 1. Even though it is important to measure the extinction as exactly as possible, it should be noted that most of the objects in Figure 1 would still appear to have excess line flux even if the Brackett line extinction was assumed to be zero. It is also possible that the observed infrared luminosity does not equal the total luminosity. This is possible if not all of the source photons are absorbed by dust and re-emitted within the beam width of the photometric observation. It is easy to imagine special geometries in which this may be the case. This special geometry argument, however, loses force in view of the large number of objects found to have excess line flux.

Errors in distance can also affect the conclusions about excess line flux. Distance errors scale both the luminosity and the line power by the square of the distance and are represented by 45° lines in Figure 1. At high luminosities the distance line is almost parallel to the ZAMS track. Most of the excess line flux objects would require distance errors of 4 to 5 to bring them back to the ZAMS. In some cases the lack of He I emission is then incompatible with the new ZAMS spectral type required at the higher distances. This review of observational problems indicates that the phenomenon is probably real and represents physical conditions present in the objects.

The excess line flux objects in Figure 1 occupy a region of the HR diagram which represents objects with radii 3 to 4 times smaller than their ZAMS counterparts with the equivalent Lyman continuum luminosity. This is not a state predicted by any current star formation theories or numerical simulations. An alternative is to abandon the concept of a single blackbody temperature for the star. The model would be a relatively low luminosity star at its normal ZAMS temperature with an added component of Lyman continuum luminosity to provide the ionized region for the emission lines. In view of the early evolutionary state of the objects it is natural to look at mass accretion processes as a possible source of energy for the excess Lyman continuum luminosity. It also appears that, whatever mechanism is responsible, it cannot compete with the high ionizing flux of stars with spectral types earlier than about O8 (Figure 1). This limitation is also consistent with mass accretion. Accretion luminosities are proportional to M_*/R_* for a given accretion rate where M_* and R_* are the mass and radius of the central object. The ratio of M/R is essentially constant over the ZAMS when compared to the very large range of the Lyman continuum luminosity.

In order to be effective the bulk of the accretion luminosity must be emitted in the Lyman continuum. This suggests that the accretion energy is deposited in a small volume which results in a high temperature emission region. Two possible mechanisms are thin shock regions for spherically symmetric accretion and the boundary layer for impact on the star of material accreted from a disk. In all such cases a restriction on the emitted power is imposed by the Eddington limit. A simplified analysis for the boundary layer accretion from a viscous disk yields a limit on the accretion luminosity L_{AC} of

$$L_{AC} \leq 4/9 \frac{c\, G\, M_*}{K_{av}} \approx 10^3\, L_\odot\, M_*/M_\odot \tag{2}$$

where K_{av} is the average opacity per unit mass and is taken equal to the electron scattering opacity to determine the numerical factor on the right hand side. This limit is consistent with the observations of the phenomena only for stars later than about O8.

Emission processes other than recombination radiation should also be considered for the production of the infrared lines. The observed flux in the Bγ line is on the order of a solar luminosity for most of the objects which implies a total line luminosity of at least $10^3\ L_\odot$ for either recombination or thermal line emission. This very high luminosity appears to be beyond the range of most chromospheric models.

The conclusion is that the line excess is the result of processes, probably accretion, occurring during the formation of the central star.

G. IR SPECTROSCOPIC SURVEYS

The increasing maturity and sensitivity of infrared spectroscopy has opened the possibility of comprehensive spectroscopic surveys of the infrared sources in star formation regions. Pioneering work in this area has been carried out by Elias (Elias 1978a, b, c) who has studied several dark clouds. Low resolution 2μm spectroscopy on most of the available objects was used to discriminate between field stars and objects associated with the cloud. The spectra also indicated the spectral type of several of the objects in the regions. No highly luminous O stars were found but the spectra were not of sufficient resolution to detect Bγ radiation from the H II regions around late O and early B stars. The infrared techniques allow much younger clouds with higher extinctions to be studied than can be handled in optical surveys. Complete surveys in both spatial extent and wavelength will allow both the energetics and evolution of star formation to be studied in dark clouds.

One of the difficulties in surveying the types of stars found in star formation regions is the lack of standard spectra by which the objects being studied can be identified. Later stars are easily identified by the strong CO absorption bands but relatively few criteria exist for stars earlier than KO. G stars have relatively few strong features but have weak hydrogen Brackett series lines and CO absorption. The Brackett lines increase in strength for F stars and the CO absorption disappears. The Brackett lines are significantly Stark broadened and tend to decrease in strength toward early B spectral types. O stars show almost no features in most cases; however, both early B and O stars show strong emission lines of hydrogen and helium if there is enough nearby material to produce an ionization bounded H II region. These characteristics have not been quantified, however, in the manner that optical spectral characteristics have been carried out such as K stars at 2μm (Ridgway 1974) and T Tauri stars at 10μm (Cohen 1980). The surveys will also be useful in determining the populations in the nuclei of other galaxies from infrared spectra as well as important in determining the nature of the objects studied.

REFERENCES AND BIBLIOGRAPHY

B. Forbidden Lines

Baluteau, J.-P., Bussoletti, E., Anderegg, M., Moorwood, A. F. M. and Coron, N.: 1976, Ap. J., 210, p. L45, ([OIII], [SIII]).
Dain, F. W., Gull, G. E., Melnick, G., Harwit, M. and Ward, D. B.: 1978, Ap. J., 221, p. L17, ([OIII]).
Dalgarno, A. and McCray, R. A.: 1972, Ann. Rev. Astron. and Astrophys., 10, p. 375.

Hefele, H. and Schulte, J.: 1978, Astron. and Astrophys., 66, p. 465, (Int. IR Lines).
Jura, M.: 1978, Protostars and Planets, Gehrels, Ed., University of Arizona Press, Tucson, Arizona, p. 165.
Lester, D. F., Dinerstein, H. L. and Rank, D. M.: 1979, Ap. J., 232, p. 139, (Int. IR Lines).
McCarthy, J. F., Forrest, W. J. and Houck, J. R.: 1979, Ap. J., 231, p. 711, ([SIII]).
Melnick, G., Gull, G. E., Harwit, M.: 1979a, Ap. J. (Letters), 227, p. L29, ([OI]).
Melnick, G., Gull, G. E., Harwit, M.: 1979b, Ap. J. (Letters), 227, p. L35, ([OIII]).
Melnick, G., Gull, G. E., Harwit, M., and Ward, D. B.: 1978, Ap. J. (Letters), 222, p. L137 ([OIII]).
Moorwood, A. F. M., Baluteau, J.-P., Anderegg, M., Coron, N. and Biraud, Y.: 1978, Ap. J., 224, p. 101, ([OIII], [SIII]).
Moorwood, A. F. M., Salinari, P., Furniss, I., Jennings, R. E., and King, K. J.: 1979, preprint, ([OIII], [OI], [NIII]).
Phillips, F. D., Huggins, P. J., Kuiper, T. B. H. and Miller, R. E.: 1980, preprint, [CI] (610μm).
Rank, D. M., Dinerstein, H. L., Lester, D. F., Bregman, J. D., Aitken, D. K. and Jones, B.: 1978, M.N.R.A.S., 185, p. 179, (Int. IR Lines).
Russell, R. W., Melnick, G., Gull, G. E. and Harwit, M.: 1980, preprint, [CII].
Simpson, J. P.: 1975, Astron. and Astrophy., 39, p. 43, (Theoretical Study of Forbidden Lines).
Storey, J. W. V., Watson, D. M., and Townes, C. H.: 1979, Ap. J., 233, p. 109, ([OIII], [OI]).

C. Continuum and Unidentified Features

Capps, R. W., Gillett, F. C. and Knacke, R. F.: 1978, Ap. J., 226, p. 863, (W33 OH).
Cohen, M.: 1980, M.N.R.A.S., in press, (T Tauri Stars).
Erickson, E. F., Tokunaga, A. T., Knacke, R. F. and Haas: 1980, preprint, (45μm Ice).
Erickson, E. F., Caroff, L. J., Simpson, J. P., Strecker, D. W. and Goorvitch, D.: 1977, Ap. J., 216, p. 404, (Sgr B2).
Erickson, E. F., Strecker, D. W., Simpson, J. P., Goorvitch, D., Augason, G. C., Scargle, J. D., Caroff, L. J. and Witteborn, F. C.: 1977, Ap. J., 212, p. 696, (KL).
Erickson, E. F. and Tokunaga, A. T.: 1980, preprint, (W51, W49).
Gillett, F. C. and Forrest, W. J.: 1973, Ap. J., 179, p. 483, (BN).
Gillett, F. C., Jones, T. W., Merrill, K. M. and Stein, W. A.: 1975, Astron. and Astroph., 45, p. 77, (VI Cyg No. 12, BN).
Herter, T., Duthie, J. G., Pipher, J. L. and Savedoff, M. P.: 1979, Ap. J., 234, p. 897, (W51 and W49).
Merrill, K. M., Russell, R. W. and Soifer, B. T.: 1976, Ap. J., 207, p. 763, (Ices and Silicates in Molecular Clouds).

Penston, M. V., Allen, D. A. and Hyland, A. R.: 1971, Ap. J. (Letters), 170, p. L33, (BN).
Pipher, J. L., Duthie, J. G. and Savedoff, M. P.: 1978, Ap. J., 219, p. 494, (Orion).
Pipher, J. L. and Soifer, B. T.: 1976, Astron. and Astrophy., 46, p. 153, (S255).
Puetter, R. C., Russell, R. W., Soifer, B. T. and Willner, S. P.: 1979, Ap. J., 228, p. 118, (K3-50, W51).
Russell, R. W., Soifer, B. T. and Puetter, R. C.: 1977, Astron. and Astrophy., 54, p. 959, (BN).
Soifer, B. T., Russell, R. W. and Merrill, K. M.: 1976, Ap. J., 210, p. 334, (W51, K3-50, NGC 7538).
Soifer, B. T., Puetter, R. C., Russell, R. W., Willner, S. P., Harvey, P. M. and Gillett, F. C.: 1979, Ap. J. (Letters), 232, p. L53, (W33).
Thompson, R. I. and Tokunaga, A. T.: 1979, Ap. J., 229, p. 153, (S140, Mon R2, R Mon).
Tokunaga, A. T., Erickson, E. F., Caroff, L. J. and Dana, R. A.: 1978, Ap. J. (Letters), 224, p. L19, (S140).
Tokunaga, A. T. and Young, E. T.: 1980, Ap. J. (Letters), in press, (3.3μm feature).
Willner, S. P.: 1976, Ap. J., 206, p. 728, (NGC 7538).
Willner, S. P.: 1977, Ap. J., 214, p. 706, (W3).

D. Molecular Absorption and Emission Lines

Gautier, T. N. III, Fink, U., Treffers, R. R. and Larson, H. P.: 1976, Ap. J. (Letters), 207, p. L129, (H_2 discovery paper, see chapter by Beckwith for further references).
Hall, D. N. B., Kleinmann, S. G., Ridgway, S. T., and Gillett, F. C.: 1978, Ap. J. (Letters), 223, p. L47, (CO absorption in BN).
Scoville, N. Z., Hall, D. N. B., Kleinmann, S. G. and Ridgway, S. T.: 1979, Ap. J. (Letters), 232, p. L121, (CO emission in BN).
Storey, J. W. V., Watson, D. M., Townes, C. H., Haller, E. E. and Hansen, W. L.: 1980, Paper presented at IAU Symposium 96.
Watson, D. M., Storey, J. W. V., Townes, C. H., Haller, E. E. and Hansen, W. L.: 1980, preprint.

E. Recombination Lines

Hall, D. N. B., Kleinmann, S. G., Ridgway, S. T. and Gillett, F. C.: 1978, Ap. J. (Letters), 223, p. L47, (BN).
Harvey, P. M. and Lada, C. J.: 1980, Ap. J., 237, p. 61, (V645 Cyg).
Joyce, R. R., Simon, M. and Simon, T.: 1978, Ap. J., 220, p. 156, (BN).
Panagia, N.: 1973, A. J., 78, p. 929, (ZAMS calculations).
Righini-Cohen, G., Simon, M. and Young, E. T.: 1979, 232, p. 782, (DR 21, W 75, K3-50).

Simon, T., Simon, M. and Joyce, R. R.: 1979, Ap. J., 230, p. 127, (W3, GL 490, Lk Hα 101, BN, Mon R2, S255, Rosette, R Mon, NGC 2264, W33, 645.1+0.1, W51, GL 2591, S140, NGC 7538).
Soifer, B. T., Russell, R. W. and Merrill, K. M.: 1976, Ap. J., 210, p. 334, (W51, K3-50, NGC 7538).
Thompson, R. I., Strittmatter, P. A., Erickson, E. F., Witteborn, F. C. and Strecker, D. W.: 1977, Ap. J., 218, p. 170, (MWC 349, Lk Hα 101, MWC 297).
Thompson, R. I. and Tokunaga, A. T.: 1978, Ap. J., 226, p. 119, (NGC 2264).
Thompson, R. I. and Tokunaga, A. T.: 1979, Ap. J., 231, p. 736, (GL 490 and GL 2591).
Thompson, R. I. and Tokunaga, A. T.: 1980, Ap. J., 235, p. 889, (W51 and K3-50).
Thronson, H. A. and Harper, D. A.: 1979, Ap. J., 230, p. 133, (Far-infrared and radio measurements).
Thronson, H. A., Thompson, R. I., Harvey, P. M., Rickard, L. J. and Tokunaga, A. T.: 1980, preprint, (W5).
Tokunaga, A. T. and Thompson, R. I.: 1979a, Ap. J., 229, p. 583 (M17).
Tokunaga, A. T. and Thompson, R. I.: 1979b, Ap. J., 233, p. 127, (S106 and S235).
Werner, M. W., Becklin, E. E., Gatley, I., Matthews, K., Neugebauer, G.: 1979, M.N.R.A.S., 188, p. 463, (NGC 7538).
Wynn-Williams, C. G., Becklin, E. E., Matthews, K., Neugebauer, G.: 1978, M.N.R.A.S., 183, p. 237, (G333.6-0.2).

F. Excess Line Flux

Jennings, R. E.: 1975, "H II Regions and Related Topics" Lecture Notes in Physics Vol. 42, Wilson, T. L.and Downes, D. Eds. Springer-Verlag Berlin, Heidelberg, New York, p. 137.

G. Infrared Spectroscopic Survey

Cohen, M.: 1980, M.N.R.A.S., in press, (T Tauri Stars).
Elias, J. H.: 1978a, Ap. J., 223, p. 859, (IC 5146).
Elias, J. H.: 1978b, Ap. J., 224, p. 453, (Ophiuchus).
Elias, J. H.: 1978c, Ap. J., 224, p. 857, (Taurus).
Ridgway, S. T.: 1974, Ap. J., 190, p. 591, (K stars).

DISCUSSION FOLLOWING PAPER DELIVERED BY R. I. THOMPSON

WERNER: It is, of course, possible that energy is escaping in some of your objects, leading to an underestimate of luminosity. Also, since we may expect anisotropies in the distribution of matter around the source, I wonder if you have looked for effects such as correlations between luminosity deficit and reddening.

THOMPSON: The fact that 7 out of 8 objects show the effect described would argue against invoking special geometries. Most of these sources have no radio flux from them, and are therefore very compact regions which must have a rather high density around them. I would suspect that their dust has not yet blown away.

ZUCKERMAN: I think that you are somewhat underestimating the power of the VLA at 1 cm. Even right now your remarks are marginal, but as the receivers on the VLA improve, the radio sensitivity should become much better.

THOMPSON: That is true, but for self-absorbed sources, for instance those which Paul Harvey has found to be optically thick right to 100 μm, the infrared method will win, while for less compact objects radio observations will be more sensitive.

SANDELL: Can you detect sources in Brackett-γ or -α which cannot be detected photometrically?

THOMPSON: No. If they cannot be seen broadband they cannot be seen narrow band.

THE IMPLICATIONS OF MOLECULAR HYDROGEN EMISSION

Steven Beckwith
Department of Astronomy, Cornell University, Ithaca, N.Y., U.S.A.

Abstract: Emission from vibrationally excited molecular hydrogen has been discovered in a variety of objects of widely differing ages and environs including molecular clouds, planetary nebulae, and a Seyfert galaxy. The observations of the H_2 spectra indicate this emission arises in hot, nearly thermalized gas. While there is still some disagreement between detailed predictions of hydrodynamic calculations and recent observations, it is generally believed that energy supplied to the interstellar gas in the form of shock waves is responsible for the observed H_2 emission.

Several of the H_2 sources are molecular clouds associated with ongoing star formation, most notably the Orion Molecular Cloud. From the intensity, strength, temperature, and velocity of the molecular hydrogen emission, it is estimated that at least 10^{48} ergs has been deposited in the cloud over the last thousand years or so in the form of bulk kinetic energy. There is no clear explanation for this process, since the energy is large and the timescale short, and it appears unlikely that we should observe such events unless they occur frequently. Among the other H_2 sources in molecular clouds, NGC 7538, DR 21, and W3 are of similar spatial extent and apparent luminosity as the Orion emission.

1. INTRODUCTION

When Gautier, Treffers, and their collaborators (Gautier et al. 1976, Treffers et al. 1976) discovered vibrationally excited molecular hydrogen in the Orion molecular cloud and in NGC 7027, they uncovered a component of interstellar gas which was almost completely unanticipated by earlier molecular observations. This component consists of dense interstellar material, primarily molecular hydrogen, which has been heated to several thousand degrees, and it contrasts with the more commonly observed molecules which indicate dense molecular gas at a few tens of degrees. The hot gas is widely believed to result from shock waves in the clouds. The unusual strength of the H_2 lines, indicating very energetic shocks, has been a surprise. Prior to the

C. G. Wynn-Williams and D. P. Cruikshank (eds.), Infrared Astronomy, 167–178.

discovery of H_2 emission and the nearly simultaneous discovery of high-velocity CO line wings, millimeter line observations showed gas velocities which were typically less than 10 km s^{-1}, not high enough to excite the vibrational lines of molecular hydrogen.

Following the Orion observations, molecular hydrogen emission was detected from a variety of other celestial objects. Beckwith, Persson, and Gatley (1978) observed H_2 toward five more planetary nebulae, and Beckwith et al. (1978a) found H_2 emission from T Tauri. Gautier (1978) discovered spatially extended H_2 emission from several molecular clouds, and, more recently, Fischer and her collaborators (Fischer, Righini-Cohen, and Simon 1980; Fischer et al. 1980) have observed H_2 in two more clouds which are sites of recent star formation. Elias (1980) has measured H_2 emission lines from seven Herbig-Haro objects in a sample of nineteen. In several cases, it can be shown that the H_2 is probably shock-heated as in Orion; in no case can it be shown that the H_2 is excited by any other process.

Several approaches have been taken to understand the implications of this emission. A variety of new lines have been observed in the shocked regions, notably Orion, to determine the structure and physical condition of the gas flows. Surveys are being made to determine the frequency with which molecular shock waves are generated in the interstellar medium. These surveys, although difficult, now suggest gas flows are relatively common in a variety of celestial objects. The analyses of those objects which show H_2 emission indicate that in some objects (for example NGC 7027), the shocked gas can be easily explained with known gas flows, whereas in others (for example Orion), it is difficult to identify a source which can provide sufficient energy to drive the shock waves. There has been some speculation and considerable controversy about the nature of the source in Orion and its relation to star formation, and it is this problem which is of greatest current interest in the study of H_2 emission.

Theoretical studies of shocked molecular hydrogen have elucidated other aspects of the problem, and the discoveries have stimulated a new interest in the radiative cooling of molecular shocks. The theoretical work of Hollenbach and Shull (1977); Kwan (1977); and London, McCray, and Chu (1977) is essential to the interpretation of the infrared observations. More recent work has gone beyond the H_2 to discuss the direction of new observations. For the sake of brevity, only those calculations which are directly applicable to existing H_2 observations are discussed in this review. The rich literature of other calculations is discussed by Hollenbach (1979) and McKee and Hollenbach (1980).

The following discussion is divided into three sections. The first section (Section 2) described the H_2 emission from the Orion nebula. Orion is the best studied of all the H_2 sources, and it provides a basis for comparison with other H_2 emission regions. The next section summarizes the new results on other H_2 sources. Most of these results have been obtained recently, and only limited data are available for discussion. In the final section, the future of the H_2 observations is briefly discussed.

2. THE ORION MOLECULAR CLOUD

Orion is the best-studied of the known H_2 emission sources. The H_2 emission was first detected by Gautier *et al.* (1976), and subsequently a wealth of observational information has been obtained by various groups (see, e.g., Grasdalen and Joyce 1976; Beckwith *et al.* 1978b; Beckwith, Persson, and Neugebauer 1979; Simon *et al.* 1979; Nadeau and Geballe 1979; Beck, Lacy, and Geballe 1979). Many different molecular hydrogen transitions have been observed in Orion. Because the molecule is homonuclear, all the transitions are electric quadrapole, and every line observed is optically thin with optical depths typically less than 10^{-4}. Thus, from the observations of the vibrational transitions, the excitation temperature, column densities, and extinction can be derived in a self-consistent way.

The relative level populations indicate a region in approximate thermal equilibrium at a temperature of 2000 K; the column density of this gas is of order 3×10^{20} cm^{-2} which is only a small fraction (<1%) of the total H_2 column density in the cloud estimated from other molecular lines. The emission region extends over an area roughly 0.2 pc across, and it is inside the molecular cloud as indicated by the 40 visual magnitudes of extinction to the emission region (see Beckwith, Persson, and Neugebauer, 1979, for a more detailed discussion). There is also H_2 emission which arises from a somewhat cooler ($\lesssim$ 1000 K) region with greater column density (Beck, Lacy, and Geballe 1979). The linewidths vary from 60 km s^{-1} FWHM with wings extending to over 90 km s^{-1} from the line centers in spectra taken near the center of the emission region to less than 30 km s^{-1} FWHM in spectra taken near the edges (Nadeau and Geballe 1979). The line centers are at 9 km s^{-1} (LSR) identical to the line centers of the radio molecular lines to within the observational uncertainties.

The temperature and velocity of the H_2 contrast sharply with most of the molecular material along the line of sight. As inferred from several molecules, notably CO, most of the gas is less than 100 K, with linewidths less than 10 km s^{-1} centered on 9 km s^{-1} LSR, and total column densities of order 10^{23} cm^{-1} or more (Zuckerman 1973). The H_2 emission thus indicates that a small fraction of this gas is quite hot and moving with supersonic velocity relative to the main cloud. Yet the apparent spatial extent of the hot H_2, $\sim$ 0.2 pc, is comparable to the extent of the cloud core, $\sim$ 1 pc. Although only a small fraction of the total molecular material in the core is hot at any time, a much larger fraction has probably been through a hot phase as the hot region grew to its present size.

Most of these observations are readily explained by assuming the H_2 is heated by shock waves driven into the molecular cloud by supersonic gas flows. The calculations of Hollenbach and Shull (1977); Kwan (1977); and London, McCray and Chu (1977) show that shocks should produce the observed column densities of molecular hydrogen at 2000 K if the ambient volume density is of order 10^7 cm^{-3} and the shock velocities range between 10 and 25 km s^{-1}. The required volume densities while high are consistent with other molecular density indicators. Furthermore, several of the most recent observations were anticipated by the first calculations such as the existence of a cooler post-shock region studied by Beck and her collaborators and the CO emission from high-rotational states ($J \sim 25$) studied by Storey et al. (1980). The observed velocities, however, are outside of the range allowed by the theory, the range of shock velocities being limited on the high end by the speed at which all the H_2 molecules are dissociated by the shock. While there has been some dispute about the exact value of this speed (e.g. Dalgarno and Roberge 1979), it is probably not far in error at the derived densities. Since this velocity is less than the observed velocity of much of the H_2, there has been some effort to bring the theory into parity with the observations (e.g. Hollenbach and McKee 1979, Kwan 1979, Draine, Roberge, and Dalgarno 1979, Chevalier 1980). None of these explanations has gained widespread acceptance, but it is generally assumed that some kind of shock-heating excites the H_2, and since we expect shocks to occur anyway from the supersonic velocities indicated by the extreme H_2 line wings, we will assume the H_2 is shock-excited.

Simultaneous with the discovery of the H_2 emission, the CO line profiles in the direction of the cloud core were shown to exhibit wings extending beyond ±50 km s^{-1} (Zuckerman, Kuiper, and Rodriguez-Kuiper 1976; Kwan and Scoville 1976). The flows which produce these velocities are presumably connected with the flows which produce the H_2 emission. There are now a variety of observations of various molecules such as CO, H_2O, and NH_3 which probe various aspects of these flows (see for example Phillips et al. 1977, Wilson et al. 1979, Genzel et al. 1980, and Storey et al. 1980). Gas flows of this type described above have rather remarkable implications for the energetics of the cloud core. Simple dimensional arguments based on either the H_2 or CO observations indicate the kinetic energy input to the cloud per unit time must be large, and the momentum input per unit time is greater than can be obtained from simple radiation pressure models by a factor of a hundred to a thousand, depending on the assumptions. The gas flows thus appear to be caused by a process such as mass loss (with extraordinary mass-loss rates) or an explosion (of energies near those of supernovae), presumably associated with star formation in the cloud core.

If we assume the typical flow velocity is 30 km s^{-1} and the radius of the flow region is 0.1 pc, then the time for the region to expand to its observed size at this velocity is about 3,000 years. For explosions or winds where the overall velocity is a decreasing function of time, this is an upper limit. The total luminosity of the H_2 emission is of order 1000 $L_\odot$, so if this emission has proceeded for the expansion time, then about 3×10^{47} ergs have been radiated. Isothermal shock waves radiate energy at about the same rate as they deposit bulk kinetic energy in a flow, so we might expect that the expanding gas now contains a few times 10^{47} ergs of kinetic energy. This estimate gives almost the same result as estimates of the kinetic energy, based on CO observations (Zuckerman, Kuiper, and Rodriguez-Kuiper 1976; Kwan and Scoville 1976). The momentum in the flow can be estimated in a similar fashion to be roughly 10^{41} g cm s^{-1}. For comparison, the total energy radiated by the infrared cluster in 3,000 years is a few times 10^{49} ergs. The total momentum which is transferred to the gas through radiation pressure is $(\tau\ L/c)\ t = 10^{39}\ \tau$ g cm s^{-1}. The factor τ is essentially the number of times a typical photon is absorbed or scattered before escaping the expanding gas. Notice that this estimate is based on the total luminosity, size, and velocity of the H_2.

These estimates show that while the kinetic energy in the gas is only 1% of the energy radiated by the infrared cluster over lifetime of the flow, the total momentum is unusually large. If the gas is driven by radiation pressure only, the factor τ must be at least 100 in the equation above. Kwan and Scoville (1976) suggested a supernova may have exploded within the cloud less than 1000 years ago and caused an expansion of the cloud core. Other authors (e.g. Genzel and Downes 1977) have favored mass loss from a star or stars in the infrared cluster as the energy input; the mass-loss rates required are of order 10^{-4} $M_{\odot}$ yr^{-1} or more, in the simplest models (Beckwith 1979). Neither supernovae nor objects with such high mass-loss rates were supposed to exist within the Orion molecular cloud prior to these observations. These observations have thus added two new pieces to the puzzle of star formation. They suggest that some short-lived energetic object is associated with at least one well-studied star forming region, and they show that the structure of the core immediately surrounding a premain-sequence association may be swept clean by one of the members of the association. The expansion time of less than 3000 years is perhaps the most striking feature of these remarks. The a priori probability that we should observe such a short-lived phenomenon is small unless the phenomenon occurs frequently.

3. SOURCES OF MOLECULAR HYDROGEN EMISSION

Molecular hydrogen emission has been observed from a variety of celestial objects including molecular clouds, planetary nebulae, Herbig-Haro objects, T Tauri stars, supernova remnants, and even the nucleus of a Seyfert galaxy. The spatial extent and luminosity of the H_2 emission varies by several orders of magnitude within the sample. The observations of each different type of object have been interpreted assuming shock-heating.

Since the 1979 review of these sources (Beckwith 1979), three significant results have been obtained. First, Fischer, Righini-Cohen, and Simon (1980) have discovered two more examples of the Orion phenomena in DR 21 and OMC 2. Additionally, Fischer et al. (1980) have re-examined the molecular hydrogen emission in NGC 7538, discovered by Gautier (1978), and support Gautier's suggestion that NGC 7538 is an example of the Orion phenomenon. Second, Elias (1980) has discovered H_2 emission from seven Herbig-Haro objects in a sample of nineteen. Shock waves have been suggested to explain earlier optical line observations of these objects, and Elias suggests his H_2 observations may be explained by shock-heated gas as well. Third, Beckwith et al. (1980) have extended the observations of NGC 7027, and they conclude that the H_2 may be heated at the outer boundary of the ionized gas by a shock wave driven by the expanding nebula. This contrasts with the earlier interpretation that the H_2 planetaries are excited in neutral clumps embedded in ionized gas (Beckwith, Persson, and Gatley 1978).

The molecular clouds contain the most extended and luminous H_2 emission of all these sources. DR 21 and NGC 7538 have apparent H_2 luminosities which are comparable to Orion. In both objects, the molecular hydrogen emission is of lower surface brightness and greater extent than the Orion emission. The arguments of the last section show that if the flow velocities in these objects are of order 50 km s^{-1}, or less, then these emission regions are older, but have total energy contents which are similar to Orion. The theoretical work mentioned earlier places a lower limit of about 10 km s^{-1} to the speed at which a shock wave can excite appreciable amounts of molecular hydrogen. An upper limit on the age of these new sources is the size divided by 10 km s^{-1}, or 10^5 years for the largest region, DR 21. Similar conclusions were reached by Gautier about W3 and S140. In all of these regions, the total continuum luminosity is of order 10^5 $L_\odot$. Therefore, the arguments concerning stellar wind power or explosive energy needed to cause the events in Orion may be applied to the newly discovered H_2 sources. Note that since the extinction has not yet been measured to the newest H_2 sources, these arguments are based only on apparent surface brightness.

Perhaps the most remarkable discovery is the observation of molecular hydrogen emission from NGC 1068 by Thompson, Lebofsky, and Rieke (1978), recently confirmed by Scoville et al. (1980). A crude estimate of the percentage of molecular clouds which contain H_2 emission sources of similar luminosity as the Orion source can be made from this observation. If we assume the total luminosity of a typical molecular cloud is the same as Orion, 2×10^5 $L_\odot$ (Werner et al. 1976), then an upper limit to the number of such clouds is obtained by dividing the total luminosity of NGC 1068, 3.7×10^{11} $L_\odot$ (Telesco, Harper, and Loewenstein 1976), by this luminosity. The limit is $(N_{clouds})\leq2\times10^6$. The same calculation applied to H_2 emission only, where the apparent luminosities of the v = 1→0 S(1) line in NGC 1068 and Orion are 3.5×10^6 and 2.5 $L_\odot$, respectively, implies that 1.5×10^6 clouds exhibit molecular hydrogen at the same strength as Orion. Thus, almost every cloud is a strong H_2 emitter! While this calculation depends upon questionable assumptions, it is difficult to escape the conclusion that a substantial amount of the molecular gas in NGC 1068 has undergone shock-heating. When better statistics become available on sources in our own galaxy, it will be possible to estimate the rate at which energy is deposited in molecular clouds by these gas flows.

Herbig-Haro objects and the star T Tauri are the other premainsequence objects which exhibit H_2 emission (Elias 1980, Beckwith et al. 1978a). The emission from the HH objects has a similar excitation temperature to Orion. Elias notes the linewidths in these objects are probably large, based on the intensity variations of the Q branch, as they are attenuated by the telluric observation. Shocks have been suggested by Schwartz (1978) to explain the optical lines, and the H_2 emission is consistent with this picture. While the observations of the H_2 in T Tauri are limited to the $v = 1 \rightarrow 0$ S(1) line, plausible assumptions about the emission show it can arise from shocks driven by a stellar wind from T Tauri. There is some controversy about the existence of such a wind, however (Kuhi 1964, Ulrich 1976).

Planetary and protoplanetary nebulae have been shown to display H_2 emission. Because the H_2 emission from NGC 6720 showed good spatial correlation with the 6300 Å line of [OI], Beckwith, Persson, and Gatley (1978) suggested the H_2 in planetaries is excited in neutral clumps embedded in the ionized nebula; this interpretation was given by Capriotti (1973) to explain the [OI] emission. Recently, Beckwith et al. (1980) have analyzed observations of the H_2 emission from NGC 7027, and on the basis of the spatial distribution and line intensity ratios they conclude the H_2 is excited by a shock wave at the outer boundary of the ionized nebula. At this time, it is not known if shock waves excite the H_2 emission seen in all planetaries or if some other excitation process is responsible (e.g. Black 1978). Measurements of the H_2 vibrational temperature can in principle answer this question as they have in Orion. If the H_2 seen in planetaries is shock-heated, planetaries should be good examples for a comparison of the theoretical predictions of shock-wave calculations with observations, since the planetaries are geometrically simple and the gas flows can be mapped with a variety of spectroscopic techniques. A crude estimate of the H_2 mass in NGC 7027 based on the H_2 luminosity gives a value of 1 to 4 $M_\odot$. This value is higher than estimates of the mass of ionized matter for NGC 7027 and for most planetaries.

Finally, molecular hydrogen emission has been detected in the supernova remnant IC 443 by Treffers (1979). In this case, shock waves had been detected by DeNoyer (1979a, 1979b) on the basis of her observations of the OH and HI velocity profiles around the source. Beckwith and DeNoyer (1981) show this emission to be extended and roughly cospatial with the CO emission with a temperature less than about 4000 K.

The sample of known H_2 sources is limited primarily by sensitivity, since sensitive searches over large areas are exceedingly time-consuming with existing instrumentation. For example, Scoville et al. (1979) were unable to find H_2 emission from ten southern hemisphere objects, in spite of diligent effort. On the other hand, Gautier and Fischer and her collaborators have found molecular hydrogen emission by searching large areas for extended emission of low-surface brightness. This emission often bears no obvious spatial relationship to other molecular or infrared emission (e.g. Fischer, Righini-Cohen, and Simon 1980) thus compounding the search problem.

4. FUTURE WORK

Perhaps the two most important unresolved issues which come out of this work are the nature of the driving source for the gas flows observed in the molecular clouds, and their overall importance to the collapse of these clouds. At least in the core of Orion, the H_2 and radio molecular line observations indicate turbulent energy is being deposited in the cloud at a high enough rate to affect the line shapes. It will be useful to understand the observed H_2 line intensities in detail from theoretical work to obtain accurate estimates of the energy in the shocks. The overall importance of the phenomena may be assessed by more sensitive surveys of many molecular clouds. As mentioned above, these searches are very time-consuming with available instrumentation, but Gautier and Fischer and her collaborators have already made progress in this important area.

Several groups are searching for other H_2 lines. Beck and her collaborators (Beck, Lacy, and Geballe 1979; Beck et al. 1980) have demonstrated the efficacy of measuring the $v = 0$, $J = 4 \rightarrow 2$ line at 12 μm. This line probes a larger, cooler portion of the shocks than the 2 μm lines and, furthermore, should be less susceptible to the extinction which plagues the near-infrared lines. Young and Knacke (1980) have observed the $v = 0$, $J = 11 \rightarrow 9$ line at 4.7 μm; they find significant differences between the observed line intensity and that predicted from the simplest theory. My colleagues and I have recently searched for the $J = 9 \rightarrow 7$, $8 \rightarrow 6$, $7 \rightarrow 5$, and $6 \rightarrow 4$ lines between 5 and 8 μm, specifically to avoid the extinction which complicates the interpretation of the near-infrared spectra. These lines provide additional probes of the shock structure. Hall, Scoville and their collaborators (in preparation) have measured several more H_2 lines from the $v = 2$ state to better determine the vibrational temperature. There are still unexplained differences between the observations and the theory, and it is crucial to extend the observations to other lines if these differences are to be understood.

There has been a long-standing interest in the longer wavelength lines at 28 μm and 89 μm, the latter being very strongly forbidden. These lines have excitation energies which are much less than those of any other mentioned in this article. If these lines can be detected, they may provide us with information about the overall abundance and distribution of molecular hydrogen in the interstellar medium (however, Drapatz and Michel [1974] pose serious doubts about the observability of these lines). Observations at these wavelengths are difficult for a variety of reasons, and it is unlikely these lines will be detected in the near future. Nonetheless, the importance of the results and the unexpected strength of the near-infrared lines emphasizes the need to make these observations whenever it is possible.

REFERENCES

Beck, S. C., Lacy, J. H., and Geballe, T. R.: 1979, Ap. J. (Letters) 234, L213.
Beck, S. C., Serabyn, E., Lacy, J. H., Geballe, T. R., and Smith, H. A.: 1980, paper presented at IAU Symposium #96 on Infrared Astronomy.
Beckwith, S.: 1979, IAU Symposium #87 on Interstellar Molecules.
Beckwith, S. and DeNoyer, L. K.: 1981, in preparation.
Beckwith, S., Gatley, I., Matthews, K., and Neugebauer, G.: 1978a, Ap. J. (Letters) 223, L41.
Beckwith, S., Persson, S. E., and Gatley, I.: 1978, Ap. J. (Letters) 219, L33.
Beckwith, S., Persson, S. E., and Neugebauer, G.: 1979, Ap. J. 227, p. 436.
Beckwith, S., Persson, S. E., Neugebauer, G., and Becklin, E. E.: 1978b, Ap. J. 223, p. 464.
Beckwith, S., Persson, S. E., Neugebauer, G., and Becklin, E. E.: 1980, Astron. J., in press.
Black, J. H.: 1978, Ap. J. 222, p. 125.
Capriotti, E. R.: 1973, Ap. J. 179, p. 495.
Chevalier, R. A.: 1980, preprint.
Dalgarno, A. and Roberge, W. G.: 1979, Ap. J. (Letters) 233, L25.
DeNoyer, L. K.: 1979a, Ap. J. (Letters) 228, L41.
DeNoyer, L. K.: 1979b, Ap. J. (Letters) 232, L165.
Draine, B. T., Roberge, W., and Dalgarno, A.: 1979, Bull. ASS 11, p. 689.
Drapatz, S. and Michel, K. W.: 1974, Astron. and Astrophys. 36, p. 211.
Elias, J. H.: 1980, preprint.
Fischer, J., Righini-Cohen, G., and Simon, M.: 1980, Ap. J. (Letters) 238, L155.
Fischer, J., Righini-Cohen, G., Simon, M., Joyce, R. R., and Simon, T.: 1980, preprint.
Gautier, T. N., III: 1978, Ph. D. Thesis, University of Arizona.
Gautier, T. N., III, Fink, U., Treffers, R. R., and Larson, H. P.: 1976, Ap. J. (Letters) 207, L129.
Genzel, R. and Downes, D.: 1977, Astron. Astrophys. 61, p. 117.
Genzel, R., Reid, M. J., Moran, J. M., and Downes, D.: 1980, preprint.
Grasdalen, G. L. and Joyce, R. R.: 1976, Bull. AAS 8, p. 349.
Hollenbach, D. J.: 1979, IAU Symposium #87 on Interstellar Molecules.
Hollenbach, D. J. and McKee, C.: 1979, Ap. J. Supp. 41, p. 555.
Hollenbach, D. H., and Shull, J. M.: 1977, Ap. J. 216, p. 419.
Kuhi, L. V.: 1964, Ap. J. 140, p. 1409.
Kwan, J.: 1977, Ap. J. 216, p. 713.
Kwan, J.: 1979, Bull. AAS 11, p. 688.
Kwan, J. and Scoville, N.: 1976, Ap. J. (Letters) 210, L39.

London, R., McCray, R., and Chu, S. I.: 1977, Ap. J. 217, p. 442.
McKee, C. F. and Hollenbach, D. J.: 1980 Ann. Rev. of Astron. 18, in press.
Nadeau, D. and Geballe, T. R.: 1979, Ap. J. (Letters) 230, L169.
Phillips, T. G., Huggins, P. J., Neugebauer, G., and Werner, M. W.: 1977, Ap. J. Letters 217, L161.
Schwartz, R. D.: 1978, Ap. J. 223, p. 884.
Scoville, N. Z., Hall, D. N. B., Kleinmann, S. G., and Ridgway, S. T.: 1980, paper presented at IAU Symposium #96 on Infrared Astronomy.
Scoville, N. Z., Gezari, D. Y., Chin, G., and Joyce, R. R.: 1979, Astron, J. 84, p. 1571.
Simon, M., Righini-Cohen, G., Joyce, R. R., and Simon, T.: 1979, Ap. J. (Letters) 230, L175.
Storey, J. W. V., Watson, D. M., Townes, C. H., Haller, E. E. and Hansen. W. L.: 1980 preprint, paper presented at IAU Symposium #96 on Infrared Astronomy.
Telesco, C. M., Harper, D. A., and Loewenstein, R. F.: 1976, Ap. J. (Letters) 203, L53.
Thompson, R. I., Lebofsky, M. J., and Rieke, G. H.: 1978, Ap. J. (Letters) 222, L49.
Treffers, R. R.: 1979, Ap. J. (Letters), 233, L17.
Treffers, R. R., Fink, U., Larson, H. P., and Gautier, T. N., III: 1976, Ap. J. 209, p. 793.
Ulrich, R. K.: 1976, Ap. J. 210, p. 377.
Werner, M. W., Gatley, I., Harper, D. A., Becklin, E. E., Loewenstein, R. F., Telesco, C. M., and Thronson, H. A.: 1976, Ap. J. 204, p. 420.
Wilson, T. L., Downes, D., and Bieging, J.: 1979, Astron. & Astrophys. 71, p. 275.
Young, E. T. and Knacke, R. F.: 1980, paper presented at IAU Symposium #96 on Infrared Astronomy.
Zuckerman, B.: 1973, Ap. J. 183, p. 863.
Zuckerman, B., Kuiper, T. B. H., and Rodriguez Kuiper, E. N.: 1976, Ap. J. (Letters) 209, L137.

DISCUSSION FOLLOWING PAPER DELIVERED BY S. BECKWITH

BECKMAN: It is important to emphasize that it could well be possible for radiation pressure alone to drive the mass flow (in Orion and similar sources); a factor of 100 is seen to be required in converting photon momentum to mass momentum. The simple formula $h\nu/c$ does not allow for the more efficient coupling of momentum by photons "bouncing around" in a rather opaque shell before their lengthening wavelength allows them to escape.

BECKWITH: That is true; by allowing photons to be multiply reflected you can amplify the momentum you may obtain from them by a factor of τ where τ is the optical depth. The problem is that, for a region like Orion, you need τ to be of order 100 and consequently for the photons to be reflected about 100 times before being absorbed; I know of no models for dust grains which have albedos of 99%.

JOHNSON: Why is it not possible for the molecular gas to be dissociated in the shock and then rapidly recombine on the grains in the high density region? Would not this give a very thin layer of atomic hydrogen for a short time?

BECKWITH: Such a mechanism has, in fact, been suggested by Hollenbach and McKee. A big advantage of the non-dissociating models is that they provide a natural thermostat mechanism which can explain why several of these sources, not just Orion, have H_2 excitation temperatures of 2000 K.

HOLLENBACH: Shocks which dissociate molecules also have a "thermostat" which produces H_2 excitation temperatures of $T \simeq 2000$ K. H_2 molecular abundances are kept low at higher temperatures because of rapid collisional dissociation.

THOMPSON: The maximum H_2 emission does not seem to correlate with the IR maximum in most protostellar sources. Is this true and does it have any physical significance?

BECKWITH: It is true, but the significance is less clear. Part of the explanation is probably that the emission we see is as much dependent on density irregularities in the surrounding medium as on the location of the source of the expanding flow. In any case, we are most sensitive to regions which are quite extended, at which point the distribution of expanding gas may well have lost some of its original symmetry.

T. L. WILSON: What is the upper limit on range of H_2 densities in the shocked gas? Can it be a factor of 10 higher than the 10^5 cm^{-3} you gave as the density required to thermalize the molecular hydrogen?

BECKWITH: Easily. In our original shock models we derived an average density of something like 3×10^7 cm^{-3}, but this, of course, refers only to a very small amount of the gas that has been shocked. To obtain the density in a larger region we need detailed shock models.

HOLLENBACH: I would like to mention that a theoretical problem arises in modeling the 2 μm intensities in Orion as postshock emission. It is difficult to prevent gas phase chemistry from producing so much H_2O that the H_2 emission is quenched by the dominant IR H_2O emission, which cools the gas.

BECKWITH: As techniques improve we ought to be able to see the H_2O vapor lines, and be able to determine exactly how much cooling is going on.

LADA: On the subject of the high velocity wings in the CO lines, I I would comment that similar wings have now been detected in other sources, such as GL 490 and Cepheus A. They are not hard to detect, but were missed in the past because they were not looked for. As you say, these high velocity phenomena are probably common.

CONTINUUM OBSERVATIONS OF THE INFRARED SOURCES IN THE ORION MOLECULAR CLOUD

G. L. Grasdalen, R. D. Gehrz, and J. A. Hackwell
Department of Physics and Astronomy
University of Wyoming
Laramie, Wyoming 82071

The details of the processes which result in the formation of new stars are still virtually entirely obscure. Surprisingly little can be said with any assurance beyond the general view that the interstellar medium collects in large clouds, involving at least tens of solar masses, which then fragment into smaller pieces which proceed to collapse into individual stars. Numerous highly detailed theoretical predictions have been generated for the sequence of events expected in the fragmentation and collapse processes. However, the meager observational evidence we have on suspected young objects is fascinatingly at variance with these predictions.

The goal is obviously to understand the formation of stars of all masses. The formation of massive stars is particularly attractive for observational analysis. The large expected luminosity of young massive stars makes them accessible to detailed observation even at relatively large distances.

Further, massive stars appear to form in relatively large complexes which have stars in a wide range of evolutionary stages. One of the brightest of the objects thought to be in the early phases of massive star formation is the Becklin-Neugebauer-Kleinmann-Low complex near the Orion nebula.

The first observations of what we now know as the infrared manifestation of the Orion molecular complex were made by Becklin and Neugebauer (1967). They discovered an infrared point source apparently embedded in the Orion Nebula. This point source is commonly referred to as the BN object. Shortly after their discovery observations at longer wavelengths by Kleinmann and Low (1967) revealed a complex extended 20 micron nebula associated with the BN object. The common designation for this region is the KL nebula. To simplify matters we will refer to the region as the BNKL complex, since it seems clear that the entire region is an active region of star formation.

C. G. Wynn-Williams and D. P. Cruikshank (eds.), Infrared Astronomy, 179–186.

The infrared complex is now understood to be intimately involved with a dense molecular cloud. Although it is commonly referred to as the center of the cloud the exact relationship of the infrared sources to the molecular cloud is not yet clear. It is prudent not to allow overly symmetric models of the region to dominate our analysis of the object.

Our new observations of the BNKL complex were obtained with the Wyoming Infrared Observatory (WIRO) 2.3 meter telescope. The method of observation differed substantially from previous infrared maps of this region. The secondary of the WIRO telescope is controlled in position via the central computer system; this allows scans in the direction of chopping to be accomplished without driving the telescope. For the BNKL observations the chopping was in the direction of declination. The throw of the secondary was either 64" or 80" for the observations reported here. Thus there were no corrections applied to the data for signal in the reference beam. The signals from the detectors, germanium bolometers, were amplified and sampled every 5 ms by the computer, which then generated estimates of the signal for each chop of the secondary mirror. This allows single declination scans to be completed quickly. At a chopping frequency of 10 Hz a segment 60" long can be sampled every 1.0" in 6.0 seconds. After a declination scan, the telescope is moved the sampling spacing in Right Ascension and another scan of the secondary initiated. From this process, which is entirely automatic and digital, an "image" of the region is recorded. This technique has been described by Grasdalen et al. (1979).

The advantages of this proceedure are 1) the data do not require complex corrections for the negative beam; they are taken in a total power mode, 2) the image is obtained rapidly; a typical time for the accumulation of a frame is less than ten minutes; thus drifts in telescope position, detector sensitivity and sky conditions are minimized, and 3) since the data are entirely digital the reduction of the material can be handled in impersonal ways.

In Figure 1 we present a contour map of the 20 micron flux from the BNKL region derived from observations with a 3".5 beam. There are a number of condensations visible in the region. The positioning of the map has been derived from offsets from θ^1C in the Trapezium region; it depends only on the position of that star given in the SAO catalog. By comparing individual frames of the BNKL region we can estimate the positions of these condensations and the expected errors in their positions. The repeatability of the positions within the BNKL region is extraordinary; relative positions within the complex are stable to a fraction of an arc second. The results of these determinations are given in Table 1. The source IRc 6 in Table 1 refers to the extended patch of 20 micron emission nearly detached from the main nebulosity.

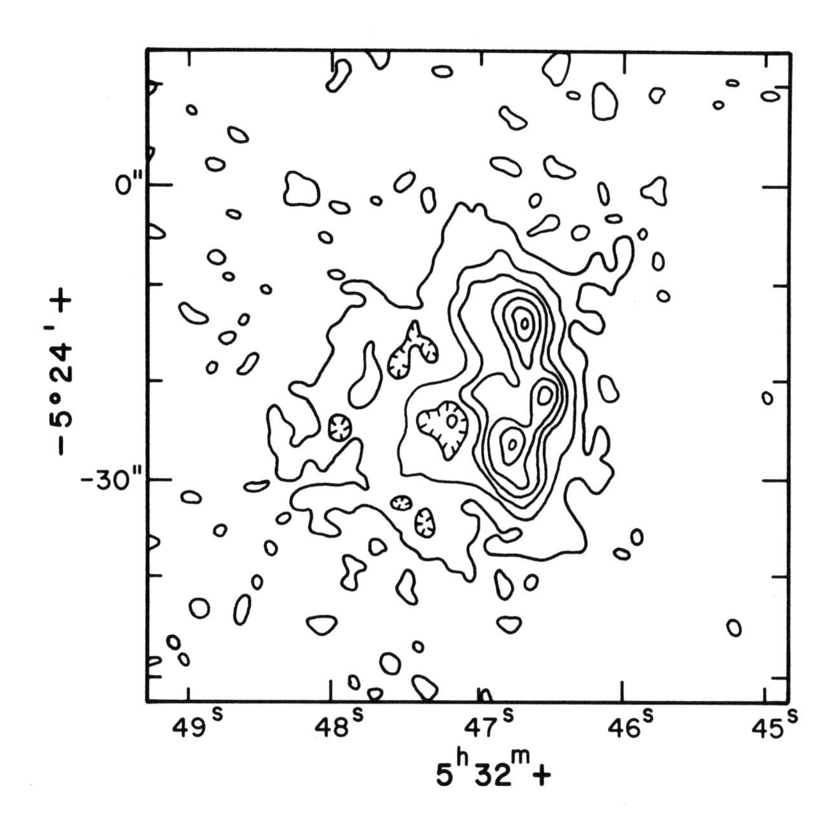

Figure 1. Contours of 20 μm emission from the BNKL complex. The beam was 3".5 in diameter.

In general the positions of the condensations agree quite well with those derived by Rieke, Low and Kleinmann (1973). In Figure 2 we have superposed their 20 micron contours on the WIRO map. However, we do not see a condensation in the position of IRc 5, the southernmost of their condensations. The reason for this descrepancy is not yet clear. The most tantalizing possibility is that this difference represents a real temporal change in the BNKL region. The region is undoubtedly related to the early phases of stellar evolution. Rapid and extreme variability is a common feature of well-studied optical pre-main-sequence objects. Thus it is reasonable to watch carefully for variability in this extreme region. During the past year we have been monitoring this region; as yet we have not seen any indications of variability. It is possible that the discrepancy with the older material is an artifact of the complex data analysis procedure required in the early days of infrared astronomy. In our maps of the BNKL region, there is a tongue of emission extending from IRc 4; this may have been displaced by Rieke,

Low and Kleinmann into the separate source IRc 5. Until this region has been monitored for several more years this situation is unlikely to be completely resolved.

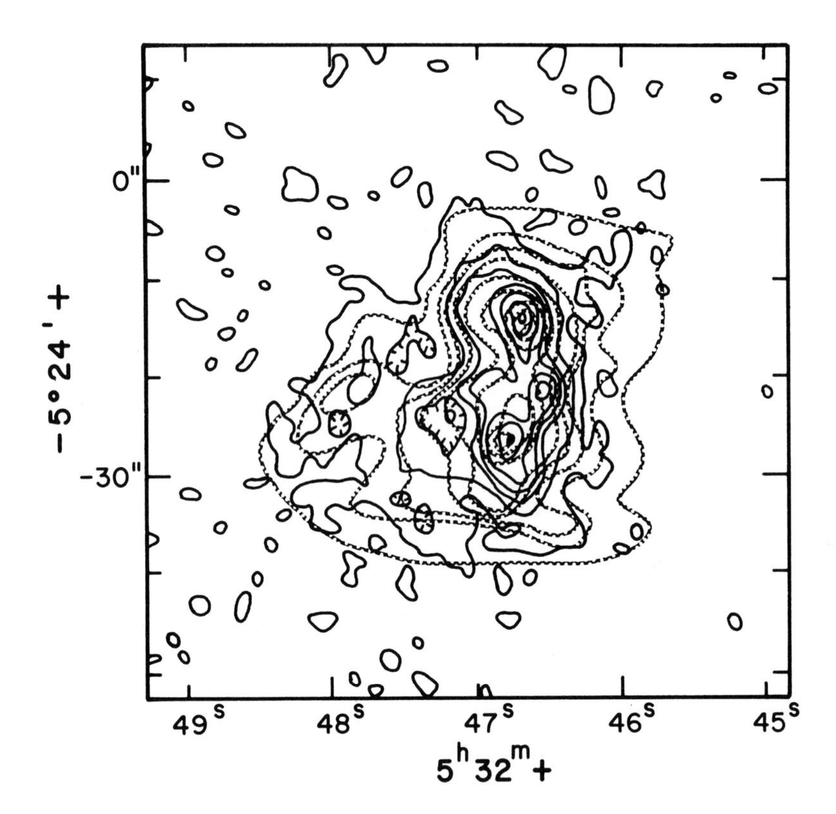

Figure 2. The WIRO 20µm observations with the 20µm contours derived by Rieke, Low and Kleinmann (1973) superposed.

Table 1
SOURCE POSITIONS IN THE BN-KL COMPLEX

	$05^h\ 32^m$ +	$-5°24'$
BN (IRc1)	46.68	15.9"
IRc2	46.98 ± 0.02	23".4 ± 0.3
IRc3	46.58 ± 0.02	23.2 ± 0.3
IRc4	46.79 ± 0.02	27.8 ± 0.3
IRc6	47.7 ± 0.07	23 ± 1

Position errors are relative uncertainties with respect to BN. The absolute uncertainty in the position of BN is 0".5 in each coordinate.

The BN object is known to exhibit a strong absorption feature near 10 microns, presumably due to silicate absorption (eg. Gillett and Forest 1973). Recently Aitken et al. (1980) have reported spectral scans of the two sources IRc 2 and 4, both of which show very strong absorption features. Rieke et al. (1973) diagnosed the presence of these features from the photometric behavior of the sources between twenty and five microns. Using an array spectrometer system we have mapped the region in six wavelength bands which cover the silicate absorption feature. The optical and detector system has been described by Gehrz, Hackwell and Smith (1976). When used on the WIRO 2.3 meter telescope the system has beam sizes of 3".5. The important feature of the instrument is that the spectral observations are made simultaneously through the same focal plane aperture. The contour maps derived from these observations are presented in Figure 3. From these maps we have derived photometry for the individual condensations. The energy distributions of the condensations are plotted in Figure 4. From this plot it is immediately obvious that the depth of the silicate feature in the BN object is relatively weak for the regions. Aitken et al. (1980) have in fact suggested that the apparently fainter sources in the KL nebula are the primary energy sources for the region. Examination of the contour maps reveals that the silicate extinction overlies the entire nebulous region. There is no evidence for an emission component with the characteristics of the Trapezium region: silicate emission peaking near 10.5 microns. Conservatively estimating the extinction to the diffuse component of emission as 30 magnitudes in the

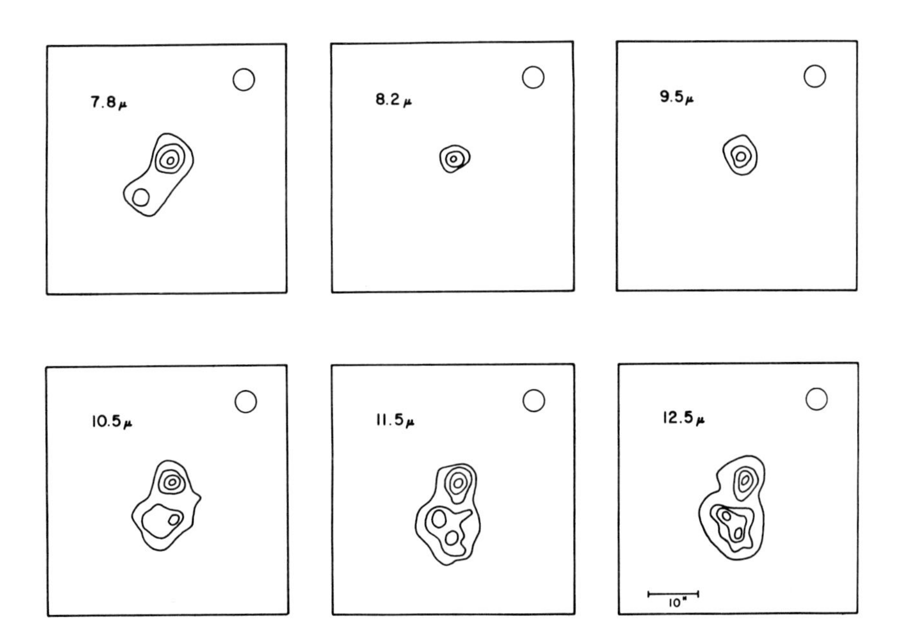

Figure 3. The array spectrometer data obtained for the BNKL region. The beam size is shown by the circle in the upper right of each panel.

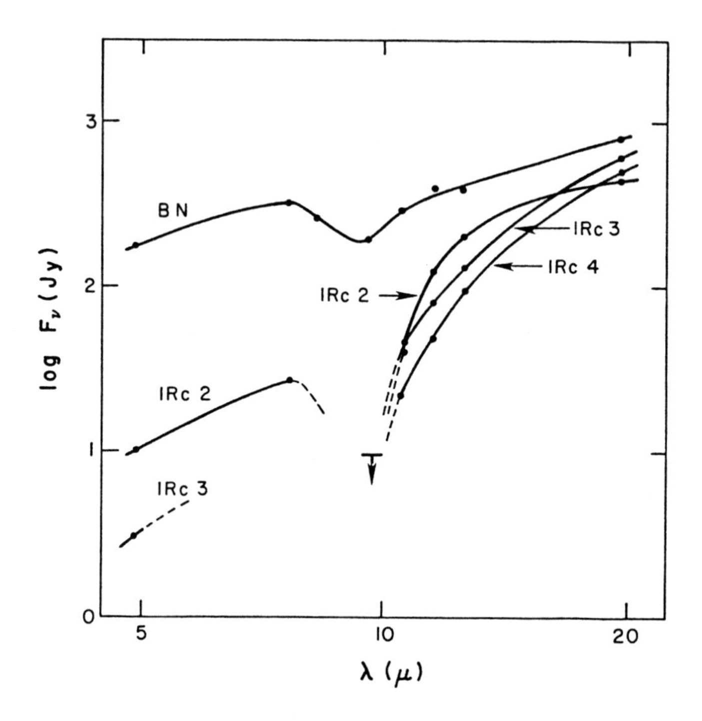

Figure 4. The mid infrared energy distribution for the four most prominent condensations in the BNKL complex.

visual we obtain a lower limit comparable to the extent of the nebula. The condensations are embedded in very dense material.

The similarity between the spectral behavior of the condensation and the surrounding nebula raises the question of the nature of the condensations. From photometric observations we cannot rule out the possibility that the condensations are not small stellar objects but dense knots in the nebula. For the BN object the angular diameter measurements of Sibille, Chelli and Lena (1979) demonstrate the object is less than 0.10" in diameter.

The region contains a number of radio molecular sources. One of the most interesting of these is the SiO maser source. The position of this source and IRc 2 agrees within the reported errors. There are also a large number of H_2O maser sources in this region (Genzel and Downes 1977, Genzel et al. 1980). A number of these appear to cluster near the condensation IRc 4 and run nearly along the tongue of 20 micron emission extending south and west from that source. As the positional uncertainty of the two wavelength regions improves we can expect further improvement in the relationship between the infrared sources and the radio sources. Currently the relationship is suggestive that the two types of observation may refer to the same physical object, but the case is not yet closed. There may only be a close relationship between the regions, not complete coincidence.

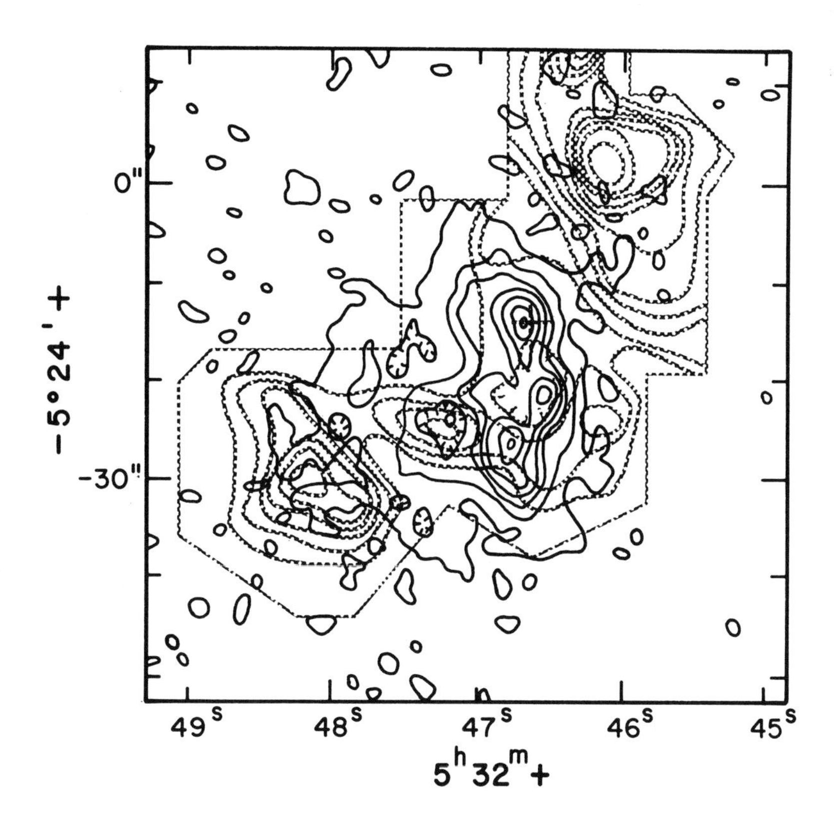

Figure 5. The WIRO 20μm observations with the contours of H_2 quadrapole emission line given by Beckwith et al. (1978) superposed. Note the strong anti-correlation of thermal dust emission and H_2 line emission.

The molecular hydrogen emission in the BNKL region has also presented a difficult case of interpretation. There have been no clear associations between the molecular hydrogen emission and other types of observation. (Gautier et al. 1976, Beckwith et al. 1978). From our 20 micron map we find a remarkable anti-correlation between the H_2 emission and the 20 micron flux. This is illustrated in Figure 5 where the contours of H_2 emission from Beckwith et al. have been superposed on the 20 micron map. The simplest interpretation would be to ascribe the reduction of intensity of the H_2 emission in regions of high 20 micron flux to the obscuration of overlying material. This would imply that the H_2 emission must lie behind the BNKL complex. Since the H_2 emission is due to shocked gas this geometry is attractive. It is difficult to see how substantial flows of energy could propagate through the dense material lying between us and the BNKL complex. If however the condensations lie on the far side of the obscuring material a relatively clear space might well exist on the far side of the complex. The situation would be almost exactly turned around from the currently accepted picture of the Trapezium region, in which the clear zone is believed to face us.

REFERENCES

Aitken, D. K., Roche, P. F., Spenser, P. M. and Jones, B. 1980, MNRAS in press.
Becklin, E. E. and Neugebauer, G. 1967, Ap. J. 147, 799.
Beckwith, S., Persson, S. E., Neugebauer, G. and Becklin, E. E. 1978, Ap. J. 223, 464.
Gautier, T. N., Fink, U., Treffers, R. R. and Larson, H. P. 1976, Ap. J. 207, L129.
Gehrz, R. D., Hackwell, J. A. and Smith, J. R. 1976, Pub. Astron. Soc. Pac. 88, 971.
Genzel, R., Becklin, E. E., Wynn-Williams, C. G. and Downes, D. 1980. Paper presented at IAU Symposium No. 96.
Genzel, R. and Downes, D. 1977, Astronomy and Astrophysics, 61, 117.
Gillett, F. C. and Forrest, W. J. 1973, Ap. J. 179, 483.
Grasdalen, G. L., Herzog, A. D., Hackwell and Gehrz, R. D. 1979, BAAS 11, 712.
Kleinmann, D. E. and Low, F. J. 1967, Ap. J. 149, L1.
Rieke, G. H., Low, F. J. and Kleinmann, D. E. 1973, Ap. J. 186, L7.
Sibille, F., Chelli, A. and Lena, P. 1979, Astron. and Astrophys. 79, 315.

Editors' Note: The record of the discussions which followed the several papers on the Orion Molecular Cloud have been combined and follow Scoville's review in this volume.

SPECTROSCOPY OF THE ORION MOLECULAR CLOUD CORE

N.Z. Scoville
Department of Physics and Astronomy
University of Massachusetts
Amherst, MA 01003

ABSTRACT

Recent infrared and radio spectroscopic data pertaining to the Orion BN-KL infrared cluster are reviewed. A new, high resolution CO map shows that the thermal structure over the central 10'(1.5 pc) in the Orion molecular cloud is dominated by energy sources in the infrared cluster and M42. Peak CO brightness temperatures of 90 K occur on KL and near the bar at the southern edge of M42.

Within the central 45" of the infrared cluster, both radio and IR data reveal a highly energetic environment. Millimeter lines of several molecules (e.g. CO, HCN, and SiO) show emission over a full velocity range of 100 km s^{-1}. These supersonic flows can be modeled as a differentially expanding envelope containing a total of ~5 $M_\odot$ of gas within an outer radius of $r \simeq 1.3 \times 10^{17}$ cm. Over the same area emission is seen from vibrationally excited molecular hydrogen at an excitation temperature of 2000 K. The high velocity mm-line emission and the NIR H_2 lines are clearly related since they exhibit similar spatial extents and line widths. Comparison of the total cooling rate for all the H_2 lines with the estimated kinetic energy and expansion time for the mm-emission region indicates that the H_2 emission probably arises from shock fronts where the expanding envelope impinges on the outer cloud.

Near IR spectroscopy also probes ionized and neutral gas closely associated with BN. Br α and Br γ emission is detected from an ultracompact HII region of mass $M_{HII} \lesssim 10^{-4}$ $M_\odot$. Full widths for the HII lines are ~400 km s^{-1}. CO bandhead emission detected in BN at $\lambda \simeq 2.3$ μm is probably collisionally pumped in a high excitation zone ($n_{H+H_2} > 10^{10}$ cm^{-3} and $T_K \simeq 3000$ K) at only a few AU from the star. The velocity of both the HII and CO emission is $V_{LSR} \simeq +20$ km s^{-1}; thus BN appears to be redshifted by 11 km s^{-1} with respect to OMC-1.

C. G. Wynn-Williams and D. P. Cruikshank (eds.), Infrared Astronomy, 187–205.

I. INTRODUCTION

In the last few years we have witnessed a remarkable change in our comprehension of the evolution of active star formation regions like the Orion nebula. Far from the initial picture of an interstellar cloud gently condensing to form stars, recent mm and near infrared spectroscopic data attest to an extraordinary level of activity at the cloud core. Within the BN-KL infrared cluster, NIR emission lines are detected from vibrationally excited H_2 and CO, indicating dense, neutral gas at temperatures of several thousand degrees, and high velocity wings are observed on many of the mm-wavelength rotational transitions (e.g. $J = 1 \rightarrow 0$, CO), indicating hypersonic gas motions over a range of 100 km s^{-1}. Similar phenomena have now been detected in almost a dozen other cloud core regions; it is therefore evident that a thorough study of the events in Orion can contribute to our understanding of cloud evolution, specifically the disruptive phases following an internal episode of star formation. The near infrared spectroscopy is also yielding important constraints on the nature of the embedded sources like BN and their immediate environs. Most helpful to observations of continuum sources has been the recent development of Fourier transform spectrometers with sufficient resolution ($\Delta K \lesssim 0.2$ cm^{-1}) to remove absorption features in the earth's atmosphere and to resolve both "stellar" and interstellar features.

In this review of spectroscopic results, our focus is upon the active central region ($r \leq 2 \times 10^{17}$ cm) of the Orion molecular cloud encompassing infrared continuum sources and the excited, high velocity molecular gas. The two phenomena are very likely to be related since several of the continuum sources are presumably very young stars and the gas disturbances may ultimately be traced back to the evolution of such stars trapped within a cloud. In studying this activity we find a convenient marriage of the complementary radio and IR data. The former generally provides an excellent view on the lower excitation, more extended gas phases; the latter is uniquely capable of sampling trace quantities of gas with high excitation conditions or the small volumes immediate to the embedded stars. To gain some perspective on the central core region we start the discussion below with a brief description of the very extended Orion molecular cloud as gained from millimeter line data.

II. THE ORION GIANT MOLECULAR CLOUD

The full extent of the Orion molecular cloud is best appreciated from maps of the $J = 1 \rightarrow 0$ (2.6-mm) CO line on account of the ubiquity of CO and the very modest requirements ($n_{H_2} \gtrsim 300$ cm^{-3}) for excitation of this transition above the 2.7 K background. The first complete, low resolution map of this emission by Kutner et al. (1977) showed the cloud strung out roughly parallel to the galactic plane (SE-NW) with dimensions of $1 \times 5^{\circ}$ or 10×45 pc. The overall mass estimated from a virial analysis and from the calculated CO column densities is about 1×10^5 $M_{\odot}$, most of which is presumably H_2 at a mean density $n_{H_2} \sim 300$ cm^{-3}.

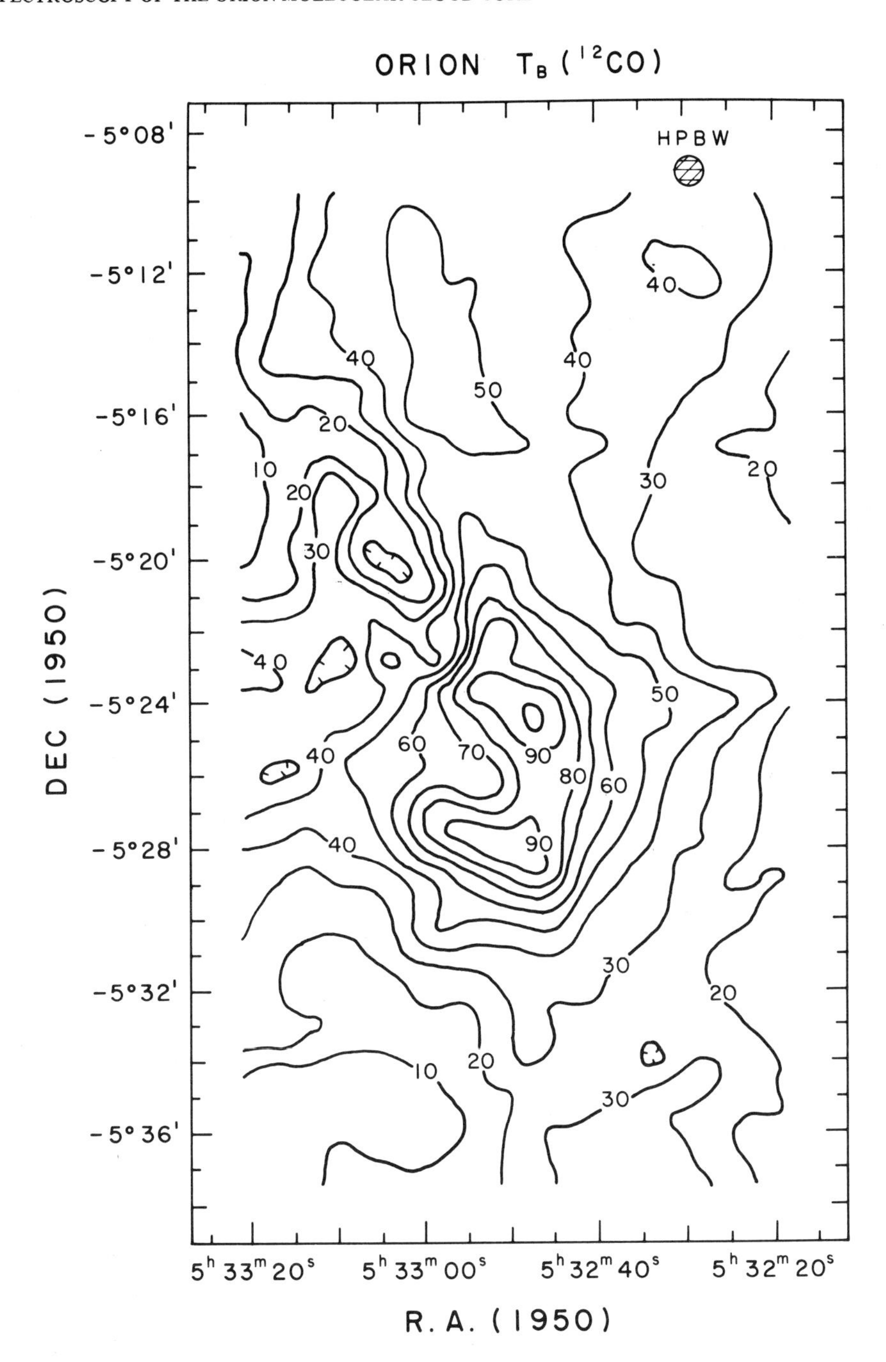

Figure 1. The CO (J = 1 → 0) brightness temperature distribution for the central 1/2° of the Orion nebula (Scoville, Schloerb, and Goldsmith 1980b). The map was densely sampled with the 45-foot University of Massachusetts antenna (HPBW = 45").

In Figure 1 a high resolution map of the CO brightness temperature is shown for the central $1/2^{\circ}$ surrounding M42 and the KL-BN cluster. The area shown in Figure 1 was fully sampled using the 45-foot mm-telescope (HPBW = 45") at the University of Massachusetts (Scoville, Schloerb, and Goldsmith 1980b). Dominant features of the emission on this scale are an obvious peaking in the line intensity near M42 and KL and the pronounced N-S elongation, especially to the north of the cloud core. The fine structure in the vicinity of KL and M42 closely follows some of the highest resolution far infrared data. In particular, a peak is seen corresponding to KL-BN and a ridge shows up for the first time at the southern ionization front in M42 similar to the FIR "bar" feature of Werner et al. (1976). It is significant that the CO brightness temperature and the 50/100 μm color temperatures agree within the calibration errors (~15%) for both the KL peak and the bar. In these regions, the CO transition is both optically thick and the levels are thermalized ($n_{H_2} >> 3000$ cm^{-3}), so we may interpret the CO brightness temperature map as a map of gas kinetic temperature. Thermal equilibrium between the dust grains and H_2 gas is expected for $n_{H_2} > 5 \times 10^4$ cm^{-3} (Goldreich and Kwan 1974) which will probably hold here.

III. THE ORION CLOUD CORE

a) Radio Frequency Lines

The remarkable characteristic in the mm-lines, setting the central 1' of the core off from the remainder of the cloud, is the velocity dispersion of the emission seen there. Outside the core the full line widths are always $\lesssim 5$ km s^{-1} and the peak velocities vary only 5 km s^{-1} over the full 5° length of the cloud. Yet within the core the emission of nearly half the observed molecules (e.g. CO, CS, SiO, SO, SO_2, and HCN) can be followed out 30 to 50 km s^{-1} with respect to the cloud rest frame at $V_{LSR} = 9$ km s^{-1}. Figure 2 shows the ^{12}CO and ^{13}CO spectra at the KL position. This high velocity gas component (the "plateau" source on account of the line shape) was first observed in $J = 1 \rightarrow 0$ CO and analyzed by Zuckerman, Kuiper, and Kuiper (1976) and Kwan and Scoville (1976). The latter authors suggested that the gas was in a differentially expanding envelope, perhaps the result of an explosive event in the cloud core.

New measurements of the high velocity 2.6-mm CO emission at the best available spatial resolution, 45" using the 45-foot University of Massachusetts antenna, find the center at $\Delta\alpha = 0 \pm 4''$, $\Delta\delta = 0 \pm 5''$ relative to BN (Solomon et al. 1980). Within the errors IRC-2 could be located within the centroid but KL which is 12" south of BN appears ruled out. The diameter of the emission region is 36" when corrected for the telescope HPBW, giving a radius of $r_{CO} = 1.3 \times 10^{17}$ cm.

From comparison of the ^{12}CO and ^{13}CO profiles in Figure 2, one sees that the opacity of the ^{12}CO high velocity wings is $\lesssim 3$ in the $J = 1 \rightarrow 0$ line if $[^{12}CO/^{13}CO] \lesssim 89$. The low opacity of the 2.6-mm line wings

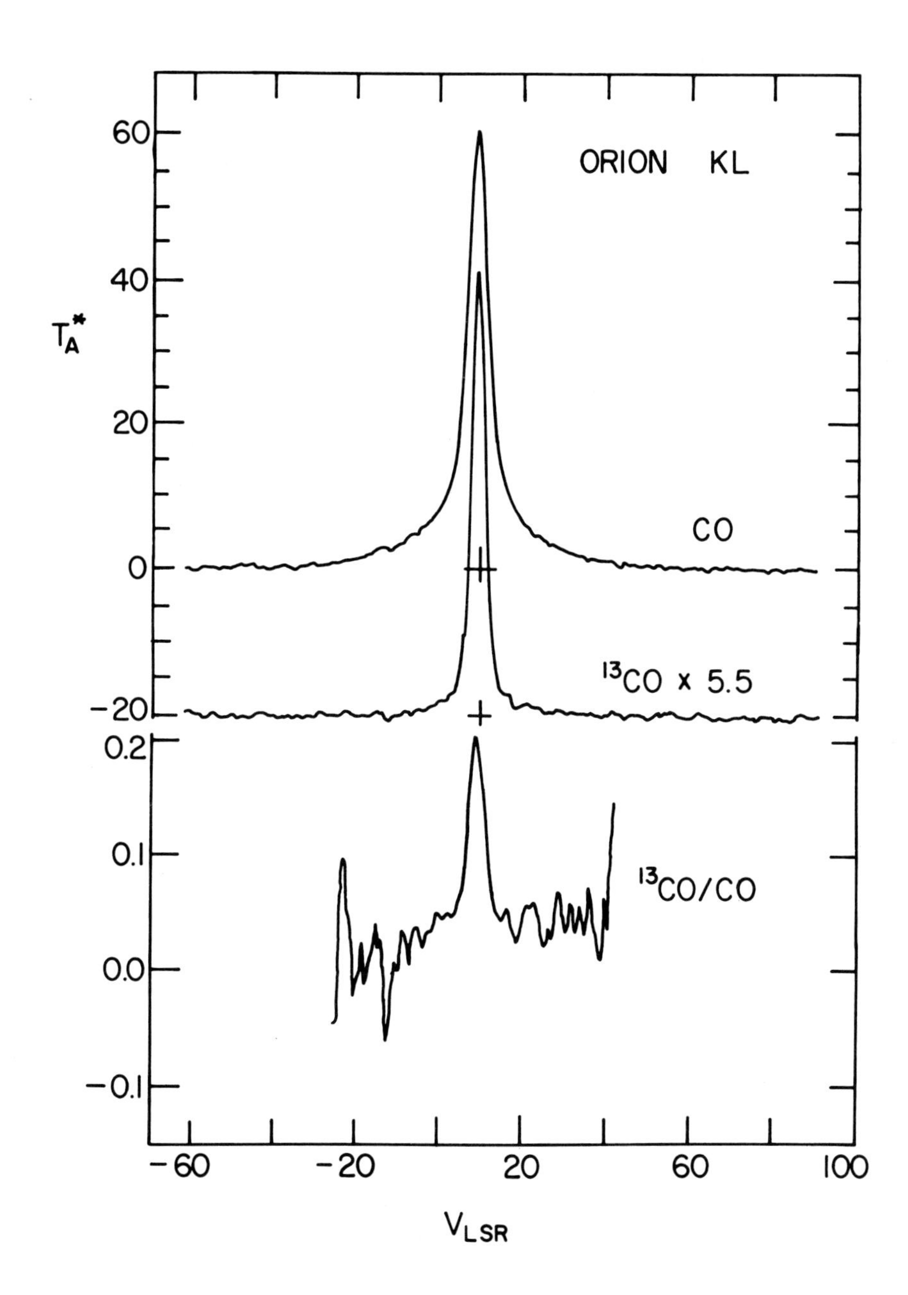

Figure 2. Spectra of CO and ^{13}CO corrected antenna temperature (T_A*) at the position of KL. Note that the high velocity emission, prominent in ^{12}CO, is hardly visible in ^{13}CO due to the low opacity of this kinematic component compared to the 9 km s^{-1} component from OMC-1.

allows one to directly use the integrated flux in the line wings to estimate the CO column density and hence the total mass of high velocity gas. For a rotational temperature of 200 K, as indicated by the higher CO lines (Phillips et al. 1977), and assuming 10% of the C is bound in CO, this mass is $M_{H_2} = 5\ M_\odot$. The total kinetic energy involved in the high velocity motions at present is therefore $\sim 1\text{-}2 \times 10^{47}$ ergs and the momentum is $\sim 200\ M_\odot$ km s^{-1}.

It should be noted that the center of the high velocity emission and its size appear different in several other molecules. Using the new mm-interferometer at Hat Creek, Baud et al. (1980) locate the ground state SiO emission ($v = 0$, $J = 2 \rightarrow 1$) at IRC-2 with a size of only 12" and from data obtained on the 100-m Effelsberg antenna, Zuckerman, Palmer, and Morris (1980) identify a "hot" NH_3 component with IRC-4 near the center of KL. Phillips et al. (1977) find the $J = 3 \rightarrow 2$ CO emission centered midway between BN and KL, only 4" from IRC-2. And from VLBI data, Genzel et al. (1979) deduce that the SiO maser emission from $v = 1$ and $v = 2$ is from IRC-2; however, the relationship of this emission to the other high velocity emission is unclear since the line shape and velocity extent ($v = -10$ to $+21$ km s^{-1}) appear rather similar to those in late-type giant stars. An outstanding question at present is which, if any, of the molecular line maps one should adopt for the most meaningful picture of the high velocity gas distribution. My own preference is for the 2.6-mm CO line: its opacity can be demonstrated to be low ($\tau < 3$); its excitation is likely to be saturated at the gas kinetic temperature and therefore the brightness translates directly into column density (rather than an emission measure $\propto n^2L$), and lastly the abundance of CO in the gas is likely to be relatively stable against perturbations resulting from passage of shock fronts. In contrast, the higher CO lines will have at least 4 times greater opacity, while SiO, SO, SO_2, and HCN probably suffer all the above difficulties to some extent.

In addition to the relatively continuous (in velocity) emission seen in the millimeter lines, the Orion cloud core shows H_2O maser emission with over 50 discrete velocity components between $v = \pm 50$ km s^{-1}. Position measurements obtained in VLBI experiments over the last three years have been recently analyzed by Genzel et al. (1980) to yield the first proper motion detections in Orion. Noticing that the proper motion vectors appear to radiate out of a common center, they attempted to fit the kinematic data with an expanding envelope and identify the center of expansion as IRC-2. A subset of this data for which proper motions could be determined in both RA and DEC is shown in Figure 3.

b) Infrared Lines of H_2 and CO

Probably the greatest surprise in the spectroscopic studies of Orion has been the discovery in 1976 by Gautier et al. of the 2 μm H_2 emission lines. Their detection was totally unanticipated since the $v = 1$ states from which the emission arises are at an energy $E/k \simeq 6800$ K above ground and the emission was found to be spatially extended.

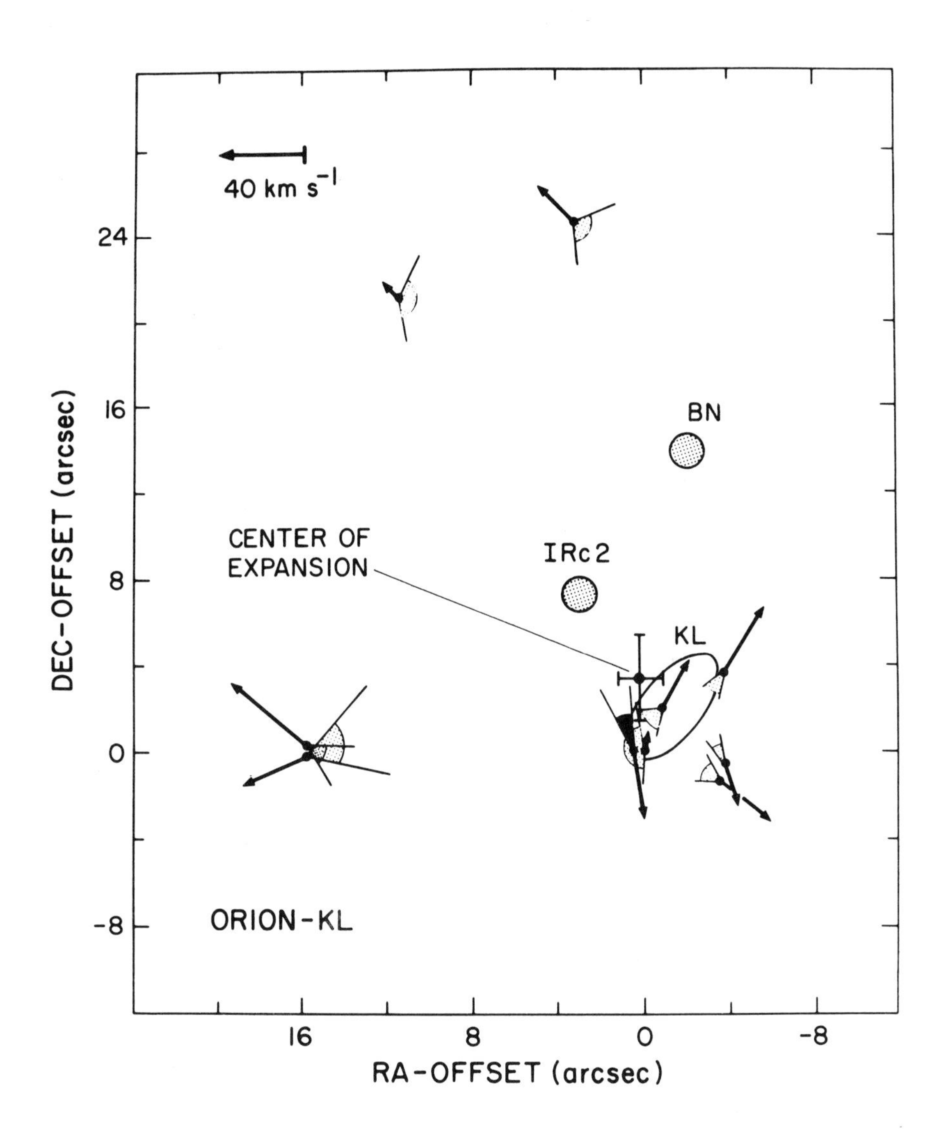

Figure 3. The proper motion vectors are shown for all H_2O masers in Orion for which motions could be measured in both RA and DEC (Genzel et al. 1980). All velocity vectors are relative to the center of expansion, that is, they have been corrected for the motion of the reference maser feature based upon the model parameters.

The initial interpretation that the H_2 emission was collisionally excited in shock heated gas has been borne out by subsequent observations and theoretical models.

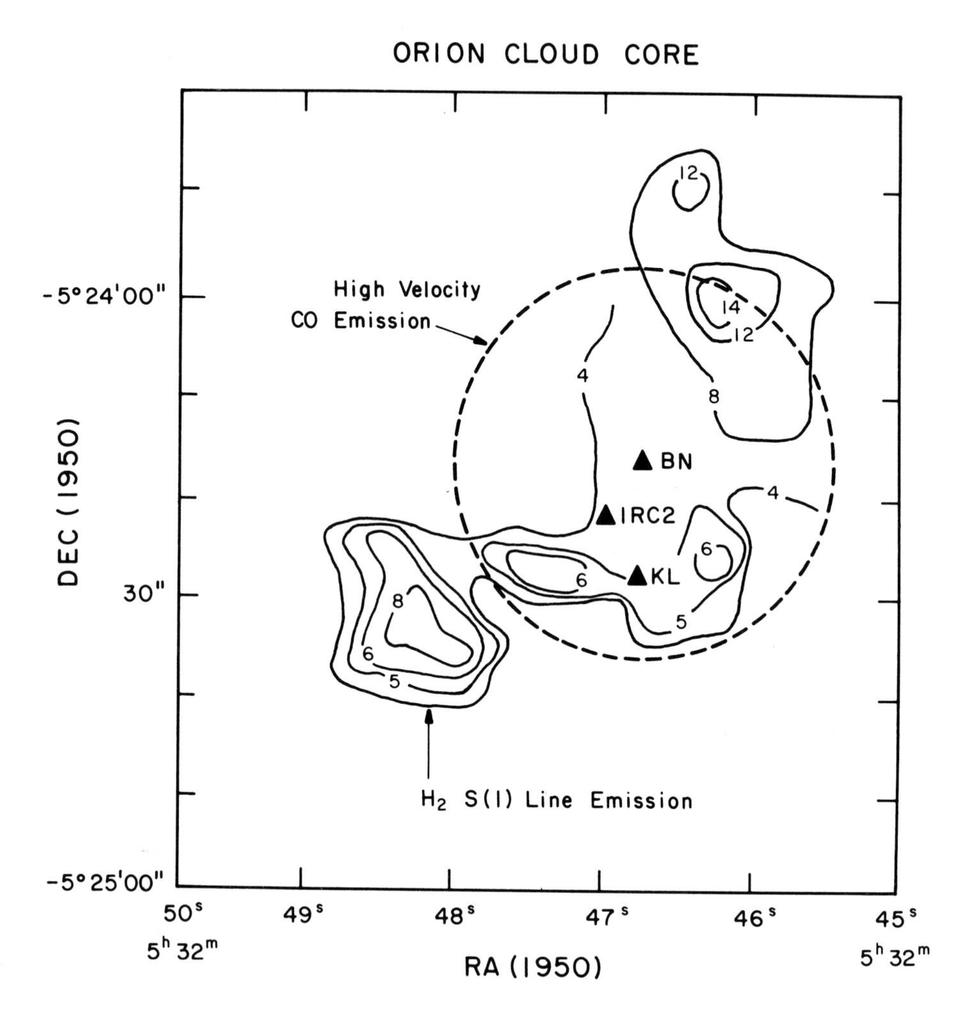

Figure 4. The H_2 S(1) line emission from the Orion nebula observed with 5" aperture (Beckwith et al. 1978). The size and position of the high velocity CO (J = 1 → 0) emission is also shown (Solomon, Huguenin, and Scoville 1980). The intensity contour units are 6.1 x 10^{-4} ergs s^{-1} cm^{-2} sr^{-1}. Peak 1 (highest contour 14) is 18" NW of BN; Peak 3 (highest contour 6) is 8" E of KL; and Peak 5 (highest contour 12) is the most northern source.

The spatial distribution of the H_2 emission appears extremely nonuniform and clumpy on the scale of 5-10", but low level emission does appear over at least 40". A high resolution map (5") in the S(1) line (v = 1 → 0, J = 3 → 1) made by Beckwith et al. (1978) is shown in Figure 4. In the map at least five distinct H_2 peaks may be seen--none of which show correlation with previously identified continuum sources.

Using the fact that several of the observed H_2 lines originate from a common upper state, for example S(1)(v = 1 → 0, J = 3 → 1) and Q(3) (v = 1 → 0, J = 3 → 3), Beckwith, Persson, and Neugebauer (1979) and Simon et al. (1979) have estimated the differential extinction in front of a couple of the H_2 peaks. Both investigations measure the 2 μm reddening and infer $A_v \simeq 40 \pm 10$ mag (or $N_{H+2H_2} \simeq 6 \times 10^{22}$ cm^{-2} for a standard gas-to-dust ratio). Since the extinction estimated for three locations was similar within the errors, it appears likely that most of the spatial structure seen in Figure 4 is actually in the H_2 source and cannot be attributed solely to a clumpy foreground extinction. When the observed S(1) line luminosity of 2.5 $L_\odot$ from the entire Orion source is corrected for the measured extinction ($A_{2\mu m} = 4 \pm 1$ mag) and the contributions of the other H_2 lines are included (a factor of 10 for $T_{vib} = 2000$ K), the total implied H_2 luminosity is 1000 $L_\odot$ (Beckwith et al. 1979).

Theoretical analyses for the emission have favored collisional excitation of the H_2 in gas which has been shock heated to several thousand degrees (Hollenbach and Shull 1977, Kwan 1977). The high extinction deduced to the H_2 suggests that these shocks cannot merely be at the interface of M42 and OMC-1. Indeed the similarity of the estimated extinction for the H_2 and that found for the HII region associated with BN (see below) implies that both regions are at an equivalent distance inside the cloud. Both the large spatial extent of the H_2 emission within a dense cloud and the relatively weak v = 2 → 1 emission, compared to v = 1 → 0, argue against the UV fluorescense excitation for H_2. The shock models in which the H_2 vibrational states are collisionally excited can give the proper luminosity (1000 $L_\odot$) and correct line ratios ($\bar{T}_{ex} = 2000$ K) with an ambient density $n_{H_2} \gtrsim 10^6$ cm^{-3} and shock velocities in the range 10-25 km s^{-1} (Kwan 1977). In this picture the emitting layer of <u>hot</u> H_2 is extremely thin (~10^{13} cm along the line-of-sight) since the observed column densities are only ~10^{20} cm^{-2}. The overall extent of the H_2 emission <u>is</u> similar to that of the high velocity CO as can be seen in Figure 4 where an outline of the CO volume is superposed. This suggests that the H_2 is excited in shock fronts where the expanding envelope seen in the mm-lines collides with the ambient cloud (Kwan and Scoville 1976, Kwan 1977).

The most extensive data at high spectral resolution to measure the H_2 velocities and map the shock front kinematics have been taken by Nadeau and Geballe (1979) using a Fabry-Perot interferometer with 10" aperture and by Scoville et al. (1980a) using the KPNO-FTS with 3".75 aperture. An extremely large velocity range ~100 km s^{-1} for the H_2 emission was first reported by Nadeau and Geballe (1979). Figure 5 shows the FTS spectra taken on three of the H_2 peaks in Figure 4 with the seven visible H_2 lines spliced together. An enlarged view of the S(1) line from Peak 1 (shown in Figure 6) demonstrates that significant emission can be seen over at least 150 km s^{-1}. Although the emission profiles for the three positions appear quite different, they do <u>all</u> exhibit a very large velocity range, 100-150 km s^{-1} for the emission. One of the most puzzling aspects of the H_2 emission is how the molecules

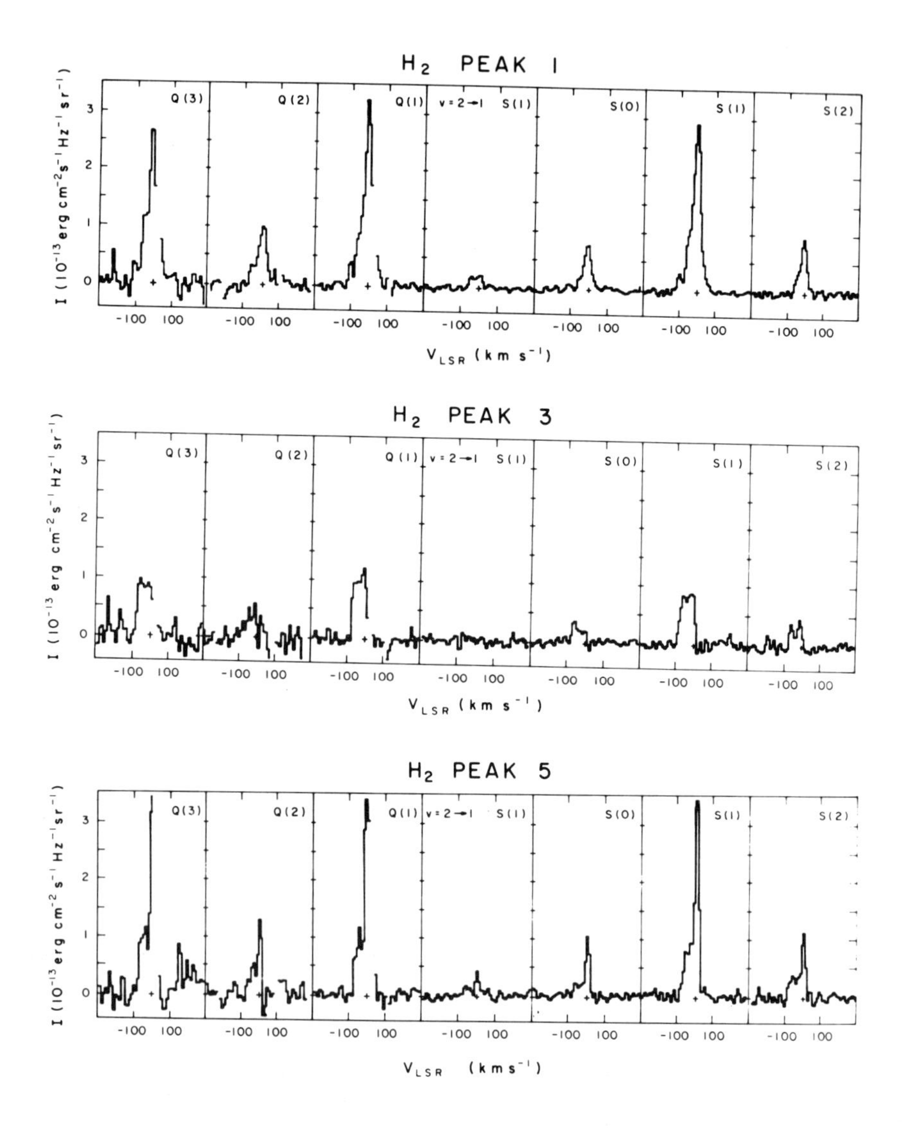

Figure 5. Seven H_2 transitions in $v = 1 \rightarrow 0$ and $2 \rightarrow 1$ bands are shown for three H_2 peaks shown in Figure 4. (The positions of these peaks are given in the Figure 4 caption.) The spectra were taken with the KPNO-FTS on the 4-m telescope using a 3″.75 aperture (Scoville et al. 1980a).

can survive acceleration to such high velocities. Theoretical models predict collisional dissociation of all the preshock H_2 for shock velocities exceeding 25 km s^{-1} if the density is above 10^5 cm^{-3} as it should be here (Kwan 1977; Dalgarno and Roberge 1979). Thus most of the

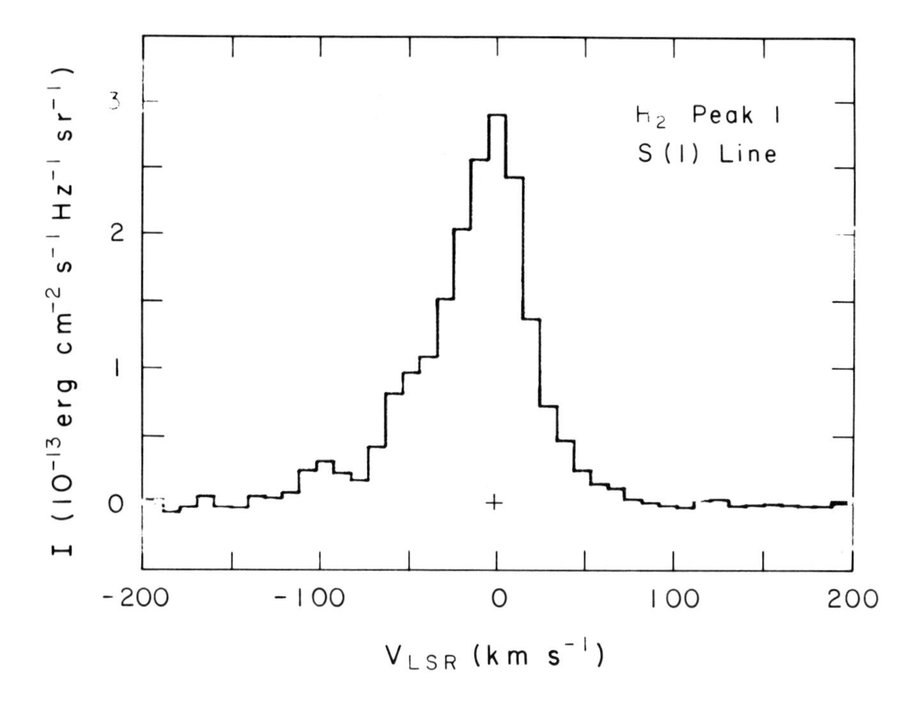

Figure 6. An expanded spectrum of the v = 1 → 0 S(1) H_2 line on Peak 1 showing emission over more than 150 km s^{-1} (Scoville et al. 1980a).

very high velocity H_2 may have been reformed downstream from the shock after the dissociated gas cools below about 4000 K. Alternatively if the original H_2 were contained largely in clumps, the gas in the center of these clumps could conceivably be accelerated without dissociation.

The H_2 kinematic data along a strip running between Peak 1 in the northwest and Peak 2 in the southeast is shown in Figure 7. The lines broaden towards the center of the region, but there is no evidence of the line splitting one expects in a symmetric, expanding shell. The absence of splitting could be due merely to inhomogeneities in the ambient medium--something which was already evident from the clumpiness of the spatial maps. It is also noteworthy that the emission profiles of S(1) and Q(3) appear similar in shape (Figure 5). Thus the different velocities seen in a given direction are not located behind grossly dissimilar columns of dust. We can rule out an expanding shell model in which the interior volume is filled with dust to the extent that one velocity component suffers twice as much extinction as the other. It does appear that the H_2 emission in the direction of IRC-2 suffers about twice as much extinction as in the other positions. Figure 7

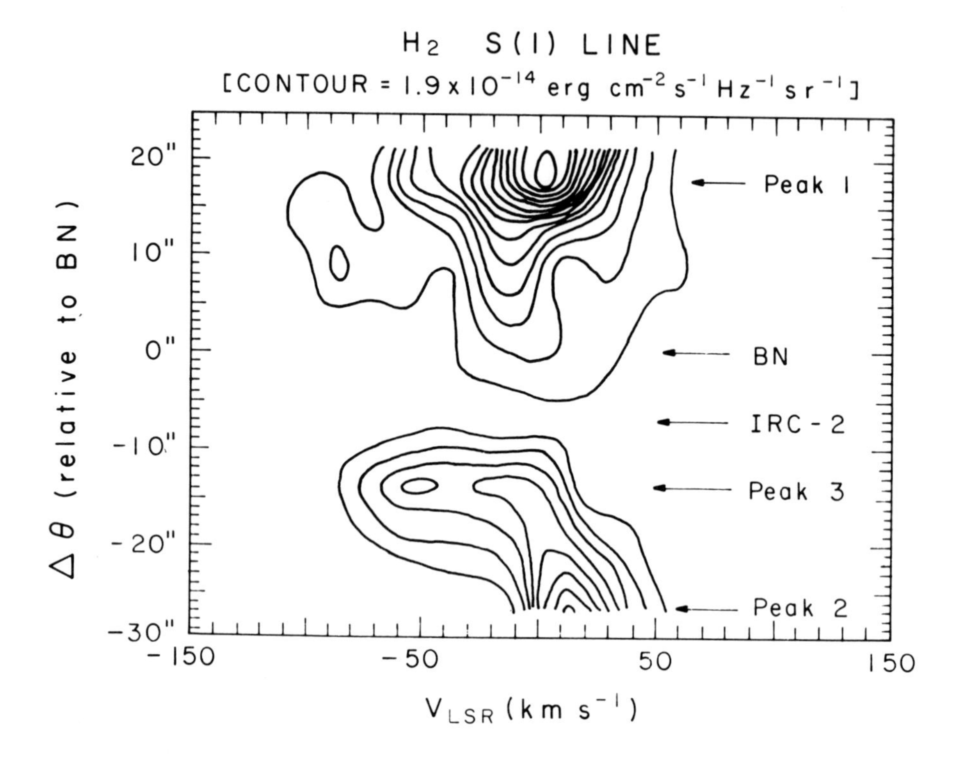

Figure 7. Spatial-velocity contour map from the FTS data at the S(1) line along a strip crossing the Orion cloud core (northwest to southeast).

shows little or no S(1) emission at IRC-2, yet we quite easily detect the Q(3) line there. The 10 μm silicate feature also is particularly deep in IRC-2 (Aitken et al. 1980).

A dynamic timescale of 1300 yrs is found for the H_2 shock front when one divides the radius of the emission region (2×10^{17}cm) by the measured half-width of 50 km s^{-1}. Assuming that the H_2 luminosity is constant over this time, we estimate that a total of 1.5×10^{47} ergs has been radiated in the vibrational lines. The equivalence of this integrated cooling rate to the kinetic energy ($1 - 2 \times 10^{47}$ ergs) estimated earlier for the high velocity CO envelope supports the earlier suggestion that the CO envelope feeds kinetic energy to the H_2 shocks where the expanding gas hits the stationary cloud, OMC-1.

IV. SPECTROSCOPY OF BN

Despite the fact that most of the high resolution spectroscopy of

BN has been in two narrow bands at λ = 2.1 - 2.4 μm and 4.5 - 4.8 μm, the initial observations already demonstrate real advantages over photometric techniques for study of bright, embedded sources. For BN the NIR photometric data indicate a color temperature $T_c \simeq 500$ K and a NIR luminosity $L_{\lambda \lesssim 20\ \mu m} \gtrsim 2 \times 10^3\ L_\odot$ (Becklin, Neugebauer, and Wynn-Williams 1973). The radius at which the observed continuum arises must therefore be greater than $r_{BB} = 4 \times 10^{14}$ cm. On the other hand, the spectroscopic data now reveal emission from CO (at λ = 2.3 μm) for which the analysis indicates a radius of 10^{13} cm and mass of just $10^{-7}\ M_\odot$. Thus, in addition to probing the physical conditions and dynamics of varied excitation regimes in both neutral and ionized gas, the spectroscopy is also capable of penetrating closer to the ultimate source of energy and detecting the small quantities of matter one expects in a primitive solar nebula.

The spectroscopic data in BN reveal at least three distinct regimes: an HII region from which Br α and Br γ are observed; a hot, dense molecular region giving rise to CO emission at 2.3 μm; and cooler molecular gas producing CO absorption at 4.6 and 2.3 μm. The first two regions are probably at a radius of a few AU; the latter is probably much further out, possibly in the high velocity mm-line envelope.

The Br α emission in BN was first detected and mapped by Grasdalen (1976) and Joyce, Simon, and Simon (1978) using a grating spectrometer. The measured flux indicated an exciting star with a Lyman continuum emission rate $7 \times 10^{46}\ s^{-1}$ (equivalent to a B0 main sequence star with $L \simeq 10^4\ L_\odot$) assuming the HII was optically thin and photoionized. Hall et al. (1978) measured a velocity of V_{LSR} = + 21 km s^{-1} for both Br α and γ and detected broad symmetric wings on the Br α line. Recent FTS data in the vicinity of Br γ is shown in Figure 8. The line here shows two components: a narrow one with ΔV_{FWHM} = 30 km s^{-1} and a high velocity one with emission out to $\pm$ 200 km s^{-1}. Based upon these kinematic properties, i.e. the 11 km s^{-1} redshift of the HII center velocity from the cloud center of mass and the large linewidth, it appears more reasonable to attribute the emission to circumstellar gas rather than a more extended HII region ($r \gtrsim 10^{15}$ cm) formed in the cloud by BN. Comparison of the Br α/Br γ flux ratio with that from optically thin, photoionized HII gives a Br α/Br γ color excess of 1.6 $\pm$ 0.1 mag. or $A_v \simeq 30$ mag. to the region (Hall et al. 1978). Figure 8 also shows the $^2S - ^2P^o$ transition of NaI with a kinematic width similar to the broad HII component. This doublet was previously detected in the F8 supergiant IRC + 10420 where the analysis indicated the emission could be pumped by absorption of UV at 3000 Å out of the ground state (Thompson and Boroson 1977).

From the upper limits to the radio free-free flux in BN, the size of the HII region must be less than 5×10^{15} cm (Grasdalen 1976). This size combined with the measured emission measure imply that $M_{HII} < 10^{-4}\ M_\odot$. It is thus clear that the present kinetic energy and momentum in the HII region are insufficient (by two orders of magnitude) to drive the flow in the high velocity CO envelope.

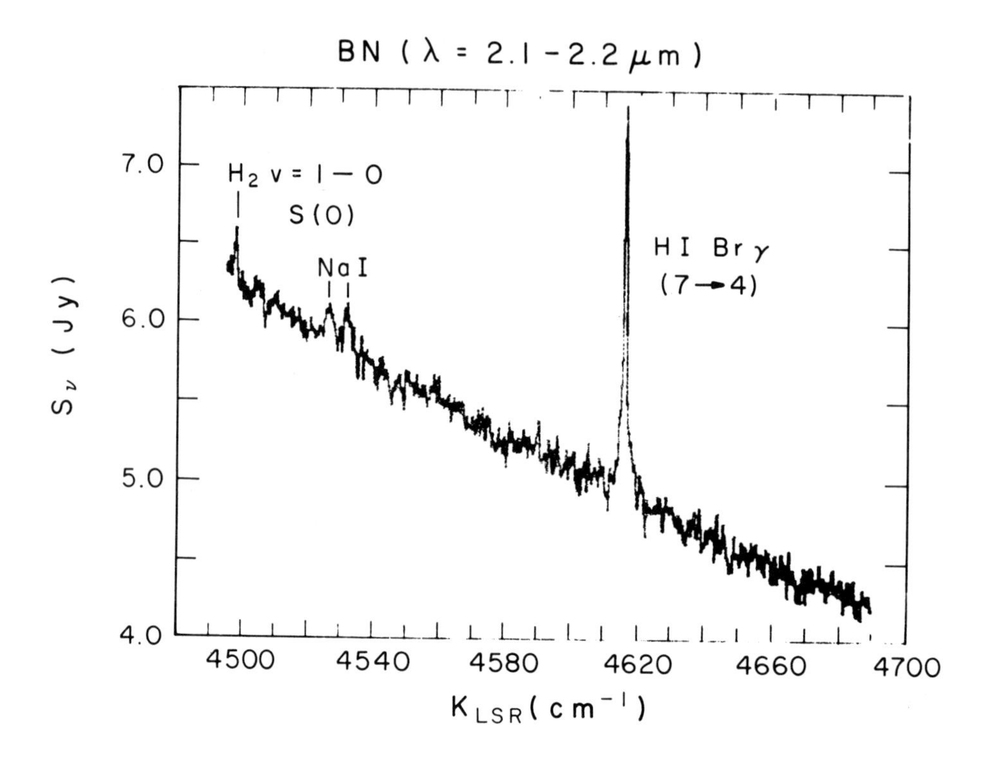

Figure 8. FTS spectrum of BN in the vicinity of the Br γ and Na I lines (Scoville et al. 1980a).

Molecular gas in BN was first observed as CO absorption in the 4.6 μm fundamental band from gas along the line of sight (Hall et al. 1978). At high resolution they found two doppler components: one at V_{LSR} = 9 km s^{-1} arising from the foreground part of OMC-1 and the other at $V_{LSR} \simeq -16$ km s^{-1}. The second system must arise from gas expanding toward us relative to BN; conceivably it is the foreground part of the "plateau" source. [Later data at 4.6 μm has revealed still a third system in emission at $V_{LSR} \simeq +20$ km s^{-1} (Scoville et al. 1980a).]

Emission from high excitation molecular gas much closer to BN is seen at λ = 2.3 μm. Figure 9 shows the CO overtone bandhead features (v = 2 → 0, 3 → 1, and 4 → 2) first detected by Scoville et al. (1979). The highest states from which the emission arises (v = 4, J ≃ 50) are at an equivalent energy E/k ≃ 19000 K and their radiative decay time is only 5 ms. Collisional excitation of the states requires $n_{H+H_2} > 10^{10}$ cm^{-3} and $T_K \gtrsim 3000$ K. From lower limits to the optical depth of the bandheads and the observed fluxes, we find that the emitting region must be only ~10^{13} cm in size. The center velocity and width of the individual lines are $V_{LSR} \simeq 25 \pm 10$ km s^{-1} and

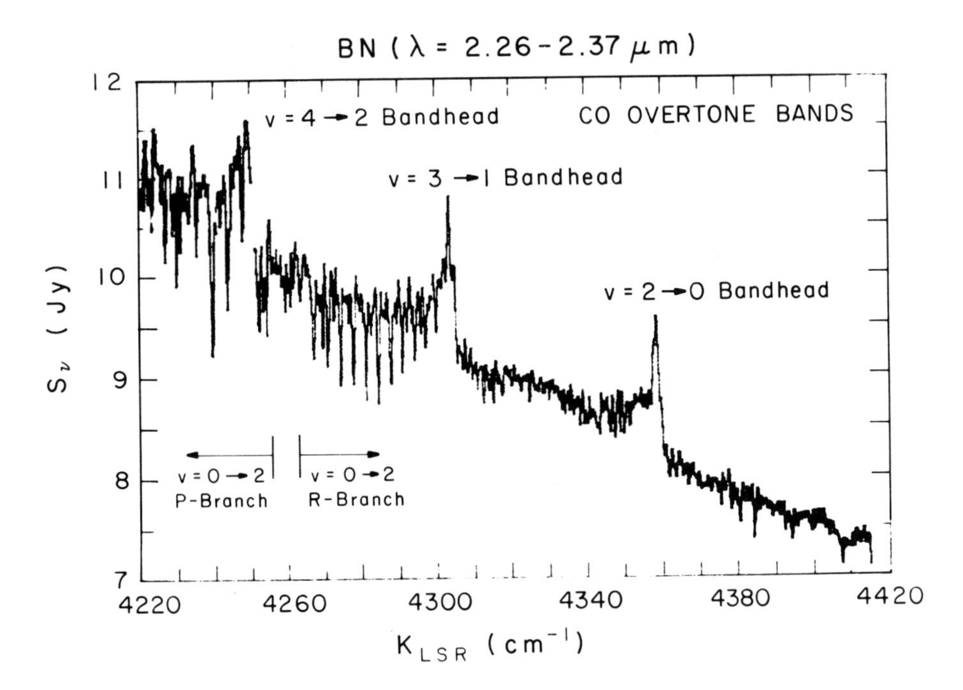

Figure 9. FTS spectrum of BN at λ = 2.3 μm showing CO absorption in the low J lines of the v = 0 → 2 overtone band and emission in the v = 2 → 0, 3 → 1, and 4 → 2 overtone bandheads (J = 50). Data was taken on the KPNO 4-m telescope with a 3".75 aperture and 0.3 cm^{-1} (20 km s^{-1}) spectral resolution (Scoville et al. 1980a).

$\Delta V \simeq 100$ km s^{-1}. Thus the CO emission probably arises from a high density, circumstellar shell--perhaps just adjacent to the HII region. Since both the HII and CO velocities are V_{LSR} = 20-25 km s^{-1} and these lines probably originate at very small radii from BN, we conclude that BN is probably moving along the line of sight at 10-15 km s^{-1} relative to OMC-1. This could represent the orbital velocity of BN within the young star cluster.

V. CONCLUSIONS

As a result of close coordination between the radio and infrared spectroscopic investigations and theoretical modeling we might now claim at least partial understanding of the physical conditions and energetics in the high velocity gas and associated shock fronts of Orion. Despite the reasonably good specification we now have for the *present state* of the active gas, the most intriguing issues such as the driving mechanism for the flow and the source of activity are still very open. A radiatively driving wind was originally discounted by Kwan and Scoville (1976) on account of the shape of the millimeter lines which can be fit well by a differential (non-constant velocity) flow and the insufficient momentum in the radiation field of KL ($P_{rad} = L_{KL}\dot{N}\ \tau_{exp}/c$). On the other hand a <u>single</u> explosive event to drive the flow can be criticized since the short time over which its effects are detectable ($\lesssim 10^4$yrs) would appear to conflict with recent data reporting high velocity mm-lines and H_2 emission in several other cloud core regions. A conceivable solution to the radiative momentum problem could occur if the IR photons in the central source were trapped and reused ~50-100 times (Phillips and Beckman 1980; Solomon, Huguenin and Scoville 1980). Alternatively the objection to an explosive event could be overcome if the energy is released in many minor events--perhaps some phase of stellar instability.

The best candidates for the origin of the activity appear to be BN and IRC-2. Future IR spectroscopy of IRC-2, similar to what has been described for BN, will be extremely helpful in defining its characteristics. Measurement of the physical conditions and gas dynamics so immediate to young stars and protostars is a unique capability of infrared astronomy.

Though most of the discussion here has relied upon the NIR spectroscopy it is clear that much will be gained in the future by a shift toward longer wavelengths to pick up the pure rotational transitions of hydrides and fine structure transitions. In Orion there have already been detections of the S(2) transition of H_2 at 12.3 μm (Beck, Lacy and Geballe 1979) and the CO $J = 21 \rightarrow 20$ and $J = 22 \rightarrow 21$ transitions at 120 μm (Watson et al. 1980). These longer wavelength lines furnish the missing probe of the excitation regimes between those sampled in the millimeter lines and in the vibrational transitions of the NIR.

This is contribution number 446 of the Five College Observatories. This research is supported in part by National Science Foundation grant AST79-19821.

REFERENCES

Aitken, D.K., Roche, P.R., Spenser, P.M., and Jones, B. 1980, M.N.R.A.S. (in press).
Baud, B., Bieging, J.H., Plambeck, R., Thornton, D., Welch, J., and Wright, M. 1980, IAU Symposium No. 87 (Interstellar Molecules), ed. B. Andrew: Reidel, Dordrecht, p. 545
Beck, S.C., Lacy, J.H., and Geballe, T.R. 1979, Ap. J. (Letters), 234, L213.
Becklin, E.E., Neugebauer, G., and Wynn-Williams, C.G. 1973, Ap. J. (Letters), 182, L7.
Beckwith, S., Persson, S.E., Neugebauer, G., and Becklin, E.E. 1978, Ap. J., 223, 464.

Beckwith, S., Persson, S.E., Neugebauer, G. 1979, Ap. J., 227, 436.
Dalgarno, A.A. and Roberge, W.G. 1979, Ap. J., 233, L25.
Gautier, T.N., Fink, U., Treffers, R.R., and Larson, H.P. 1976, Ap. J. (Letters), 207, L129.
Genzel, R., Moran, J.M., Lane, A.P., Predmore, C.R., Ho, P.T.P., Hansen, S.S., and Reid, M.J. 1979, Ap. J. (Letters), 231, L73.
Genzel, R., Reid, M.J., Moran, J.M., and Downes, D. 1980, Ap. J. (submitted).
Goldreich, P. and Kwan, J. 1974, Ap. J., 189, 441.
Grasdalen, G.L. 1976, Ap. J. (Letters), 205, L83.
Hall, D.N., Kleinmann, S.G., Ridgway, S.T., Gillett, F.C. 1978, Ap. J. (Letters), 223, L47.
Hollenbach, D.J. and Shull, J.M. 1977, Ap. J., 216, 419.
Joyce, R.R., Simon, M., and Simon, T. 1978, Ap. J., 220, 156.
Kutner, M.L., Tucker, K.D., Chin, G., and Thaddeus, P. 1977, Ap. J., 215, 521.
Kwan, J. 1977, Ap. J., 216, 713.
Kwan, J. and Scoville, N.Z. 1976, Ap. J. (Letters), 210, L39.
Nadeau, D. and Geballe, T.R. 1979, Ap. J. (Letters), 230, L169.
Phillips, J.P. and Beckman, J.E. 1980, M.N.R.A.S. (in press).
Phillips, T.G., Huggins, P.J., Neugebauer, G., and Werner, M.W. 1977, Ap. J. (Letters), 217, L161.
Scoville, N.Z., Hall, D.N.B., Kleinmann, S.G., and Ridgway, S.T. 1979, Ap. J. (Letters), 232, L121.
Scoville, N.Z., Kleinmann, S.G., Hall, D.N.B., and Ridgway, S.T. 1980a (in preparation).
Scoville, N.Z., Schloerb, F.P., and Goldsmith, P.F. 1980b (in preparation).
Simon, M., Righini-Cohen, G., Joyce, R.R., and Simon, T. 1979, Ap. J. (Letters), 230, L175.
Solomon, P.M., Huguenin, G.R., and Scoville, N.Z. 1980, Ap. J. (Letters), in press.
Thompson, R.I. and Boroson, T.A. 1977, Ap. J. (Letters), 216, L75.
Watson, D.M., Storey, J.W.V., Townes, C.H., Haller, E.E., and Hansen, W.L. 1980, Ap. J. (in press).
Werner, M.W., Gatley, I., Harper, D.A., Becklin, E.E., Lowenstein, R.F., Telesco, C.M., and Thronson, H.A. 1976, Ap. J., 204, 420.
Zuckerman, B., Kuiper, T.B.H., and Kuiper, E.N.R. 1976, Ap. J. (Letters), 209, L137.
Zuckerman, B., Palmer, P., and Morris, M. 1980, B.A.A.S., 12, 483.

DISCUSSION FOLLOWING PAPERS ON THE ORION MOLECULAR CLOUD

ZUCKERMAN (TO GRASDALEN): It has been suggested by some astronomers studying the VLBI maps of H_2O emission that IRc4 is simply a hot spot in a cloud rather than a bona fide star or protostar like BN or IRc2. Could you comment on that?

GRASDALEN: I deliberately used the word condensation to describe the peaks in the KL nebula, because I do not believe we yet have enough information to determine the fundamental problem of the source of energy.

THOMPSON (TO SCOVILLE): What do you think the total luminosity of BN is and does it power KL?

SCOVILLE: This is basically an infrared photometric problem which I do not feel qualified to answer. You need small aperture far infrared measurements to look for symmetry around BN, when integrated over all wavelengths. There are other sources in the complex and I see no reason to suggest that BN is the only powerful source. The luminosity of BN would have to be a little greater than 10^4 $L_\odot$ if one accounts for the Brα and γ emission with photoionization by a main sequence star (B0-B1).

RIEKE: Unpublished maps at 34 μm show that the Orion region is basically double, centered on BN and IRc4. This implies IRc4 is a luminosity source in its own right.

AITKEN: IRc4 shows a silicate extinction significantly different from that to most heavily obscured sources. The minimum is shifted to shorter wavelengths which can be explained by strong temperature gradients within the beam. This implies that a source of luminosity is present within IRc4.

GENZEL: Using the IRTF, Downes, Becklin, Wynn-Williams, and I have shown that IRc2 has a deeper silicate absorption feature than any of the other peaks, and, consequently, that it is a substantial source of luminosity for the BNKL region. Its position coincides within 1.5 σ with the proper motion expansion center of the H_2O masers in Orion (Genzel, Reid, Moran, and Downes, preprint). IRc2 is already known to coincide with the SiO maser in Orion, and we now propose that large-scale (10^{-4} or 10^{-3} $M_\odot$ yr^{-1}) outflow from IRc2 is the cause of both the maser features and the "plateau" source in Orion.

ROWAN-ROBINSON: IRc2 sounds very like one of those extreme M6 supergiants like VY CMa or NML Cyg. Is this a possibility?

GENZEL: The mass loss rate needed for IRc2 is a factor of 100 higher than that attributed to M supergiants.

GRASDALEN: The error bars for the center of H_2O expansion embrace IRc4, but are outside of IRc2. Why do you reject your error bars and pick IRc2 as the center of expansion, particularly in view of Aitken's result that the silicate extinction is stronger in IRc4 than IRc2?

GENZEL: Firstly the positional discrepancy is only 1.5 σ. Secondly, the H_2O masers have features which are identical to the SiO maser features, and we know that the SiO maser is well identified with IRc2. A 5 km s^{-1} proper motion of IRc2 could also account for the discrepancy. Our data with a 3.8 arcsec beam and narrow band filters show IRc2 to have a much deeper silicate feature than IRc4.

AITKEN: 10-μm spectroscopy with the AAT and a 3.4 arcsec beam indicate deeper absorption in IRc2 than IRc4. However, when you take into account radiative transfer effects in order to fit the shift of minimum wavelength in IRc4 you come up with a much larger silicate optical depth to IRc4 than to IRc2. We find roughly similar luminosities of around 5000 $L_\odot$ for BN, IRc2, and IRc4.

BECKLIN: Since the discovery by Hall *et al*. that the BN object has many of the properties of a B star, there has been the nagging question as to why it and other infrared sources in regions of star formation are so bright at near infrared wavelengths ($\lambda < 10$ μm). The answer could come from the recent work by Genzel and others that at least one such object shows evidence for large mass loss. If this mass loss is a general property of all such objects, then the infrared emission could result from dust formed out of the material lost from the stars. Such processes would be similar to those seen in many novae.

SPECTROPHOTOMETRY OF DUST

D. K. Aitken
Anglo-Australian Observatory and University College London

I INTRODUCTION

In recent years there have been a number of extensive review papers and books on interstellar dust (e.g. Savage and Mathis, 1979, and references therein). This short review deals only with the spectral properties of dust within the wavelength range 3-25μm, with particular emphasis on recent results.

Infrared spectral features provide information on the chemical composition and structure of dust grains, by comparison with likely laboratory materials and the expected vibrational frequencies of atomic groups in the solid state.

The first observation of a broad infrared emission feature at 10μm in oxygen rich giants and supergiants was made by Woolf and Ney (1969), and attributed to the stretching resonance of the Si-O bond. Since then a total of 13 infrared bands, resolved with moderate resolving power and attributed to solid state transitions, have been discovered in the wavelength regions 3-13.5 and 16-25μm. Plausible laboratory counterparts exist for some of these features but in no case is there a completely unambiguous identification.

II INDIVIDUAL FEATURES

A. The 9.7μm Feature

This feature, with a full width at half maximum of 2-3μm, is seen in emission in the circumstellar shells of oxygen rich giants and supergiants (e.g. Merrill and Stein, 1976a) and also in emission, absorption or a combination of these in a variety of stellar objects (Russell et al 1975; Cohen, 1980), compact H II regions and the interstellar medium (Forrest et al, 1978), molecular clouds (Capps et al, 1978), some planetary nebulae (Aitken et al, 1979a), and presumed progenitors of planetary nebulae (Puetter et al, 1978, Aitken et al, 1980b) and comets (Hackwell, 1971; Merrill, 1974). It has not been

C. G. Wynn-Williams and D. P. Cruikshank (eds.), Infrared Astronomy, 207–221.

observed in the spectra of carbon stars or novae. For brevity references are given only to recent work or if not referenced in earlier review articles.

The commonly accepted interpretation of this feature is that it is due to the stretching vibration of the Si-O bond in grains of disordered silicates. Crystalline forms of silicate minerals such as quartz, olivine and orthopyroxene can be ruled out on the basis of positional mismatch, lack of structure and independently by polarisation observations between 8-13μm of the BN object and the galactic centre (Capps and Knacke, 1976; Dyck and Beichman, 1974). The band strength required by the polarisation observations are an order of magnitude smaller than in terrestrial silicates, but are consistent with those of amorphous silicates. Amorphous silicates produced in the laboratory (Day and Donn, 1978; Day, 1979; Stephens and Russell, 1979) in fact provide a very good fit to the smooth astronomical feature as also do the structurally disordered hydrous silicates found in type I carbonaceous chondrites (e.g. Penman, 1976; Rose, 1979). The material in the meteorites has not been subjected to temperatures in excess of 500K since their formation and contains material from early in the history of the solar system. The correspondence between laboratory measurements of the optical properties of these materials and the astronomical feature must be regarded as very strong evidence for the correctness of the silicate identification.

Silicates also are expected to have band structure in the 20μm region due to bending of the Si-O-Si bond, and such a feature near 18μm has been observed in emission and absorption (Treffers and Cohen, 1974; Forrest _et al_, 1978; Forrest _et al_, 1976; Forrest and Soifer, 1976).

The feature can also be fit arguably well by a number of other materials or mixtures: a) Mixtures of particles of the diatomic oxides Si-O and Mg-O have been suggested by Duley _et al_, 1979. This model accounts for the 9.7μm and 18μm features, though not so convincingly as disordered silicates, and also the 2200Å feature. b) Mixtures of high molecular weight organic compounds can also be constructed to fit the data (Hoyle and Wickramasinghe, 1979; Khare and Sagan, 1979). Probably the strongest arguments against such a mixture being representative of a significant part of the interstellar medium is the observed appearance of the feature in oxygen rich but not carbon rich stars, and that it requires a correlation which is not observed between the absorption feature at 3.07μm and the 9.7μm feature.

B. The 11μm Broad Feature

This feature is characterised by emission increasing fairly sharply from near 10.5μm to a broad maximum in the 11μm region with a less well defined long wavelength turn off at around 12.7μm. These wavelengths correspond respectively to the longitudinal and transverse optical phonon frequencies in silicon carbide, which define the well known forbidden gap for propagation of electromagnetic waves. The spectral shape of the feature depends on the shape and size distribution of the grains.

This broad feature is seen in emission in circumstellar shells around carbon stars (Merrill and Stein, 1976a) and some planetary nebulae (Willner et al, 1979a; Aitken et al, 1979a). The feature is rarely seen in absorption (Jones et al, 1978), and in the general interstellar extinction $\tau_{SiC}/\tau_{sil} < 0.1$. Band strengths for silicon carbide grains are likely to be somewhat greater ($\beta m \simeq 1.4 \times 10^4 cm^2 g^{-1}$, Dorschner et al, 1977) than for amorphous silicates ($\beta m \simeq 3 \times 10^3 cm\ g^{-1}$, Penman, 1976); the ratio of silicon carbide to silicates in the interstellar medium must be less than a few percent by mass.

C. Phenomenology of the Oxygen-Rich and Carbon-Rich Features

Irrespective of arguments about the precise nature of these grain materials, observationally they are present in circumstellar environments according to whether the abundance ratio of carbon to oxygen atoms is greater or less than some number close to unity and in the range 1.0-1.8. In addition to often displaying the silicon carbide feature, carbon stars also show a featureless excess emission with characteristic temperature near 1000K which is usually attributed to graphite grains.

Thus on the basis of 8-13μm spectroscopy it is possible to categorise objects according to whether grains formed in an oxygen or carbon rich environment. This can be of particular value in nebular studies where optical and UV data relevant to the important ionic species of carbon may be scarce and the interpretation complicated by temperature dependence and theoretical uncertainties. For example the bright planetary nebula IC418 has recently been demonstrated to have N(C)/N(O) = 1.8±0.8 from optical and UV studies (Harrington et al, 1980), compared with N(C)/N(O) = 0.6 for the sun. In IC418, NGC6572 (Willner et al, 1979a); NGC6790 (Aitken et al, 1979a) and IC5117 (Grasdalen, 1979) the 11μm silicon carbide emission feature is readily seen and it can be inferred that the nebular composition in these objects is enriched in carbon due to nucleosynthesis. Some other planetary nebulae show the 9.7μm feature in emission (Aitken et al, 1979a) and at least the formation of grains must have taken place in an oxygen rich environment.

The gas phase of the interstellar medium and H II regions has N(C)/N(O) close to solar values (e.g. Salpeter, 1977) and it is perhaps gratifying to find the silicate feature ubiquitous in these regions. The archetype of this feature is often taken to be that of the Trapezium region of the Orion nebula. Here complications due to a photospheric component are avoided because the nebular feature can be spatially separated from the hot stars in the Trapezium. A simple two component dust model in which emission from warm isothermal dust is attenuated by cold Trapezium material has been remarkably successful in matching the observed 8-13μm spectra of H II regions, protostellar objects and the galactic centre sources, and it can be asserted that in some cases the underlying emission is from optically thin Trapezium like material. This is true for a number of 'moderately' obscured sources such as BN and some of the galactic centre sources. When the depth of the absorption feature approaches or exceeds an order of magnitude the match becomes insensitive to the form of the underlying emission and introduces an uncertainty in the derivation of the extinction optical depth to the source. Nevertheless the good fits obtainable for a wide variety of sources and optical depths is evidence for the universality of this extinction curve at 10μm.

Estimates of the ratio $A_V/A_{9.7}$ are difficult because few sources have well determined values of A_V large enough to give significant optical depths at 10μm. Table I gives values of $A_V/A_{9.7}$, obtained for three sources.

TABLE I Comparison of visual and infrared extinction

Source	A_V	$\tau_{9.7}$	$A_V/A_{9.7}$	N(Si)/N(H)	
G333.6-0.2	20±1	1.15	17.4±1	3.9×10^{-5}	Rank *et al*, 1978
VI Cyg #12	10	0.65	15	4.6×10^{-5}	Gillett *et al*, 1975b
IRS 7 in SgrA	30	3.5	8.5±3	8.1×10^{-5}	Becklin *et al*, 1978
				3.3×10^{-5}	Solar value; Allen, 1973

The value of $\tau_{9.7}$ = 1.15 used for G333.6-0.2 differs from that used by Rank *et al*, 1978 and is more appropriate to the Bα and Hα beam sizes used to determine A_V since it has been found that the silicate optical depth is a function of beam size in this source. The range of about a factor two between the various determinations of $A_V/A_{9.7}$ may in part be due to the assumptions made in the derivations. The required silicon abundance is calculated assuming that 3/4 of $A_{9.7}$ is due to silicates with mass absorption coefficient $3\text{x}10^3\text{cm}^2\text{g}^{-1}$ and that the hydrogen column density is $2\text{x}10^{21}\text{cm}^{-2}\,A_V^{-1}$. The required abundance is rather high, as noted by Hong and Greenberg (1978), especially for the galactic centre for which the extinction is predominantly interstellar (Becklin *et al*, 1978).

A few heavily obscured sources refuse to yield good fits with the two component model. One of these, OH 0739 (Gillett and Soifer, 1976), exhibits a broad long wavelength wing to the extinction and this is presently without adequate explanation. The source IRC4 in BNKL has its minimum shifted to near 9.3μm, but this is due to failure of the two component model in a source with strong temperature gradients (Aitken *et al*, 1980c).

The Trapezium spectrum is more dilute than the excess observed in oxygen rich giants. In particular this is seen for μ Cep (Russell et al, 1975) and compared with that of the Trapezium and other oxygen rich giants in Fig 1. It seems that H II regions and the interstellar

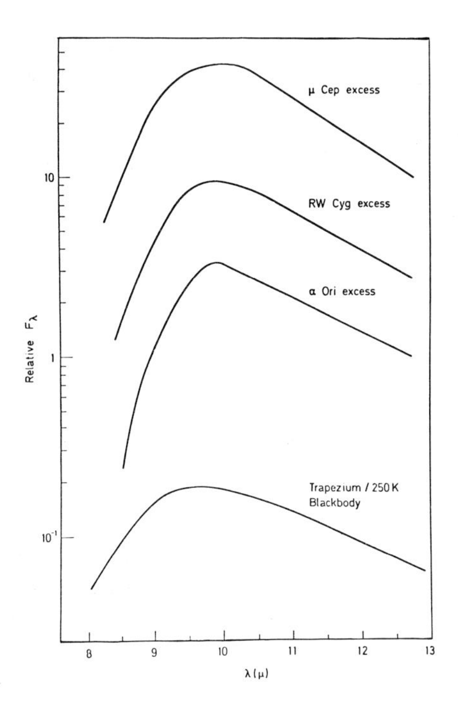

Figure 1. Comparison of the 8-13μm excess from μ Cep, RW Cyg and α Orion with that of the Orion Trapezium region. Data from Merrill et al, 1976; Russell et al, 1975; Forrest et al, 1975.

medium require an additional featureless component to that observed in the oxygen rich giants, and dust injection to the medium from carbon stars, planetary nebulae and novae is a likely cause. This difference also suggests that observationally one might be able to distinguish generally between oxygen rich regions in which grain formation has or is taking place and emission from H II regions. Certainly the eruptive variables HM Sge and V1016 Cyg, often considered to be protoplanetary nebulae, show 8-13μm spectra typical of recent grain formation (Puetter et al, 1978; Aitken et al, 1980b) in an oxygen rich environment.

D. The 3.07μm Absorption Feature

The presence of an absorption feature near 3.07μm due to interstellar water ice has been found in sources deeply embedded within molecular clouds but not in the general interstellar medium. The observed feature is narrow with a width at half maximum of about 0.3μm and while agreeing qualitatively with calculations using Mie theory and laboratory optical constants for H_2O ice, is different in details. In particular the observed feature has more extinction shortwards of 3.07μm and an additional long wavelength wing than required by theory. These difficulties can be reconciled by allowing a range of sizes and an admixture of NH_3 ice to produce the shorter wavelength extinction (Merrill et al, 1976a) and hydrocarbon ices (Hagen et al, 1980) to produce the long wavelength wing. The presence of C-H bond contamination has also been suggested by Soifer et al, (1979). A further difficulty exists however in that the broad feature expected in the 10μm ice spectrum (Bertie et al, 1969) is not observed even in sources showing very deep 3.07μm features.

Water ice also has a feature near 45μm and this has been searched for in KL. Detection of this band is reported by Papoular et al. (1978) in emission in a 4' beam, and in absorption in a 50" beam by Erickson et al. (1980). A temperature gradient in KL can explain this apparent discrepancy; however, these observations are extremely difficult and may be subject to systematic uncertainties.

The 'ice' feature is not to be confused with an absorption at the same wavelength seen in carbon stars and which has been definitively identified in high spectral resolution observations as molecular band absorption by hydrogen cyanide and acetylene (Ridgway et al, 1978). At the lower resolution typical of filter wheels the two features may be distinguished observationally with some care since the carbon star feature is slightly narrower and lacks the long wavelength wing. In practice distinction between the two features is often made on the basis of prior knowledge of the nature of the source.

In molecular clouds the ice and silicate features are essentially uncorrelated with $0.2 < \tau_{ice}/\tau_{sil} < 2$; in the interstellar medium ice absorption is not observed and $\tau_{ice}/\tau_{sil} < 0.04$ for VI Cyg No. 12 (Gillett et al, 1975b). Using a band strength $\beta_m = 1.4 \times 10^4 cm^2 g^{-1}$ for ice implies a molecular cloud ratio of mass of ice/silicate in the range 4-40%, well short of abundance constraints. Greenberg (1976), has suggested that ambient starlight in the medium converts H_2O to free OH radicals.

III OTHER EMISSION FEATURES

In 1973 Gillet et al reported the observation of two narrow but resolved features in the 8-13μm spectrum of the planetary nebula NGC7027. Since then other features have been observed and presently there are 7 narrow ($\Delta\lambda$<1μm) emission features at the wavelengths 3.28, 3.4, 3.5, 6.2, 7.7, 8.6 and 11.25μm. None of these, with the possible exception of that at 3.4μm, have ever been observed in absorption, and all save the 3.5μm feature were first discovered in NGC7027 (Fig. 2). The features have been seen in a wide range of objects encompassing a range of chemical abundances and evolutionary status (Table II) with the one apparent common feature of a plentiful supply of UV photons. The features do not break up into lines or bands at higher resolution (Tokunaga and Young, 1980; Bregman, 1977)

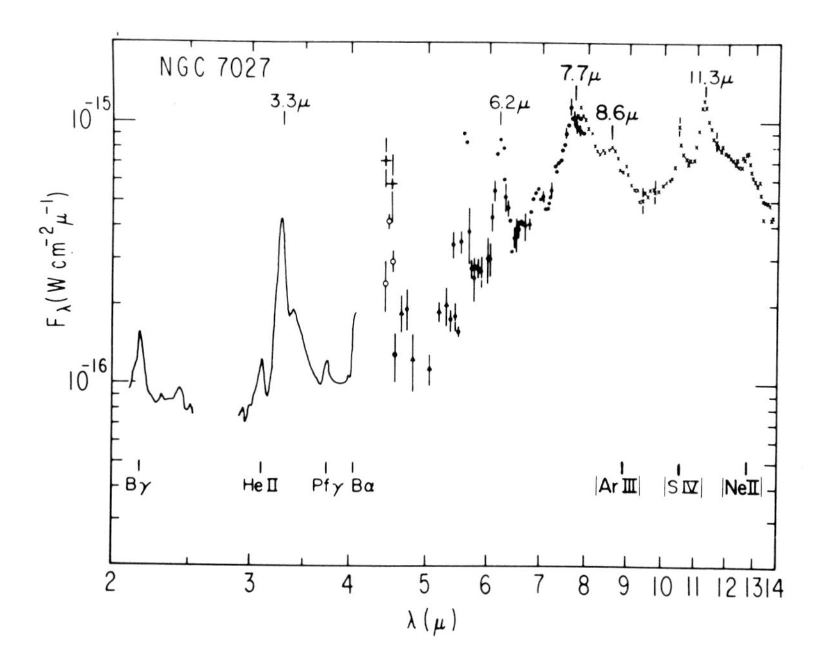

Figure 2. 2-14 μm spectrum of NGC 7027 showing the unidentified emission features together with atomic transitions. From Russell et al, (1977b).

and it is therefore assumed that the features arise from solid grains or grain mantles.

Apart from the 3.4 and 3.5μm feature there is a general tendency for the features to occur together; this is a qualitative statement, there being clear differences in the ratio of the strengths of the features although this is never very large (Table II). So far there is no agreed identification for any of the features, but while they do not necessarily arise from the same material or process it is clear that the environment which produces one of them also favours the production of others. The table reveals that there is no significance to the non-observance of the 8.6 and 11.25μm features in most H II regions.

Observations of the ionization front near θ^2 A in the Orion nebula indicate that the 11.25μm feature (Aitken et al, 1979b) and the 3.28μm feature shown in Fig 3 (Aitken and Roche, unpublished) arise from just outside the H II region. Maps of Orion (Sellgren, 1980) in the 3.28μm feature and 3.5μm continuum show that the feature distribution is different and more extended than that of the continuum and that the line to continuum ratio has a maximum close to the ionization front. Wynn-Williams et al (1980) show that in AFGL 437 the feature is similarly more extended than its H II region. All

TABLE II NARROW EMISSION FEATURES

	Band Centre and Width (μm)							
	3.28	3.4	3.5	6.2	7.7	8.6	11.25	
Source	0.05	0.08	0.08	0.2	0.5	0.3	0.25	References
PLANETARY NEBULAE								
NGC 7027	21	5	x	150	300	45	140	1,2,3,4
IC 418	1.2	x	x	-	-	-	<3	2,21
BD $+30^{\circ}3639$	4.0	1	<0.1	√	√	7	18	2,3,4,5
NGC 6790	0.4	x	x	-	-	<0.5	1	22,27
SwSt1	0.4	?	x	-	-	<0.5	<0.5	22,27
M1-11	0.5	?	x	-	-	1	1	24
NGC 6572	0.5	<0.2	x	-	-	-	<1.5	21
H II REGIONS								
NGC 7538	√	?	x	-	-	x	x	2
M17	√	√	x	-	-	-	-	17
Orion Bar 7"	2	0.5	x	√	√	4	9.5	8,9,10,28
GL 3053	(1)	(0.3)	<(.1)	(22)	(55)	(4.5)	(4.5)	11,12
G 333.6-0.2	2	x	x	-	-	<20	<20	28
G 45.1+0.1	√	x	x	-	-	<10	<10	28
STELLAR OBJECTS								
He 2-113	4	<0.6	1?	-	-	9	20	25,26
CPD$-56^{\circ}8032$	8	4	2?	-	-	<9	20	25,26
γ^2 Vel	√	x	x	-	-	-	-	26
MWC 922	√	x	x	-	-	?	?	19
HD 97048	1	3.5	6	-	-	?	3.3	20,24
EXTRA GALACTIC								
NGC 253	1	0.2	x	-	-	1.5	2	4,15
M 82	4	2	x	32	72	8	9	15,16
3C 273	.05	-	-	-	-	-	-	23
UNCLASSIFIED								
HD 44179	60	<6	<2	160	500	90	90	4,6,7,18
AFGL 437	2	0.4	<.2	-	-	8	11	13
GL 4029	√	√	√	√	√	√	√	14

Table is representative rather than comprehensive.
Approximate intensities are given in units of 10^{-18} W cm^{-2}.
Approximate intensities relative to 3.28μm given in brackets.
√ feature observed x not observed - no observations

REFERENCES:
1 Merrill *et al* 1975
2 Russell *et al* 1977a
3 Gillett *et al* 1973
4 Russell *et al* 1977b
5 Russell *et al* 1980
6 Russell *et al* 1978
7 Cohen *et al* 1975
8 Sellgren 1980
9 Soifer *et al* 1980
10 Aitken *et al* 1979b
11 Russell 1979
12 Merrill 1977
13 Kleinmann *et al* 1977
14 Willner *et al* 1980
15 Gillett *et al* 1975a
16 Willner *et al* 1977
17 Grasdalen & Joyce 1976
18 Tokunaga & Young 1980
19 Merrill & Stein 1976b
20 Blades & Whittet 1980
21 Willner *et al* 1979a
22 Aitken *et al* 1979a
23 Allen 1980a
24 Aitken & Roche 1980
25 Aitken *et al* 1980a
26 Allen 1980b
27 Jones 1979
28 Aitken & Roche, unpubl.

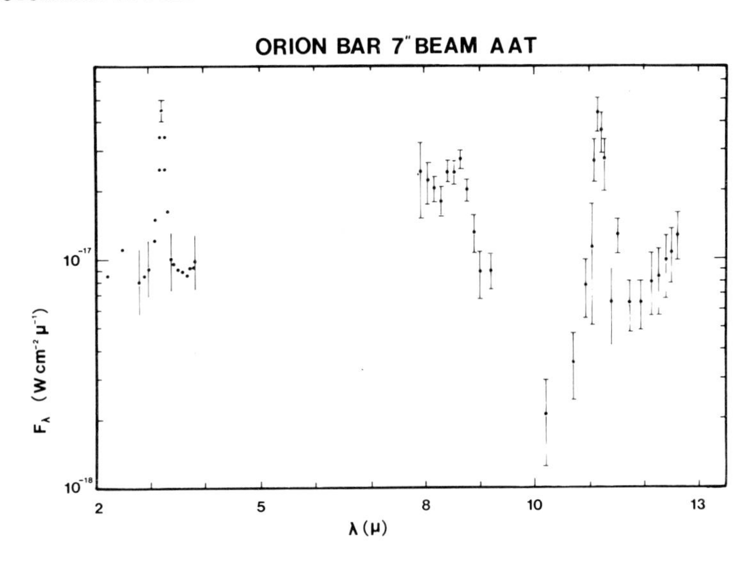

Figure 3. 3-4 μm and 8-13 μm spectrum of position 4 in the Orion bar taken with a 7" dia beam. From Aitken et al, 1979b, and Aitken and Roche, unpublished.

other observations appear to be consistent with the conclusion that the features arise just outside an H I — H II interface.

The distance from the ionizing source in Orion to the ionization front, and in AFGL 437 to the outer bound of the observed 3.28μm emission, is sufficiently large to rule out heating of normal sized grains to temperatures required to give the observed feature fluxes.

Presently there are two main ideas regarding the excitation mechanism of the features:

A Infrared Fluorescence

Allamandola et al (1979) have suggested that in sufficiently cold grains or grain mantles, UV radiation can lead to excited vibrational states which decay radiatively, i.e. IR fluorescence. They associate some of the observed bands with ices of H_2O, NH_3, CH_4 and C_2H_2. A high efficiency of conversion of UV photons to IR feature photons is required in this model and needs to approach unity for some of the features. The efficiency to all IR bands apparently exceeds unity for the Orion bar source, for instance. Such seemingly implausible efficiencies are not a fundamental problem, but a detailed theory of infrared fluorescence will need to explain them, and clearly laboratory measurements will be of value. This model naturally accounts for emission occurring from outside an H II region and may even require this constraint since grains in H II regions may not be cold enough to suppress thermal relaxation of the lattice. Absence of the features in absorption is explained since the emission process is efficient and the optical depth need only be small.

B UV Heating of Very Small Grains

Dwek et al (1980) have proposed that the features arise by thermal emission from a population of very small grains (a $\sim$ 0.01μm) which are heated by UV radiation. They show that grains of radius $\sim$ 0.01μm would reach temperatures $\sim$ 300K within the ionization front in Orion, and that the temperature reached by the grains is a strong function of the UV content of incident radiation. Essentially the ratio of cross sections in the UV to the IR is kept large by making the grains small and on the order of the characteristic wavelength of the UV field from an early O star. The relatively small variation of the observed flux ratios (in features widely separated in wavelength) from source to source implies that the features are emitted only over a narrow range of temperatures. If the grain mantles are volatile an upper bound to the emission temperature will exist, but would not explain why the 11.3μm is never observed without the 3.28μm feature. The features do not appear in absorption because the emission process is efficient due to relatively high temperature and only a small optical depth of emitting material is required.

The 3.5μm feature, unresolved with resolving power of 50, has so far been observed in only one object, HD 97048, a likely pre-main-sequence star of spectral type B9-AO Ve in the Chameleon dark cloud (Blades and Whittet, 1980), which has a variable circumstellar shell on a time scale of months. The 3.28μm and 3.4μm features are also evident but with band strengths remarkably reversed from other sources, and the 3.5μm feature is dominant. The feature is very tentatively identified with formaldehyde following Allamandola and Norman (1978) and it is pointed out that the polymerised form of formaldehyde, polyoxymethylene, retains the 3.5μm signature and additionally shows infrared activity at 10.37μm and 11.76μm (Whittet et al, 1976). The 8-13μm spectrum (Aitken and Roche, to be published) shows no evidence of such activity although the 11.25μm feature is seen.

The sequence of emission features 7.7, 8.6 and 11.25μm can mimic the shape of the silicate absorption as can be seen in Figs. 2 and 3 where the silicate optical depth is expected to be small or negligible. Thus care is needed in evaluating extinction corrections to objects in which these emission features are seen or where observations have been made by narrow band photometry.

IV OTHER ABSORPTION FEATURES

Narrow but resolved absorption features are seen at 4.61μm, 6.0μm and 6.8μm, the latter two being observed only in molecular cloud sources. These are typified by the predominantly interstellar absorption seen to the galactic centre (Fig 4a, from Willner et al, 1979b)

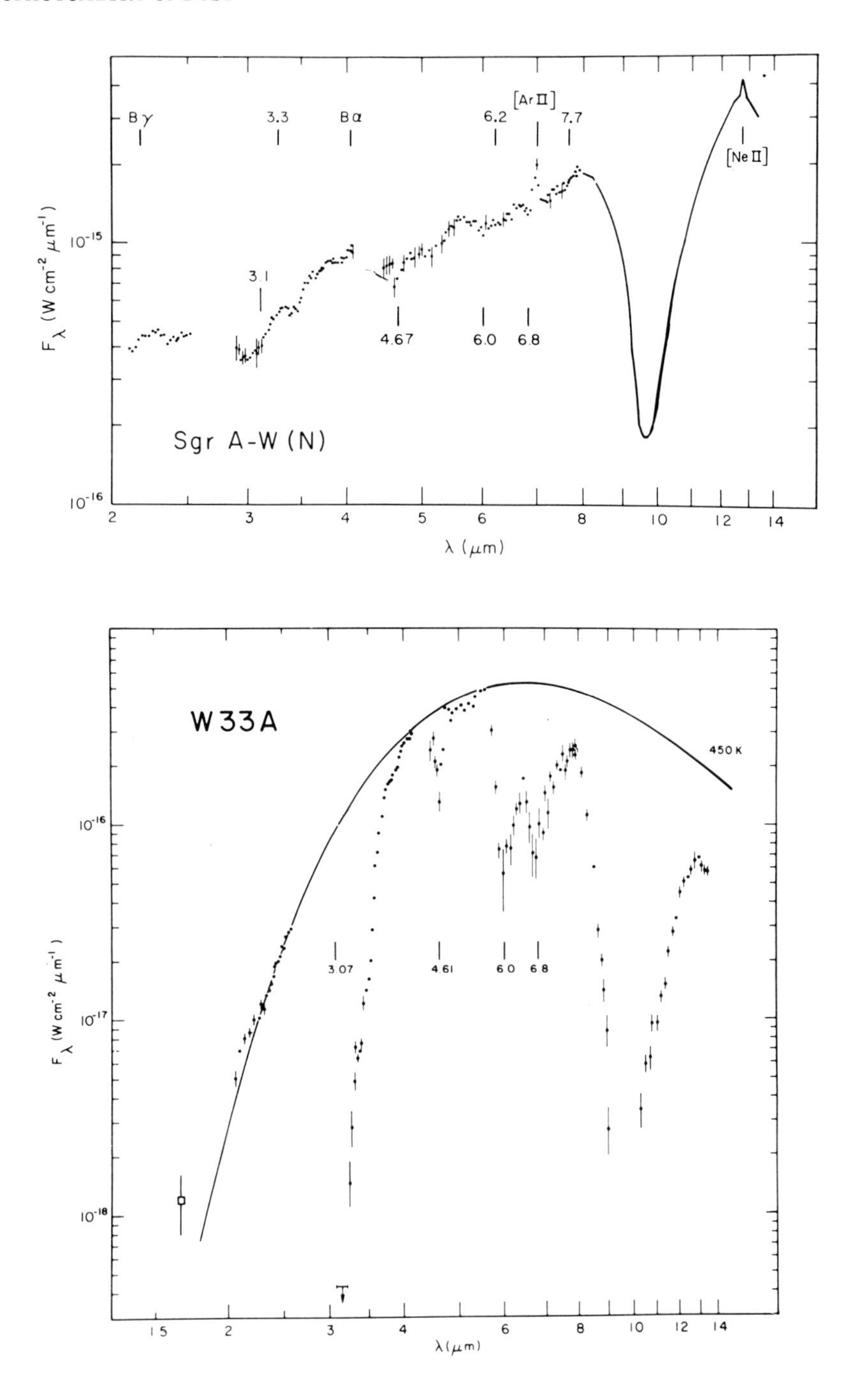

Figure 4. a) Absorption features in the interstellar medium to Sgr A; b) molecular cloud absorption features to W33A. Data from a) Willner et al, 1979a, and b) Soifer et al, 1979.

and the heavily obscured molecular cloud source W33A (Fig 4b, from Soifer _et al_, 1979).

The 4.67μm fundamental vibration band of CO has been suggested to be associated with the 4.61μm feature, although there are some admitted problems with width and wavelength shift of the observed feature (Soifer _et al_, 1980).

The strength of the 6.0 and 6.8μm absorption feature implies they must be due to abundant elements. Refractory materials appear to be ruled out as the features only appear in sources in molecular clouds and the presence of hydrated materials is unlikely for a number of reasons (Soifer _et al_, 1980). Although hydrated minerals do exhibit a band between 6.1 and 6.2μm this is significantly different from the 6.0μm band observed, which does not in any case occur in other situations where hydrated minerals would be expected. Hagen _et al_ (1980) have shown that condensates from the gas phase of a mixture of H_2O, CH_3OH, NH_3 and CO at 10 K exhibit spectra which can account for all the absorption features seen between 3 and 8μm.

The stretching vibration of the C-H band expected between 3.3-3.5μm has been suggested as the reason for the skewed shape of the 3.07μm ice feature to larger wavelengths. It has also been searched for in absorption in the interstellar medium where the ice feature is not observed. Soifer _et al_ (1976) observed a feature in the spectrum of the galactic centre, but it is not clear whether this is an emission feature at 3.3μm or the expected absorption near 3.4μm. Allen and Wickramasinghe (1980) have recently observed the galactic centre source IRS7; their data favours an absorption feature at 3.4μm with an optical depth of 0.2. An estimate of the column density of hydrocarbon material required to give this feature depends on the chemical class of the organic compounds, but if this result and interpretation is correct hydrocarbon material is not an insignificant part of the interstellar medium.

REFERENCES

Aitken, D. K., Barlow, M. J., Roche, P. F., and Spenser, P. M., 1980a, M.N.R.A.S., in press.
Aitken, D. K., Roche, P. F., Spenser, P. M., and Jones, B., 1979a, Ap.J., 233, 925.
Aitken, D. K., Roche, P. F., Spenser, P. M., and Jones, B., 1979b, Astron. and Astrop., 76, 60.
Aitken, D. K., Roche, P. F., and Spenser, P. M., 1980b, M.N.R.A.S., in press.
Aitken, D. K., Roche, P. F., Spenser, P. M., and Jones, B., 1980c, M.N.R.A.S., submitted.
Aitken, D. K., and Roche, P. F., 1980, in preparation.
Allamandola, L. J., and Norman, C. A., 1978, Astron. and Astrop., 63, L23.
Allamandola, L. J., Greenberg, J. M., and Norman, C. A., 1979, Astron. and Astrop., 77, 66.
Allen, C. W., 1973, Astrophysical Quantities, p. 31.
Allen, D. A., 1980a, Nature, 284, 323.
Allen, D. A., 1980b, private communication.
Allen, D. A., and Wickramasinghe, D. T., 1980, in preparation.
Becklin, E. E., Matthews, K., Neugebauer, G., and Willner, S. P., 1978, Ap.J., 220, 831.
Bertie, J. E., Labbe, H. J., and Whalley, E., 1969, J.Chem.Phys. 50, 4501.
Blades, J. C., and Whittet, D. C. B., 1980, M.N.R.A.S. 191, 701.
Bregman, J. D., 1977, Pub.A.S.P. 89, 335.
Capps, R. W., and Knacke, R. F., 1976, Ap.J. 210, 76.
Capps, R. W., Gillett, F. C., and Knacke, R. F., 1978, Ap.J. 226, 863.
Cohen, M., 1980, M.N.R.A.S., 191, 499.
Cohen, M., Anderson, C.M., Cowley, A., Coyne, G.V., Fawley, W.M., Gull, T.R., Harlan, E.A., Herbig, G.H., Holden, F., Hudson, H.S., Jakoubek, R.O., Johnson, H.M., Merrill, K.M., Schiffer, F.H., III, Soifer, B.T., and Zuckerman, B., 1975, Ap.J. 196, 179.
Day, K. L., 1979, Ap.J. 234, 158.
Day, K. L., and Donn, B., 1978, Ap.J.Lett. 222, L45.
Dorschner, J., Friedemann, C., and Gurtler, J., 1977, Astron.Nachr. 298, 279.
Duley, W. W., Millar, T. J., and Williams, D. A., 1979, Astrop. and Sp.Sci. 65, 69
Dwek, E., Sellgren, K., Soifer, B. T., and Werner, M. W., 1980, preprint.
Dyck, H. M., and Beichman, C. A., 1974, Ap.J. 194, 57.
Erickson, E. F., Knacke, R. F., Tokunaga, A.T., and Haas, M. R., 1980, preprint.
Forrest, W. J., Gillett, F. C., and Stein, W. A., 1975, Ap.J. 195, 423.
Forrest, W. J., Gillett, F. C., Houck, R. J., McCarthy, J. F., Merrill, K. M., Pipher, J. L., Puetter, R. C., Russell, R. W., Soifer, B. T., and Willner, S. P., 1978, Ap.J., 219, 114.
Forrest, W. J., Houck, J. R., and Reed, R. A., 1976, Ap.J.Lett., 208, L133.
Forrest, W. J., and Soifer, B. T., 1976, Ap.J., 208, L129.
Gillett, F. C., Forrest, W. J., and Merrill, K. M., 1973, Ap.J., 183, 87.
Gillett, F. C., Kleinmann, D. E., Wright, E. L., and Capps, R. W., 1975a, Ap.J.Lett., 198, L65.
Gillett, F. C., Jones, T. W., Merrill, K. M., and Stein, W. A., 1975b, Astron. and Astrop., 45, 77.
Gillett, F. C., and Soifer, B. T., 1976, Ap.J., 207, 780.
Grasdalen, G. L., 1979, Ap.J., 229, 587.
Grasdalen, G. L., and Joyce, R. R., 1976, Ap.J.Lett., 205, L11.

Greenberg, J. M., 1976, Astrophys. Sp.Sci., 39, 9.
Hackwell, J. A., 1971, Observatory, 91, 33.
Hagen, L. J., Allamandola, L. J., and Greenberg, J. M., 1980, preprint.
Harrington, J. P., Lutz, J. H., Seaton, M. J., and Stickland, D. J., 1980, M.N.R.A.S., 191, 13.
Hong, S. S., and Greenberg, J. M., 1978, Astron. and Astrop., 69, 341.
Hoyle, F., and Wickramasinghe, N. C., 1979, Astrop. and Sp.Sci., 66, 77.
Jones, B., 1979, private communication.
Jones, B., Merrill, K. M., Puetter, R. C., and Willner, S. P., 1978, Astron.J., 83, 1437.
Khare, B. N., and Sagan, C., 1979, Astrop. and Sp.Sci., 65, 309.
Kleinmann, S. G., Sargent, D. G., Gillett, F. C., Grasdalen, G. L., and Joyce, R. R., 1977, Ap.J.Lett., 215, L79.
Martin, P. G., 1971, Astrop.Lett., 7, 193.
Merrill, K. M., 1974, Icarus, 23, 566.
Merrill, K. M., 1977, in R. Kippenhahn, J. Rahe, and W. Strohmeier (eds.) "I.A.U. Colloq. 42, Proceedings," Veröff. Remeis-Sternw. Bamberg, Astron. Inst. Univ. Erlangen-Nürnberg, Band XI, Nr. 121, pp.446-494.
Merrill, K. M., and Stein, W. A., 1976a, Pub.A.S.P., 88, 285.
Merrill, K. M., and Stein, W. A., 1976b, Pub.A.S.P., 88, 874.
Merrill, K. M., Russell, R. W., and Soifer, B.T., 1976, Ap.J., 207, 763.
Merrill, K. M., Soifer, B. T., and Russell, R. W., 1975, Ap.J.Lett. 200, L37.
Papoular, R., Lena, P., Marten, A., Rouan, D., and Wijnbergen, J., 1978, Nature, 276, 593.
Penman, J. M., 1976, M.N.R.A.S., 175, 149.
Puetter, R. C., Russell, R. W., Soifer, B. T., and Willner, S. P. 1978, Ap.J., 223, L93.
Rank, D. M., Dinerstein, H. L., Lester, D. F., Bregman, J. D., Aitken, D. K., and Jones, B., 1978, M.N.R.A.S., 185, 179.
Ridgway, S. T., Carbon, D. F., and Hall, D. N. B., 1978, Ap.J., 225, 138.
Rose, L. A., 1979, Astrop. and Sp.Sci., 65, 47.
Russell, R. W., 1979, PhD., Thesis, University of California at San Diego.
Russell, R. W., Soifer, B. T., and Forrest, W. J., 1975, Ap.J.Lett., 198, L41.
Russell, R. W., Soifer, B. T., and Merrill, K. M., 1977a, Ap.J., 213, 66.
Russell, R. W., Soifer, B. T., and Willner, S. P., 1977b, Ap.J.Lett., 217, L149.
Russell, R. W., Soifer, B. T., and Willner, S. P., 1978, Ap.J., 220, 568.
Russell, R. W., Puetter, R. C., Soifer, B. T., and Willner, S. P., 1980, in preparation.
Salpeter, E. E., 1977, Ann.Rev.Astron.Astrop., 15, 267.
Savage, B. D., and Mathis, J. S., 1979, Ann.Rev.Astron.Astrop., 17, 73.
Sellgren, K., 1980, preprint.
Soifer, B. T., Puetter, R. C., Russell, R. W., Willner, S. P., Harvey, P. M., and Gillett, F. C., 1979, Ap.J.Lett., 232, L53.
Soifer, B. T., Russell, R. W., and Merrill, K. M., 1976, Ap.J., 207, L83.
Soifer, B. T., Willner, S. P., Puetter, R. C., and Russell, R. W., 1980, in preparation.
Stephens, J. R., and Russell, R. W., 1979, Ap.J., 228, 780.
Tokunaga, A. T., and Young, E. T., 1980, preprint.
Treffers, R., and Cohen, M., 1974, Ap.J., 188, 545.

Whittet, D. C. B., Dayawansa, I. J., Dickinson, P. M., Marsden, J. P., and Thomas, B., 1976, M.N.R.A.S., 175, 197.

Willner, S. P., Jones, B., Puetter, R. C., Russell, R. W., Soifer, B. T., 1979a, Ap.J., 234, 496.

Willner, S. P., Russell, R. W., Puetter, R. C., Soifer, B. T., and Harvey, P. M., 1979b, Ap.J., 229, L65.

Willner, S. P., Soifer, B. T., Russell, R. W., Joyce, R. R., and Gillett, F. C., 1977, Ap.J.Lett., 217, L121.

Willner, S. P., Puetter, R. C., Soifer, B. T., Russell, R. W., and Smith, H. E., 1980, in preparation.

Wolff, N. J., and Ney, E. P., 1969, Ap.J.Lett., 155, L181.

Wynn-Williams, C. G., Becklin, E. E., Beichman, C. A., Capps, R., and Shakeshaft, J. R., 1980, preprint.

DISCUSSION FOLLOWING PAPER PRESENTED BY D. K. AITKEN

RIEKE: The scatter in quoted values of $A_V/\tau_{silicates}$ in the literature is partly due to different definitions of τ_{sil}: if it is measured relative to a continuum joining the spectral points at 8 and 12 μm, VI Cyg No. 12 has $A /\tau_{sil} \simeq 23$; if a Trapezium-type spectrum is fitted, the fitted curve requires a significant optical depth at 8 and 12 μm, which gets added to τ_{sil} as it would be defined the other way, then VI Cyg No. 12 has $A /\tau_{sil} \simeq 15$. Similar discrepancies would be expected for other sources.

AITKEN: For the first two examples in Table 1, Trapezium-like emission was assumed. For IRS 7 in Sgr A something very similar was taken, so that this source of discrepancy is avoided in the table. We have used $\tau_{9.7}$ to refer to the total Trapezium-like optical depth since this may not be due solely to silicates.

RUSSELL: The Cornell group, using Houck's spectrometer and an amorphous silicate sample (Stephens and Russell, Astrophys. J. 228, 780, 1979) have obtained a fit to the Trapezium emission from 16 to 30 μm with a $\chi^2 = 1.4$.

MOORWOOD: Is the silicon carbide identification well accepted?

AITKEN: It has not been seriously challenged, but it cannot be taken as a definite identification.

HILDEBRAND: The Chicago group, in collaboration with Gatley, Sellgren, and Werner, has measured the ratio of near ultraviolet (0.25 μm) and far infrared (125 μm) extinction efficiencies of the reflection nebula NGC 7023. This is the subject of Stan Whitcomb's thesis. The measured ratio is much too low to correspond to bare graphite, silicate, or silicon carbide grains, but, on the basis of Aanestad's calculations, is in agreement with expected ratios for grains with ice mantles having mantle/core volume ratios $\lesssim 1$.

POLARIMETRY OF INFRARED SOURCES

H. M. Dyck and Carol J. Lonsdale
University of Hawaii
Honolulu, Hawaii 96822

1. INTRODUCTION

Polarization at infrared wavelengths has been detected from a number of different objects within the Galaxy. These include young sources associated with molecular clouds and H II regions, cool stars with thick circumstellar shells, bi-polar nebulae, and normal stars suffering interstellar polarization. Typical levels of polarization detected at 2.2 μm are up to 25% for the molecular cloud sources, less than ~5% for the cool stars, around 30% for some bi-polar nebulae and less than ~2% for interstellar polarization. For the latter three types of source the origin of the polarization is basically understood: it results from scattering of stellar radiation off small particles in the surrounding shell or nebula in the cool stars and bi-polar nebulae and by transfer of flux through a foreground medium of aligned dust grains for the interstellar polarization. The phenomenon of large infrared polarization in the young stellar and pre-stellar sources is less well understood, and it is to this problem that we address ourselves in this review.

More than 40 such young molecular cloud sources have now been surveyed in the near-infrared for linear polarization (Capps, 1976; Dyck and Capps, 1978; Kobayashi et al., 1978; Dyck and Lonsdale, 1979). About 30% of the samples show more than 10% polarization at 2.2 μm, and about 50% show more than 5%.

One can immediately discount the possibility that the polarization arises by the normal interstellar mechanism along the line-of-sight from the Sun, because the levels of polarization observed are often much higher than the maximum expected levels of interstellar polarization of ~1-2% at 2.2 μm (Hall, 1958; Wilking et al., 1980). In fact, large variations in the degree of polarization observed among sources within the same molecular cloud complex indicate that the polarization arises in the close vicinity of the infrared source, and not over a long pathlength in the interstellar medium. A good example of this phenomenon is the Orion Molecular Cloud (OMC 1) in which are embedded

C. G. Wynn-Williams and D. P. Cruikshank (eds.), Infrared Astronomy, 223–236.

BN (IRS 1) and Hilgeman's source (IRS 2), polarized at 2.2 μm at 17.5% and 10.8% respectively (Dyck and Lonsdale, 1979; Johnson, 1979), though they are separated by only ~0.05 pc in the plane of the sky. The trapezium stars, believed to lie at the front surface of the molecular cloud, are nearly unpolarized by comparison (Hall, 1958). Thus the high polarization of BN and Hilgeman's source must arise in the dense material of OMC 1 itself. Given the large thermal infrared fluxes of the young sources, the polarization can be attributed to processes involving dust grains. Understanding the mechanism of polarization may lead to information about the molecular cloud dust grains and to physical conditions within the cloud.

In the following sections we will review the possibilities that the polarization of the molecular cloud sources is due to (a) an enhanced interstellar mechanism within the dense cloud material, or (b) to scattering off dust grains in a shell or nebula around the young objects. In Section 2 we present detailed polarimetric data which are available for two of these sources, BN and GL 2591, and in Section 3, we discuss the general properties of the whole sample. In Section 4 we discuss the consequences of the conclusions reached in Sections 2 and 3 for the conditions which must exist within the molecular cloud complexes.

2. THE POLARIZATION PROPERTIES OF BN AND GL 2591

The BN source in OMC 1 was the first of its kind to be observed polarimetrically (Loer et al., 1973) and has been extensively studied in the interim. Our most up-to-date understanding of the source is that it is a B0 or B1 star undergoing mass loss (Hall et al., 1978) with a surrounding thick layer of dust which obscures it optically and which radiates the infrared spectrum. GL 2591 is a less well-studied source, with similar observational characteristics to BN. In Figure 1 we present a summary of the available photometric and polarimetric data for BN between 1.6 and 20 μm. The two prominent absorptions in the flux spectrum are the familiar "ice" and "silicate" bands. In addition, there are several noteworthy features associated with the polarization spectra.

a. The linear polarization spectrum exhibits two prominent maxima approximately coincident with the ice and silicate absorptions (Capps, 1976; Capps et al., 1978; Kobayashi et al., 1980). In the vicinity of those bands the polarization tends to be correlated with the absorption optical thickness within the band. Note also that the polarization rises again toward 20 μm (Knacke and Capps, 1979).

b. In the continuum, the linear polarization generally declines with increasing wavelength according to

$$P \propto \lambda^{-2.4}.$$

3. GENERAL PROPERTIES OF THE ENTIRE SAMPLE OF MOLECULAR CLOUD SOURCES

Linear polarization data at 2.2 μm exist for approximately 40 sources other than BN and GL 2591. Circular polarization measurements exist for approximately 25% of these. Statistical correlations between the polarization of these sources and other observable characteristics may improve our understanding of the polarization mechanism, if it is the same for most of the sources. The sample is strongly biased because polarization measurements are possible only for the brightest among the known observed molecular cloud/H II region infrared sources, whose discovery itself is subject to strong selection effects. However all known cool stars, as evidenced by CO first-overtone bands at 2.3 μm, have been excluded, and it is likely that the majority of the remaining sources are indeed very young objects.

a. Polarization Versus Optical Depth. If the polarization arises by the aligned grain mechanism, then it is possible that it may be statistically correlated with extinction in the foreground material containing the aligned grains. No clear dependence of polarization upon optical depth in either the ice or silicate bands has been demonstrated (Lonsdale, 1980), though there is a tendency for the very highly polarized sources to have deep ice bands (Joyce and Simon, 1979). This result is not surprising since such a correlation could be easily masked by variations in degree and orientation of alignment with respect to the line of sight in the different sources: indeed, correlations between optical interstellar polarization and extinction are seen only in particular clusters or clouds (Hall and Serkowski, 1963) and not amongst stars observed in different directions in the galaxy. In the molecular clouds extinction estimates are made unreliable by the possibility of residual silicate emission and the non-correlation of ice and silicate optical depths (Merrill et al., 1976).

b. The Relationship Between Interstellar and Molecular Cloud Polarization. We pointed out above that the most likely interpretation for the BN polarization data is that light from the source passes through a medium of aligned grains. Dyck and Beichman (1974) and Dennison (1977) have argued that the alignment mechanism is the galactic magnetic field compressed to a strength of several milligauss during the evolution of the Orion Molecular Cloud. Kobayashi et al. (1978) and Dyck and Lonsdale (1979) have pointed out that if one accepts this interpretation as generally valid for molecular cloud sources then one might expect a correlation between the field directions inside the cloud and in the vicinity of the cloud. The best way to look for such a relationship is to compare the position angle of polarization for a particular cloud source with the position angles of visual interstellar polarization delineated by OB stars near the cloud. From a study of 28 sources Dyck and Lonsdale (1979) showed that the infrared position angles for 70% of the sample lay within 30° of the mean interstellar polarization direction in the immediate vicinity of those sources. Since that study three additional sources have been observed (NGC 7538/IRS 4, L 1630 #41 and GL 961) which also have interstellar polari-

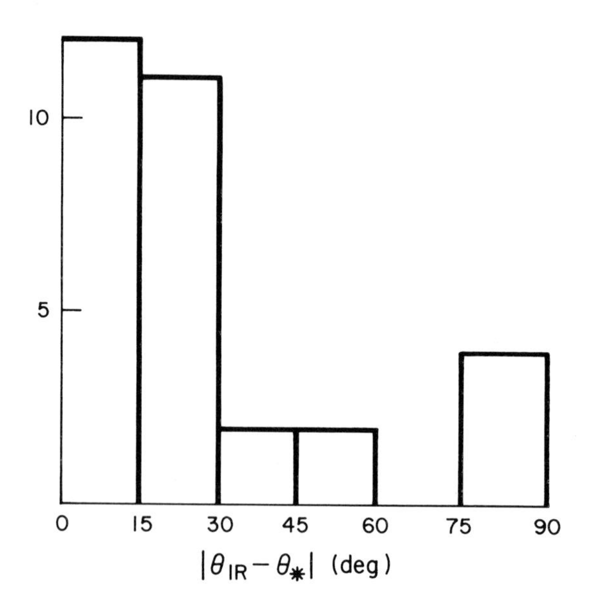

Figure 4 - A plot of the distribution of differences between infrared and interstellar polarization position angles.

zation data available, bringing the total to 31. A plot of the absolute value of the difference between the mean interstellar and the infrared polarization directions is shown in Figure 4. The position angles for the three new sources lie within 30° of the mean interstellar direction bringing the fraction to 75% of the 31 which show close agreement between the two directions. There is a strong correlation between the two sets of position angles, and therefore, justification for seeking a common mechanism for interstellar and molecular cloud polarization. To the extent that one believes that the galactic magnetic field is responsible for aligning grains which produce interstellar polarization then one must believe that it is responsible for grain alignment in molecular clouds. We conclude that this evidence demonstrates that the galactic magnetic field plays an important role in molecular cloud evolution, remaining trapped for a significant fraction of the lifetime of the cloud.

c. The Circular Polarization. Circular polarization has been searched for at 2.2 μm in nine of the molecular cloud sources including BN and GL 2591, and detected in five. It has been interpreted in terms of the aligned grain model (Lonsdale et al., 1980). As noted above, circular polarization has not been considered in the scattering model. In Figure 5 we have plotted all the available circular polarization data against linear polarization for the seven molecular cloud sources, and also for GL 2688, an object known to be a bi-polar nebula. Within

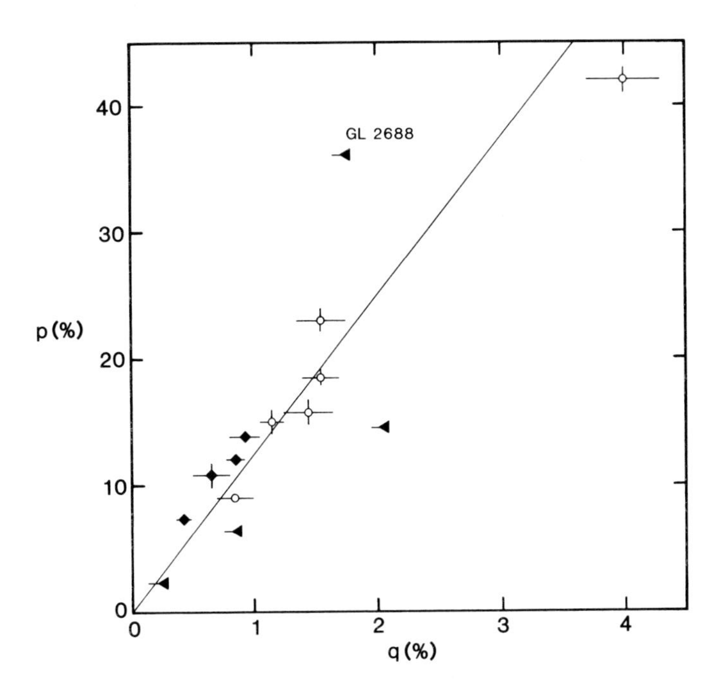

Figure 5 - Infrared circular (q) versus linear (p) polarization for a number of molecular cloud sources and for the bi-polar nebula GL 2688. The open circles are data for BN between 1.6 and 3.5 μm. The data were taken from Serkowski and Rieke (1973), Johnson (1979), Lonsdale et al. (1980), and from unpublished observations of GL 2688 made at Mauna Kea. The solid line, $q = 0.08p$, is shown for reference only. The triangles are 3σ upper limits to q.

the uncertainties of the measurements, GL 2688 shows a significantly lower ellipticity than the molecular cloud sources.

Aligned grain models which account for the circular polarization predict rotations or twists in the direction of grain alignment along the line-of-sight to the sources of the order of 20-80° (Lonsdale et al., 1980; Martin, 1980, private communication).

4. DISCUSSION

From the foregoing discussion the following conclusions arise:

I. The most plausible explanation for the polarization of BN is that it arises by transfer of radiation through a medium of aligned grains within OMC 1.

II. Since the position angles of polarization of the majority of the remaining sources in the sample correlate with those of the neighboring interstellar polarization, it is likely that the polarization mechanism is the same for these sources as for BN.

If the aligned grain model is adopted as correct, then some interesting consequences result.

a. Martin (1975), Capps (1976), and Capps and Knacke (1976) have inferred that the behavior of the ratio of polarization to optical depth within the 10 μm silicate band indicates that the interstellar silicates have a lower bandstrength than terrestrial crystalline silicates. This may indicate that they are amorphous.

b. The aligned grain models for BN and GL 2591 are crude and non-unique because of observational and computational ambiguities. For Rayleigh particles composed of pure ices and silicates the computed polarization spectra do not match the observations exactly. In order to improve the match the models may require either significantly larger particles (i.e., ones with a significantly larger scattering cross-section) or a change of composition (Dyck and Lonsdale, 1980; Lonsdale et al., 1980).

c. Gull et al. (1980), in their search for far infrared polarization, expected to observe polarization in emission from the grains which polarize by absorption at the shorter wavelengths. Such polarization would have a position angle 90° from that of the absorption polarization. The detection of polarization at 75-115 μm with a position angle similar to that of the shorter wavelength data would indicate that the polarizing region is still optically thick at these wavelengths.

d. The galactic magnetic field must have a strong influence inside the molecular clouds. Also the grain alignment mechanism must operate successfully within the clouds, and rotations or twists may exist in the direction of alignment through the cloud. Such rotations are not inconsistent with the interstellar field-infrared source position angle correlation: models would predict the two to agree within one half of the rotation angle (Martin, 1974), i.e., of order 10-40°, if there are no significant variations in grain composition along the line of sight in the cloud.

These conclusions leave us with several questions to answer. How is grain alignment effected within the dense molecular clouds? How strong do the magnetic fields have to be to produce the necessary alignment? What causes the twists in the grain alignment? To aid in answering these questions, and of independent great interest to theorists, would be the knowledge of where in the cloud does the polarization arise: do the variations in polarization between individual members of a complex arise because the objects are situated at different depths in a fairly uniform region of aligned grains throughout the cloud; or are the polarizing regions associated more closely with individual sources? Careful polarimetric work is required to answer these questions from an observational standpoint.

REFERENCES

Breger, M., and Hardorp, J.: 1973, *Astrophys. J. Letters* 183, pp. L77-L79.

Capps, R. W.: 1976, Ph.D. Thesis, Univ. of Arizona.

Capps, R. W., and Knacke, R. F.: 1976, *Astrophys. J.* 210, pp. 76-84.

Capps, R. W., Gillett, F. C., and Knacke, R. F.: 1978, *Astrophys. J.* 226, pp. 863-868.

Dennison, B.: 1977, *Astrophys. J.* 215, pp. 529-532.

Dyck, H. M., and Beichman, C. A.: 1974, *Astrophys. J.* 194, pp. 57-64.

Dyck, H. M., and Capps, R. W.: 1978, *Astrophys. J. Letters* 220, pp. L49-L51.

Dyck, H. M., and Lonsdale, C. J.: 1979, *Astron. J.* 84, pp. 1339-1348.

Dyck, H. M., and Lonsdale, C. J.: 1980, *Astron. J.* (to be published).

Elsässer, H., and Staude, H. J.: 1978, *Astron. Astrophys.* 70, pp. L3-L6.

Foy, R., Chelli, A., Sibille, F., and Lena, P.: 1979, *Astron. Astrophys.* 79, pp. L5-L8.

Gull, G. E., Russell, R. W. Melnick, G., and Harwit, M.: 1980, *Astron. J.*, submitted.

Hall, D. N. B., Kleinmann, S. G., Ridgway, S. T., and Gillett, F. C.: 1978, *Astrophys. J. Letters* 223, pp. L47-L50.

Hall, J. S.: 1958, *Publ. U.S. Naval Obs.* XVII, Part VI.

Hall, J. S., and Serkowski, K.: 1963, *Stars and Stellar Systems*, Vol. III, pp. 293-319.

Johnson, P. E.: 1979, Ph.D. Thesis, Univ. of Washington.

Joyce, R. R., and Simon, T.: 1979, *Bull. Am. Astron. Soc.*, No. 11, p. 644.

Knacke, R. F., and Capps, R. W.: 1979, *Astron. J.* 84, pp. 1705-1708.

Kobayashi, Y., Kawara, K., Maihara, T., Okuda, H., and Sato, S.: 1978, *Publ. Astron. Soc. Japan* 30, pp. 377-383.

Kobayashi, Y., Kawara, K., Sato, S., and Okuda, H.: 1980, *Publ. Astron. Soc. Japan* (to be published).

Loer, S. J., Allen, D. A., and Dyck, H. M.: 1973, *Astrophys. J. Letters* 183, pp. L97-L98.

Lonsdale, C. J.: 1980, Ph.D. Thesis, Univ. of Edinburgh.

Lonsdale, C. J., Dyck, H. M., Capps, R. W., and Wolstencroft, R. D.: 1980, *Astrophys. J. Letters* 238, pp. L31-L34.

Martin, P. G.: 1974, *Astrophys. J.* 187, pp. 461-472.

Martin, P. G.: 1975, *Astrophys. J.* 202, pp. 393-399.

Merrill, K. M., Russell, R. W., and Soifer, B. T.: 1976, *Astrophys. J.* 207, pp. 763-769.

Oishi, M., Maihara, T., Noguchi, K., Okuda, H., and Sato, S.: 1976, *Publ. Astron. Soc. Japan* 28, pp. 175-176.

Serkowski, K., and Rieke, G. H.: 1973, *Astrophys. J. Letters* 183, pp. L103-L104.

White, R. L.: 1979, *Astrophys. J.* 230, pp. 116-126.

Wilking, B. A., Lebofsky, M. J., Martin, P. G. Rieke, G. H., and Kemp, J. C.: 1980, *Astrophys. J.* 235, pp. 905-910.

DISCUSSION FOLLOWING PAPER DELIVERED BY H. M. DYCK

T. L. WILSON: Do you have any specific mechanism for aligning the grains by which you can make an estimate of the size of the magnetic field?

DYCK: I do not believe that this problem can yet be properly answered; naive calculations using the Davis-Greenstein mechanism give unreasonably large magnetic fields, of order 10^{-3} Gauss for the Orion cloud, which I am reluctant to believe.

T. L. WILSON: In any case, the Davis-Greenstein mechanism could not work in the Orion cloud because observations show that the gas and dust have the same temperature.

ALLAMANDOLA: I spoke with Ted Simon last week at JILA and he has data which show some evidence for a correlation between polarization and strong absorption at 3.1 μm, i.e., the ice band. We believe that the ice band is an indication of the presence of a mantle, although mantles do not have to have a prominent ice band. In our experiments in Leiden we are irradiating these complex molecular mixtures and find that radicals—that is, molecular sub-units which have unpaired electrons—are formed and stored. Having unpaired electrons in grains will alter the magnetic properties significantly increasing the magnetic susceptibility. Alignment via a Davis-Greenstein mechanism should then be possible without requiring unreasonably strong magnetic fields.

OKUDA: As for the linear polarization of the BN object at 3 μm band, Kobayashi has observed a spectral dependence of the polarization and found a small shift in the peak to longer wavelengths compared with the peak of the extinction. The shift can be explained well by theoretical calculation, adopting optical properties of water ice.

ELSASSER: In former publications a relation between the compactness of infrared sources and polarization was found. Is there anything new about this phenomenon which would be an essential hint to the mechanism?

DYCK: I would refer to this effect as a tendency rather than a strong correlation. Sources with a diameter smaller than about 10^{17} cm tend to be highly polarized, but there are notable exceptions such as IRS 1 in Mon R2 which is more than 20% polarized, but which is apparently extended.

T. JONES: Do you attach any significance to the $\lambda^{-2.4}$ power law in the BN source?

DYCK: Simply that the exponent in the power law is different in the BN source and in GL 2591.

BECKLIN: You used the fact that in Orion there is no radial symmetry around BN in the polarization vectors to argue against the reflection nebula hypothesis. However, the Orion cloud contains several sources of luminosity. Could it not be that each of these has an anisotropic dust distribution around it, but that there is some large-scale factor in the region, perhaps angular momentum, which aligns the asymmetries of the different objects?

DYCK: I agree that is possible, but such a mechanism would not lead to the relationship we find between the directions of polarization in infrared sources and in nearby stars. Additionally, we find that in general the direction of infrared polarization does not correlate with Galactic rotation in any easy way.

ELSASSER: The model of Staude and Elsässer was constructed because we believe, together with other authors, that it is difficult to understand the high degree of polarization found in different sources by magnetic alignment of elongated particles (Davis-Greenstein mechanism). Our model offers several tests and the new observations mentioned by Dr. Dyck show that it cannot be applied to the BN source in the way we tried. On the other hand, new results of Schulz and Lenzen of our institute on highly obscured M17 stars with polarization degrees up to nearly 30% pronounce again the difficulties of the Davis-Greenstein mechanism.

HARWIT: We have measured the 100 μm polarization of the Kleinmann-Low Nebula and find it to be $\lesssim$ 1%. This is puzzling. We had hoped to see linear polarization in emission and to find it oriented perpendicular to the direction seen in absorption in the near infrared. It suggests that the interior of the nebula might lack conditions conducive to alignment; perhaps either the alignment directions are jumbled, or there is thermal equilibrium between the gas and dust.

BECKWITH: Have you looked for correlations between overall rotation of the molecular clouds and the direction of the polarization? If, as you have suggested, the magnetic field causes grain alignment, then the correlation between the interstellar visual polarization and the infrared polarization suggests the field energy dominates the internal energy in the cloud, and you can obtain a lower limit on the field energy.

DYCK: In most cases there are not enough radio observations to establish the rotation direction of the molecular clouds associated with these infrared sources. It would be extremely valuable if radioastronomers could provide us with these data.

ZUCKERMAN: Don't hold your breath waiting for radioastronomers to solve that problem.

T. L. WILSON: The rotation of molecular clouds is difficult to measure. In Orion, there are 2 cold clouds ($T_k \simeq 70$ K) near BN/KL. To the measurement accuracy, these clouds are not rotating, and have apparently different shapes. Hence the similarity of the polarization over the Orion Molecular Cloud seems not to be simply connected to angular momentum in the clouds.

KRISCIUNAS: Your histogram of the difference of position angles of polarization of sources versus the general field polarization [Figure 4] implies a variation of grain shapes from source to source. Do you feel the number of sources you observed is large enough to validly conclude this, and, if so, are you trying to model _particular_ grain shapes in these sources?

DYCK: I think that to draw conclusions about differences in grain shape from source to source we need to draw on other kinds of data than those in Figure 4.

EMISSION LINE OBSERVATIONS OF H II REGIONS

J. H. Lacy,
Department of Physics, California Institute of Technology,
Pasadena, California 91125

I. INTRODUCTION

Optical observations provide a wealth of information on atomic abundances, the excitation (level of ionization), the kinetic temperature, and the density of the gas in H II regions. However, obscuration by dust limits optical observations of galactic H II regions to those which are nearby (within a few kpc) and which are unobscured by the molecular clouds out of which they are formed (and so generally evolved). Radio observations are unhampered by extinction but provide much more limited information. The abundance of H^+, He^+ and occasionally C^+ can be determined as well as the kinetic temperature of the gas, but no direct information on the abundance of other species or the excitation is available.

Infrared observations are in several respects intermediate to optical and radio measurements. Extinction is typically small enough not to prevent observations completely, while large enough to have an important effect on measured fluxes. The infrared spectrum contains lines of several abundant ions, so that some information can be obtained about atomic abundances and relative ionic abundances, but only a few of these lines lie in portions of the spectrum which can be observed from the ground.

Because of extinction, infrared and radio observations are most important in studies of young, compact H II regions and of variations in abundances in the Galaxy. These types of studies will be emphasized in this review.

II. OBSERVATIONS

The various infrared emission lines and the sources in which they have been observed are listed in Table 1. The first five rows of this table give the species, wavelength, wave number, ionization potential range and deexcitation density for each line.

C. G. Wynn-Williams and D. P. Cruikshank (eds.), Infrared Astronomy, 237–245.

Table 1

Infrared Emission Line Observations of H II Regions

	He I	Bγ	Bα	Pfα	Ar II	Ar III	S IV	Ne II	S III	O III	N III	O I	O III	C II
λ(μm)	2.06	2.17	4.05	7.46	6.98	8.99	10.51	12.81	18.7	51.8	57.3	63.2	88.4	156.3
ν(cm^{-1})	4850	4616.5	2467.7	1340.5	1431.6	1112.22	951.43	780.42	534.39	192.99	174.39	158.	113.18	64.0
IP(eV)	24.6-54.4	13.6-	13.6-	13.6-	15.8-27.6	27.6-40.7	34.8-47.3	21.6-40.9	23.4-34.8	35.1-54.9	29.6-54.9	0-13.6	35.1-54.9	11.3-24.4
n_c(cm^{-3})					$6.0x10^5$	$3x10^5$	$7.1x10^4$	$3.6x10^5$	$3x10^4$	3000	800	104	300	30
W3			23		16,17	17,18,48	17,18,48	17,18,48	17,30				10,43	
M42						24	24	24	5,26,30	27,30		28,43	5,10,43	39
N2024									26					39
MonR2			23											
G298.2-0.3			38			38	38	38						
G333.6-0.2	54	54	38			38		1,4,11,38,51	13	32			32	
N6334										32				
N6357										32			10,32	
SgrB2													10	
SgrA		33				19,20		2,3,19,20,49,52,53	26	46			10	
M8										6			10	
W33					17	17	17	17	17					
M17									26,31	31,32	31,32	43	10,31,32,43,45	
G29.9-0.0					17	17,41		17,41	17					
W49									6	6			6,43	
G45.1+0.1			23		17	15,17	15,17	15,17						
W51	44	42,44	42						26	32	32		10,32,43	
S88		36	36			36	36	36						
K3-50	44	42,44	25,40,42			25	25	25						
G75.8+0.4			23,35		17	17,37	17,37	17,37	17					
S106			23					14,23						
DR21			40					40				43		
S156			23											
N7538		42	42		17	17	17	17,47	6,17	6			6,43	

Note: for references see end of paper

The remaining rows give references to observations of various H II regions.

III. PROBES OF NEBULAR CONDITIONS

A. Density

Infrared and radio observations provide means of determining both the rms density and the clumpiness of the ionized gas. The rms density is given by $n_{rms} = (EM/\ell)^{1/2}$ where the emission measure, EM = $\int n_e^2 d\ell$, can be determined from radio or infrared observations of the H II emission lines or free-free emission.

The density in the regions which dominate the line emission, or the clump density, can be determined from the ratio of two lines of a multiplet where one transition is more easily thermalized than the other. If $n_{c1} < n_e < n_{c2}$, transition 1 will be nearly thermalized with $I \underset{\sim}{\alpha} \int n_{ion}\, d\ell$, whereas transition 2 will depend on n_e and $I_2 \underset{\sim}{\alpha} \int n_{ion}\, n_e\, d\ell$. n_{c1} and n_{c2} are the densities at which the transitions are thermalized. Their ratio then measures the density in the regions dominating the line emission,

$$n_{clump} = \frac{\int n_{ion}\, n_e\, d\ell}{\int n_{ion}\, d\ell} \underset{\sim}{\alpha} \frac{I_2}{I_1}$$

The lines of S III (18,33 μm) and O III (52,88 μm) have been used to measure the density in the line emitting regions. Both of these line pairs have been observed by Moorwood et al. (1980a) and O III (52,88 μm) has been observed by Baluteau et al. (1980), Dain et al. (1978) and Melnick et al. (1978). The O III line ratio is most sensitive to densities $n_e = 10^3 - 10^4\ cm^{-3}$ and S III is most sensitive to $n_e = 10^4 - 10^5\ cm^{-3}$. At low densities ($n_e < n_{c1}, n_{c2}$) the ratio depends only on the ratio of collision rates and at high densities ($n_e > n_{c1}, n_{c2}$) on the ratio of radiative decay rates. As Moorwood et al. point out, a lower value of n_e is derived from O III than S III because lower density regions dominate the O III emission. If, as in the case of O III in M17, both lines of a pair are deexcited at n_{rms}, the derived density can be smaller than n_{rms}. If only one line is deexcited, $n_{clump} > n_{rms}$. Moorwood et al. find that a three-component model can explain the observed line ratios in M17.

A mean density can be derived from lines such as O III (88 μm) and N III (57 μm) if n_e is large enough to completely thermalize the observed transition and if an assumption is made about the ionic abundance.

B. Excitation

The excitation, or level of ionization, of a region is best

determined by comparison of lines of several ionization states of an element. O III (52 or 88 μm) and O IV (26 μm), S III (18 or 33 μm) and S IV (10.5 μm), and Ar II (7.0 μm) and Ar III (9.0 μm) can be used for this purpose. Herter *et al*. (1980b) have observed S III, S IV, Ar II, and Ar III in 6 HII regions, but use their fluxes primarily to study the elemental abundances (see below).

Because the lines of Ne II (12.8 μm), Ar III (9.0 μm), and S IV (10.5 μm) are observable from the ground, the ratios of their fluxes have been used to study the excitation of H II regions, assuming that the relative atomic abundances are solar. Lacy *et al*. (1979) use these three lines to show that the excitation of Sgr A is remarkably low. Similarly, Rank *et al*. (1978) derive excitation temperatures for G333.6 - 0.2 and G298.2 - 0.3, and Lacasse *et al*. (1980) discuss the excitation of W3.

C. Elemental Abundances

In a few cases, the observed ions are likely to represent a large fraction of their elemental abundances. S^{++} and S^{+3} are excited by photons of 23.4 - 47.3 eV, Ar^{+} and Ar^{++} by 15.8 - 40.7 eV and Ne^{+} by 21.6 - 40.9 eV. Reasonably accurate elemental abundances should be obtainable for T_{*} = 30 - 40,000K for S, 25 - 35,000K for Ar, and 30-35,000K for Ne.

Herter *et al*. (1980b) measure abundances of these three elements in six H II regions and find evidence of overabundance in G29.9 - 0.0, possible underabundance in W33, and near-solar abundances in the other sources. Lester (1979) and Lester *et al*. (1979a,b) find a possible overabundance of Ar in W3 and near-solar abundances elsewhere. They did not observe Ar^{+} or S^{++} in G29.9 - 0.0, but found Ne^{+} and Ar^{++} to have nearly solar abundances.

D. Motion and Distribution

If an emission line can be resolved, its lineshape can be used to study the dynamics of the ionized gas. The distribution in velocity and position of the line emission has been studied most extensively in Sgr A West by Lacy *et al*. (1979 and 1980a). They use the Ne II line to measure the mass distribution in the central parsec of the Galaxy. Beck *et al*. (1978 and 1979) measure the rotation curves in the central regions of M82 and NGC 253 with the Ne II line.

Lacasse *et al*. (1980) study the spatial distribution of fine-structure line radiation from W3 and find a distribution quite similar to the text-book description of an H II region; S^{+3} is concentrated toward an outer (He I, H II) shell. In contrast, Lacy *et al*. (1980b) find a more complicated, clumpy structure in many H II regions, with each clump containing all three ions.

E. Uncertainties

Although the extinction to galactic H II regions is not usually large enough to prevent infrared measurements, it can significantly affect the measured fluxes. Unfortunately, the infrared extinction curve is not well known, especially beyond 5 μm. Hydrogen recombination lines can be used to measure the extinction, but are available at only a few wavelengths. Lester and Rank (1980) have determined the 4 μm extinction to a number of H II regions, from the ratio of the Bα (4.0 μm) line flux to the radio free-free flux. Bγ (2.2 μm) and Pf α(7.5 μm) have also been observed and can be used in the same way. H 6α (12.3 μm), 7α (18 μm), and 7β (11.5 μm) are all quite weak, but would provide useful probes of the extinction if measured. The 10 μm silicate absorption has been determined by fitting the absorption of continuum radiation with an opacity law which fits the emission from the hot dust in the trapezium region. The underlying emission is assumed to be either a black body (model I) or a trapezium emission spectrum (model II) (Gillett *et al.*, 1975). Unfortunately, we have no way of knowing the spectrum of the underlying emission. As a result, 10 μm extinction estimates are uncertain by a factor ~ 2. Lester and Rank (1980) show that $\tau_{4\ \mu m} \propto \tau_{10\ \mu m}$, but the constant of proportionality depends on the underlying spectrum assumed to derive $\tau_{10\ \mu m}$. Uncertainties and simplifications in the models for nebulae also introduce substantial uncertainties in the derived nebular parameters. Absorption of ionizing radiation by dust, density inhomogeneities and charge exchange reactions can affect emission line fluxes and are not often included in models. The maps of Lacy *et al.* (1980b) demonstrate that many nebulae are quite clumpy. The effects of dust are discussed below.

IV. ABUNDANCE AND IONIZATION GRADIENTS

Gradients of the ionized helium abundance, y^+, and the electron temperature, T_e, in the Galaxy have been studied extensively at radio wavelengths. Churchwell *et al.* (1974) first noted a decrease in y^+ toward the galactic center and a correlation of the decrease of y^+ below cosmic y with the infrared excess over that attributable to absorption of Lyα radiation by dust. They conclude that both effects could be explained by selective absorption of helium-ionizing photons by dust. Churchwell and Walmsley (1975) found evidence for a decrease in T_e toward the galactic center which they attribute to increased cooling resulting from increased heavy elements in the ionized gas. Several subsequent studies (e.g. Mezger *et al.*, 1980 and Wilson *et al.*, 1979) confirm the gradients in both T_e and y^+, although individual objects show a large scatter about the mean trend.

Panagia and Smith (1978) show that the correlation of y^+ with the infrared excess can be explained by absorption by dust with a cross section ratio $\sigma(\lambda<504) / \sigma(504<\lambda<912) = \sigma_{He} / \sigma_H = 4 \pm 1$. However,

Panagia (1980) concludes that the y^+ gradient can be explained by a gradient in the typical temperatures of the ionizing stars resulting from the effects of heavy elements in the stars. He also finds that the lowered stellar temperatures can explain the observed infrared excesses and in fact provide a better fit to the data than does selective absorption of helium ionizing photons.

Infrared emission line observations should be ideally suited to resolving these questions about abundances and excitation of distant H II regions. Herter et al. (1980b) have observed emission lines of He II, Ar II, Ar III, S III, and S IV from six galactic H II regions. They find an overabundance of all three elements in G29.9 - 0.0 at 5 kpc from the galactic center, as expected if an abundance gradient exists. However, W33 at 6 kpc appears to be underabundant. The primary uncertainties in this study are in the extinction correction, beam size corrections in comparisons of infrared and radio measurements, and the small size of the sample studied. As is discussed above, the extinction correction is likely to continue to be a serious problem.

No infrared survey looking for excitation gradients has yet been made. Sgr A West is ionized by radiation characterized by $T_{eff} \lesssim 31{,}000K$ (Lacy et al., 1980a), in agreement with the trend predicted by radio observations of y^+, but Sgr A may be a special case with some exotic source of ionizing radiation.

Lester (1979) and Lacasse et al. (1980) discuss the effect of dust on the relative abundances of the ions of Ne, S, and Ar. Lester uses [S IV] / [Ne II] as a measure of the excitation of the nebulae. He finds a correlation between [S IV] / [Ne II] and the free-free luminosities of H II regions, indicating that much of the variation in [S IV] / [Ne II] is due to the masses of the exciting stars. He also finds that model H II regions without dust fit the observed luminosity-excitation dependence better than do models with a λ^{-1} dust opacity in the ionizing ultraviolet.

Lacasse et al. (1980) study the fine-structure line emission from W3A. They conclude that the excitation of the nebula is lower than expected for a main-sequence star which produces the required ionizing flux. By including dust with $\sigma_{He} / \sigma_H = 5$, they were able to fit the observed ionization structure.

The differences between the conclusions of Lester and of Lacasse et al. appear to be due primarily to different interpretations of similar data. Both studies show a tendency for the required ionizing fluxes of the exciting stars of H II regions to be larger than expected for main-sequence stars of the required temperatures. This discrepancy could be due to the effects of dust, if σ_{He} / σ_H is large enough. Alternatively, it could be due to multiple ionizing stars, pre- or post- main sequence stars, or the effects of heavy element enrichment in ionizing stars. Detailed studies of the ionic distributions in nebulae combined with improved models may help resolve this question.

REFERENCES

1. Aitken, D. K. and Jones B.: 1973, Mon. Not. R. Astr. Soc. 165, p.363.
2. Aitken, D. K., Griffiths, S., and Jones, B.: 1976a, Mon. Not. R. Astr. Soc. 176, p.73P.
3. Aitken, D. K. Griffiths, J., Jones, B., and Penman, J. M.: 1976b, Mon. Not. R. Astr. Soc. 174, p.41P.
4. Aitken, D. K., Griffiths, J., and Jones, B.: 1977, Mon. Not. R. Astr. Soc. 179, p.179.
5. Baluteau, J.-P., Bussoletti, E., Anderegg, M. Moorwood, A. F. M., and Coron, N.: 1976, Ap. J. 210, p.L45.
6. Baluteau, J.-P., Moorwood, A. F. M., Biraud, Y., Coron, N., Anderegg, M., and Fitton, B.: 1980, preprint.
7. Beck, S. C. Lacy, J. H., Baas, F., and Townes, C. H.: 1978, Ap. J. 226, p.545.
   Beck, S. C. Lacy, J. H., and Geballe, R. R.: 1979, Ap. J. 231, p.28.
8. Churchwell, C., Mezger, P. G., and Huchtmeyer, W.: 1974, Astr. Ap. 32, p. 283.
9. Churchwell, C. and Walmsley, C. M.: 1975, Astr. Ap. 38, p.45.
10. Dain, F. W., Gull, G. E., Melnick, G. Harwit, M., and Ward, D. B.: 1978, Ap. J. Letters 221, p.L17.
11. de Vries, J. S., Wander Wal., P. B. and Andriesse, C. D.: 1980, Astr. Ap. in press.
12. Gillett, F. C., Forrest, W. J., Merrill, K. M., Capps, R. W. and Soifer, B. T.: 1975, Ap. J. 200, p.609.
13. Greenberg, L. T., Dyal, P., and Geballe, T. R.: 1977, Ap. J. Letters 213, p.L71.
14. Hefele, H. and Hölzle, E.: 1980, preprint.
15. Hefele, H. and Schulte in den Bäuman, J.: 1978, Astr. Ap. 66, p.465.
16. Herter, T., Pipher, J. L., Helfer, H. L., Willner, S. P., Puetter, R. C., Rudy, R. J., and Soifer, B. T.: 1980a, preprint.
17. Herter, T., Helfer, H. L., Pipher, J. L., Forrest, W. J., McCarthy, J., Houck, J. R., Willner, S. P., Puetter, R. C., Rudy, R. J., Soifer, B. T., and Gillett F. C.: 1980b, preprint.
18. Lacasse, M. G., Herter, T., Krassner, J., Helfer, H. C., and Pipher, J. L.: 1980, preprint.
19. Lacy, J. H., Baas, F., Townes, C. H., and Geballe, T. R.: 1979, Ap. J. Letters 227, p.L17.
20. Lacy, J. H., Townes, C. H., Geballe, T. R., and Hollenbach, D. J.: 1980a, Ap. J. in press.
21 Lacy, J. H., Beck, S. C., Townes, C. H., and Geballe, R. R.: 1980b, in preparation.
22. Lester, D. F.: 1979, Ph.D. Thesis, Univ. of Cal, Santa Cruz.
23. Lester, D. F. and Rank, D. M.: 1980, preprint.
24. Lester, D. F., Dinerstein, H. L., and Rank, D. M.: 1979a, Ap. J. 229, p.981.
25. ______,: 1979b, Ap. J. 232, p.139.
26. McCarthy, J. F., Forrest, W. J., and Houck, J. R.: 1979, Ap. J. 231, p.711.

27. Melnick, G., Gull, G. E., Harwit, M., and Ward, D. B.: 1978, Ap. J. Letters 222, p.L137.
28. Melnick, G., Gull, G. E., and Harwit, M.: 1979, Ap. J. Letters 227, p.L29.
29. Mezger, P. G., Pankonin, V. Schmid - Burgk, J. Thum, C. and Wink, J.: 1980, Astr. Ap. in press.
30. Moorwood, A. F. M., Baluteau, J.-P., Anderegg, M., Coron, N., and Biraud, Y.: 1978, Ap. J., 224, p.101.
31. Moorwood, A. F. M., Baluteau, J.-P., Anderegg, M., Coron, N., Biraud, Y., and Fitton, B.: 1980a, Ap. J. in press.
32. Moorwood, A. F. M., Salinari, P., Furniss, I., Jennings, R. E., and King, K. J.: 1980b, preprint.
33. Neugebauer, G., Becklin, E. E., Matthews, K., and Wynn - Williams, C. G.: 1978, Ap. J. 220, p.149.
34. Panagia, N. and Smith, L. F.: 1978, Astr. Ap. 62, p.277.
35. Panagia, N.: 1980, in P. A. Shaver (ed.), Radio Recombination Lines, D. Reidel Publ. Co., Dordrecht, Boston, London, p. 99.
36. Pipher, J. L., Sharpless, S., Savedoff, M. P., Krassner, J., Varlese, S., Soifer, B. T., and Zeilik, M.: 1977, Astr. Ap. 59, p.215.
37. Pipher, J. L., Soifer, B. T., and Krassner, J.: 1979, Astr. Ap. 74, p.302.
38. Rank, D. M., Dinerstein, H. L., Lester, D. F., Bregman, J. D., Aitken, D. K., and Jones, B.: 1978, Mon. Not. R. Astr. Soc. 185, p.179.
39. Russell, R. W., Melnick, G., Gull, G. E., and Harwit, M.: 1980, preprint.
40. Righini - Cohen, G., Simon, M., Young, E. T.: 1979, Ap. J., 232, p.782.
41. Soifer, B. T. and Pipher, J. L.: 1975, Ap. J. 199, p.663.
42. Soifer, B. T., Russell, R. W., and Merrill, K. M.: 1976, Ap. J. 210, p.334.
43. Storey, J. W. V., Watson, D. M., and Townes, C. H.: 1979, Ap. J. 233, p.109.
44. Thompson, R. I. and Tokunaga, A. T.: 1980, Ap. J. in press.
45. Ward, D. B., Dennison, B., Gull, G. E., and Harwit, M.: 1975, Ap. J. 202, p.L31.
46. Watson, D. M. Storey, J. W. V., Townes, C. H., and Haller, E. E.: 1980, preprint.
47. Willner, S. P.: 1976, Ap. J. 206, p.728.
48. _______,: 1977, Ap. J. 214, p.706.
49. _______,: 1978, Ap. J. 219, p.870.
50. Wilson, T. L. Bieging, J., Wilson, W. E.: 1979, Astr. Ap. 71, p.205.
51. Wollman, E. R.: 1976, Ph.D. Thesis, Univ. of Calif., Berkeley.
52. Wollman, E. R., Geballe, T. R., Lacy, J. H., Townes, C. H. and Rank, D. M.: 1976, Ap. J. Letters 205, p.L5.
53. _______,: 1977, Ap. J. Letters 218, p.L107.
54. Wynn - Williams, C. G., Becklin, E. E., Matthews, K., and Neugebauer, G.: 1978, Mon. Not. R. Astr. Soc. 183, p.237.

DISCUSSION FOLLOWING PAPER DELIVERED BY J. H. LACY

T. L. WILSON: Mezger no longer appeals to "selective absorption" of He ionizing photons by dust. This change of attitude is due to the new stellar atmospheres of Kurucz (Astrophys. J. Suppl. 40, p.1, 1979).

LACY: Does he explain the variation in strength of the helium recombination lines within the Galaxy by these stellar models?

T. L. WILSON: Yes, this plus a variation in the relative abundance of oxygen, nitrogen, etc. The variation in the He to H ratio with distance from the Galactic Center is influenced by noise and systematic effects. These cause scatter in the trends (in addition to any intrinsic source-to-source scatter). The gradient in electron temperature is rather more securely established, however.

JOSEPH: In this context I would comment that S. A. Morris and I at Imperial College have shown that the notion of a "geometric effect" to account for variations in the He^+/H^+ radio recombination line intensity ratios is apparently inconsistent with the data itself, and that the proposed correlation between He^+/H^+ and IR excess has very little statistical significance.

T. L. WILSON: Measurements of radio recombination linewidths in W33 at Bonn (Bieging et al., Astron. Astrophys. 64, p. 341, 1979) show T_e must be less than 6000 K in this source. The claims of Herter et al. that heavy ions are underabundant in W33 are therefore surprising.

LACY: Herter et al. state that there are problems with their W33 data, so that the apparent overabundance may not be real.

THE LARGE SCALE INFRARED EMISSION IN THE GALACTIC PLANE—OBSERVATIONS

H. Okuda
Kyoto University, Kyoto, Japan

I. INTRODUCTION

For a long time, the inner region of our Galaxy has been veiled by strong interstellar extinction in visible light. The situation has been greatly improved by recent exploration in the infrared region, where the interstellar extinction becomes practically negligible.

Infrared radiation is deeply involved in a variety of matters and processes in the galaxy. Near infrared radiation is predominantly emitted by late type stars which include the major part of the mass in the Galaxy and hence govern its dynamics. Short wavelength radiation (UV and visible) emitted from early type stars is easily absorbed by dust around the stars themselves or by interstellar dust, and reemitted in middle or far infrared regions. A variety of emission lines, fine structure lines of neutral and ionized heavy elements, as well as many molecular lines are also clustered in the middle and far infrared regions.

Since their line intensities are generally very weak, and, moreover, spectroscopic observations demand relatively difficult techniques in their detection, the surveys so far done have been limited mostly to continuum emission. Here we shall try to compile them and discuss briefly their implications to the structure of our Galaxy in its inner region.

II. SURVEY OF THE INFRARED SURVEYS

During the past decade, numbers of observations were devoted to measurements of the brightness distribution in the galactic plane, mostly either in the near infrared or far infrared regions plus a few in middle infrared. They are listed in Table 1, with their observational parameters.

C. G. Wynn-Williams and D. P. Cruikshank (eds.), Infrared Astronomy, 247–260.

In the near infrared surveys made at balloon altitudes, the observable wavelength is severely restricted within a narrow band centered at 2.4 μm, a unique band gap of the OH airglow emission which was found to show patchy distribution with rapid variation in space and time, in the course of the first trial of near IR survey (Sugiyama et al. 1973). The surveys have been made extensively in the northern sky by various authors (ref. 4, 5, 6, 7, 8, 12, 13, 14). Recently they were extended to the southern sky (ref. 15).

Observations outside the OH-band gap were also tried but were fatally influenced by the irregular change in the OH emission, and the results are restricted to limited regions near the galactic center (ref. 7, 15).

Rocket observations allowed investigators to choose their observing wavelength freely and expand it from near infrared to middle and far infrared regions (ref. 2, 3, 12, 19, 22).

Most of the far infrared surveys were conducted by balloons (ref. 1, 9, 10, 11, 16, 17, 18, 21, 23, 24) and some by airplane (ref. 20). The early works were mostly single band surveys in fragments of galactic longitudes, but the current surveys have widened the longitude range (ref. 9) and developed into two-dimensional mappings (ref. 16, 21, 24). The observing wavelength was also extended to submillimeter range (ref. 17). A complete set of two color longitude profiles was just reported recently (ref. 23).

III. GENERAL FEATURES OF THE RESULTS

1) Near Infrared Distribution

The near infrared mappings in the northern sky made by various authors in different methods and resolutions show excellent agreement with one another. Combining the recent observations in the southern sky with theirs, Hayakawa et al. (1980b) provided a synthesized map of the Milky Way between ℓ = 290° and 70°, which is reproduced in Figure 1. One can easily recognize there a familiar isophoto of an edge-on galaxy; indeed our Galaxy is the nearest and the largest edge-on galaxy!

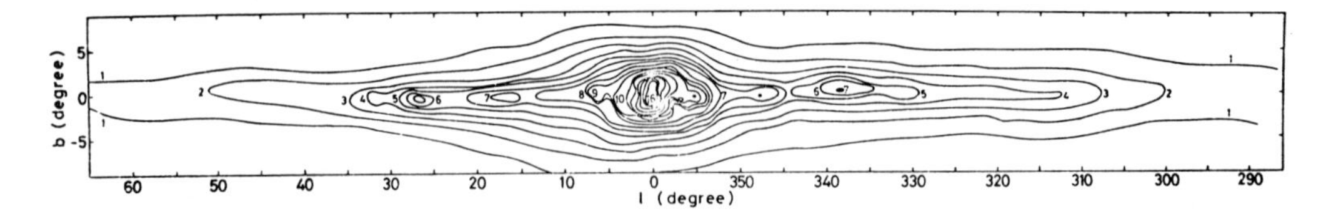

Fig. 1. The synthesized map of 2.4 μm emission in the galactic plane (Hayakawa et al. 1980b).

The central concentration, the so-called bulge, extends $\pm 15°$ in longitude and $\pm 7.5°$ in latitude, corresponding to 3 kpc and 1.5 kpc in linear size, if 10 kpc is adopted for the distance of the galactic center.

The brightness distribution in the central region of the bulge is modified by strong interstellar extinction. In fact the effect is more clearly revealed in the detailed mapping with higher resolution (Oda et al. 1979) as shown in Figure 2.

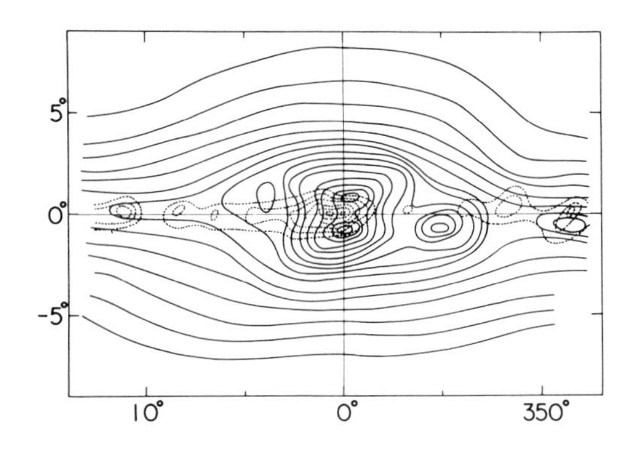

Fig. 2. The detailed map of 2.4 μm emission in the galactic central region (Oda et al. 1979), overlapped with the far infrared contours (Maihara et al. 1979).

The magnitude of the extinction is estimated to be as large as 2.7 mag for 2.4 μm at the center by assuming that the intrinsic brightness distribution follows the universal $r^{1/4}$ law (Maihara et al. 1978). About 1/3 of the extinction is considered to originate in the nuclear region (Oda et al. 1979).

Another conspicuous feature is the flat thin wing extending outward from the bulge, which is abruptly terminated near $\ell = 30°$. This behavior has been noticed commonly in various constituents of the Galaxy, such as the radio continuum, CO-molecular line, as well as the far infrared emissions; this relationship will be discussed later in more detail. In Figure 3, longitude profiles of the relevant constituents are compiled. The shoulder near $\ell = 30°$ is sometimes referred to as the 5 kpc ring (e.g. Burton 1976). In this respect, however, it is interesting to see that the distribution of the near infrared radiation is apparently asymmetrical between the northern and southern hemispheres. This may indicate that the feature is much more concerned with spiral arms rather than the ring. In fact, another shoulder at $\ell = 310°$ seems to be paired with the $\ell = 30°$ shoulder and they correspond well to the Scutum-Crux arm (Georgelin and Georgelin 1976).

A slight deviation of the brightness ridge from the galactic equator is also noticed (Hayakawa et al. 1979); the ridge shifts about 0.5° to negative latitude in the northern sky, while it lies almost on the galactic equator in the southern sky.

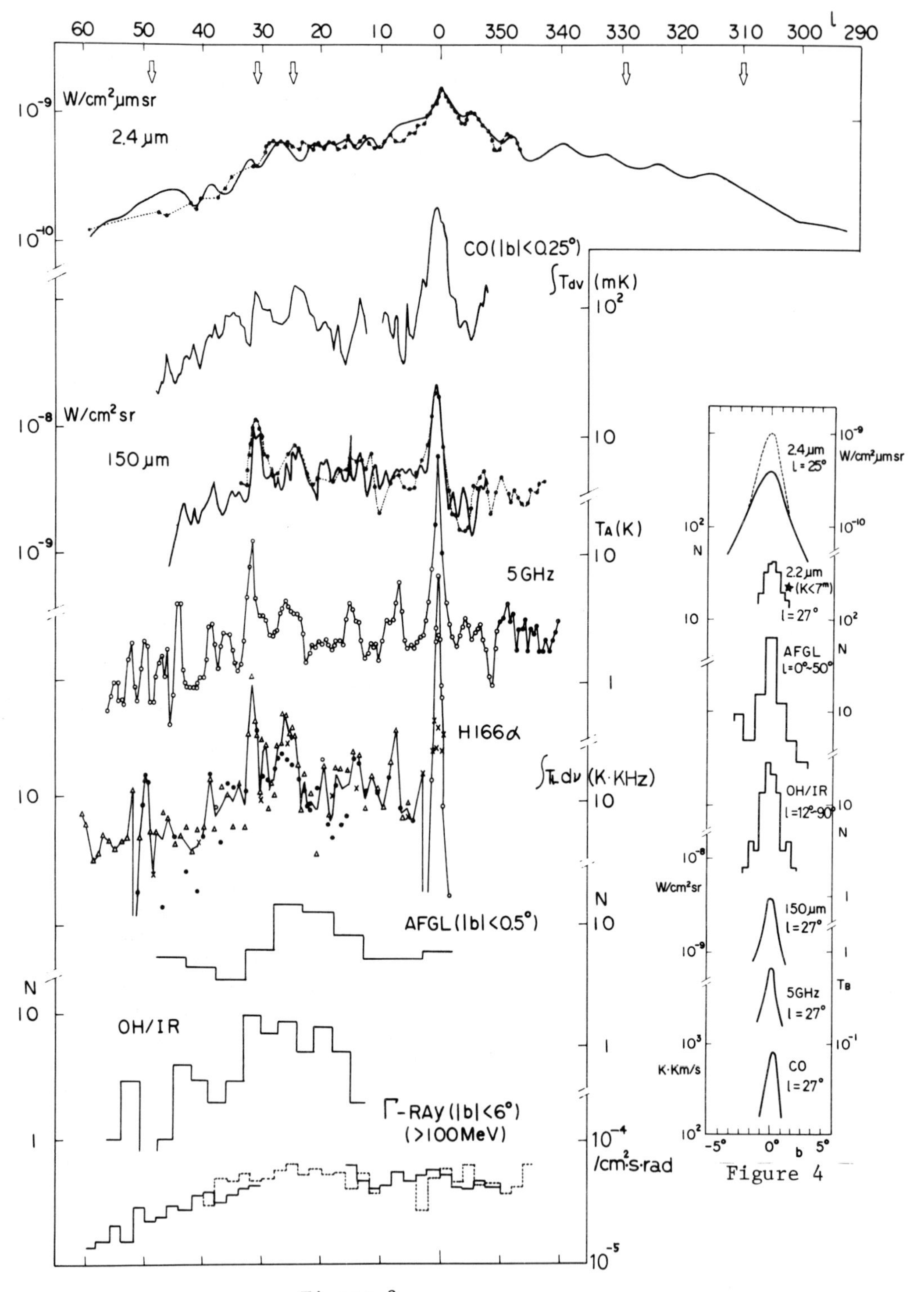

Figure 3

Figure 4

Fig. 3. Galactic longitude profiles of the 2.4 μm emission (—: Hayakawa et al. 1980b, •····• : Oda et al. 1979), CO-emission (Cohen et al. 1980 and Bania 1977), 150 μm emission •····• : Maihara et al. 1979, —: Nishimura et al. 1980), 5GHz continuum emission (Attenhoff et al. 1970), H166α line (• : Lockman 1976, ▵ : Hart and Pedlar 1976), AFGL sources (Price and Walker 1976), OH/IR sources (Bowers, 1978) and cosmic γ-rays (Bennett et al. 1977).

Fig. 4. Galactic latitude profiles corresponding to Fig. 3.

In the smaller amplitude variations in the longitude profile, one can see a clear anti-correlation with the profiles of the CO-emission and the far infrared emission. This is easily understood as an effect of the interstellar extinction, since both the CO- and the far infrared emissions are closely related to dust. The anomaly at ℓ = 355°, b=-1° was supposed to be of the same origin (Oda et al. 1979). It may be a hole in interstellar extinction through which we are looking into the deep space in the central region of the Galaxy. It is interesting to note that the area has the lowest level of the far infrared emission as is mentioned later.

The most striking feature to be pointed out in regard to the wing component is its extreme thinness. If we take into account the effect of interstellar extinction, the intrinsic width would become much thinner, 2-3° in FWHM as shown by a dotted line in Figure 4. This corresponds to a scale height of 100-140 pc even if it is located at the tangential point of the 5 kpc arm. A much smaller value of 50 pc is obtained as an extreme case (Hayakawa et al. 1980b). This is significantly smaller than the scale height of 300 pc or 400 pc for K or M-giant stars, which are presumably the most effective contributor to the near infrared emission.

In order to resolve the contributions to individual sources, a deep sky survey is now being conducted by means of a multicolor photometer (IHKL bands) attached to the groundbased telescopes with finer angular resolution (Kawara et al. 1980). The survey was made in narrow strips or bands across the galactic plane selectively sampled between ℓ = 350° and 45°. The analysis is rather preliminary as yet; however, some results are given in Figure 5. The source distribution behaves quite similarly to the brightness distribution obtained in the balloon observations, i.e., abrupt decrease in number density near ℓ = 30°, and very thin latitudinal distribution. It is particularly interesting that the latitudinal concentration is seen only in the profiles at ℓ = 26-27° and 28°. A similar concentration is found in the distributions of the AFGL sources as well as in the OH/IR sources (see Fig. 4).

Table I. Large scale IR surveys

Observers	(year)	l	b	$\lambda_{\mu m}$	f.o.v.	ref.
Hoffmann et al.	(1971)	358°-2°	±2°	100	12'	1
Houck et al.	(1971)	354°-7°	0°	100	0.25°×1°	2
Pipher	(1973)	2.5°	a profile	100	0.25°×1°	3
Hayakawa et al.	(1976)	25°-70°	±10°	2.4	3°	4
Okuda et al.	(1977)	350°-30°	±10°	2.4	1°	5
Ito et al.	(1977)	350°-33°	±10°	2.4	2°	6
Hofmann et al.	(1977)	348°-10°	±10°	2.4	2°	7
		358°-0.5°	±10°	3.4	2°	
Hofmann et al.	(1978)	352°-8°	±8°	2.45	1°	8
Rouan et al.	(1977)	28°	a profile	71-91	0.7°	9
				114-196	0.7°	
Low et al.	(1977)	348°-32°	0°	60-300	0.25°	10
				150-300	0.25°	
Serra et al.	(1978)	36°-55°	0°	71-91	0.7°	11
				114-196	0.7°	
Hayakawa et al.	(1978)	182°	-9°	2.3	4°	12
Hofmann et al.	(1978)	357°-10°	±10°	2.45	1°	13
Oda et al.	(1979)	345°-32°	±10°	2.4	0.5°	14
Hayakawa et al.	(1979)	288°-23°	±10°	2.4	0.5°,0.8°,1.7°	15
				3.4	2°	
Maihara et al.	(1979)	340°-32°	±3°	100-300	0.6°	16
Owens et al.	(1979)	345°-45°	0°	400-1000	1.6°	17
				1000-3000	1.6°	
Serra et al.	(1979)	26°-40°	0°	71-91	0.5°	18
				114-196	0.5°	
Price	(1979)	0°-30°	±5°	4	0.8°	19
		0°-320°		11	0.8°	
		0°-320°		20	0.8°	
		40°-85°		27	0.8°	
Viallefond et al.	(1980)	27.5°	a profile	114-196	6.3'	20
Nishimura et al.	(1980)	350°-45°	±2°	100-300	0.5°	21
Hayakawa et al.	(1980a)	30°-65°	0°	2.0	1.5°	22
				2.8	1.5°	
				4.8	1.5°	
Boissé et al.	(1980)	0°-85°	0°	71-95	0.4°	23
				114-196	0.4°	
Hauser et al.	(1980)	0°-63°	±3°	106	10'	24
		81°-87°		238	10'	
		129°-135°		270	10'	
		261°-285°				

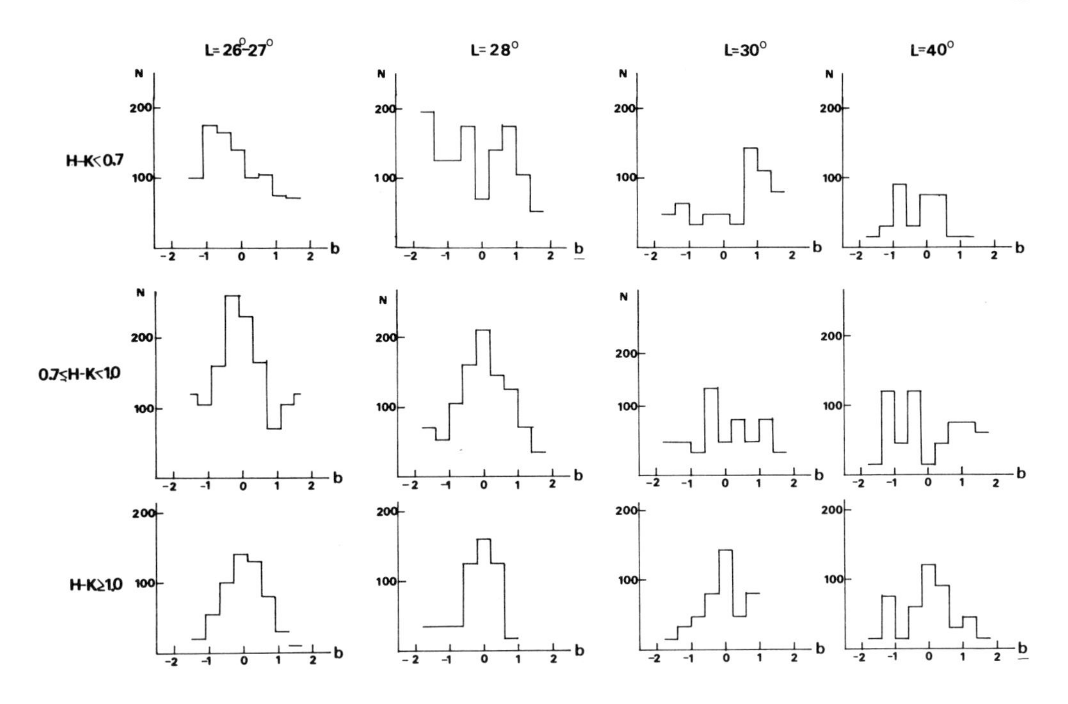

Fig. 5. The number density distribution of the infrared sources detected in the deep IR surveys (Kawara et al. 1980).

From the rocket observations the energy spectrum between 2.0-4.2 μm is found to fit a black body spectrum of 3300 K (Hayakawa et al. 1980a), but a strong enhancement was detected in the AFGL survey at 10-20 μm range (Price 1980).

2) Far Infrared Distribution

The two dimensional maps provided by Maihara et al. (1979) and by Nishimura et al. (1980) were obtained in essentially the same spectral range (100 - 300 μm) but with slightly different resolutions. They cover almost the same area in the galactic plane and show an excellent agreement with each other. The map obtained by Nishimura et al. is shown in Figure 6.

The concentration of the brightness toward the galactic center is prominent, but the distribution is more extended than previously observed in a smaller field of view by Hoffmann et al. (1971). The region is clearly silhouetted as extinction against the near infrared background (see Figure 2). The brightness peaks at Sgr B2 rather than the exact center, Sgr A, while the peak in the shorter spectral range (50-300 μm) coincides with the latter (Hoffmann et al. 1971, Low et al. 1977). This indicates that cold dust is distributed more extensively in the galactic center. In this respect, it is interesting to note that the extent of the CO clouds (Bania, 1977) is closely matched with the far infrared emission, while the thermal radio continuum is, on the other hand, concentrated strongly into the Sgr A region.

Besides the central concentration, the far IR emission is distributed along the galactic plane in a narrow band with almost constant width. As is noticed in Figure 6, the general patterns in the brightness distribution correspond strikingly well with those of the 5GHz radio map (Altenhoff et al. 1970) The relation is, however, not necessarily a priori, for the infrared emission depends on the dust abundance as well as on the heat sources. Several sources are identified with CO-cloud complexes as suggested by Nishimura et al. (1980).

Cohen et al. (1980) have made an extensive survey of CO-emission in the galactic plane and recently supplied a first complete two dimensional map in $\ell = 12°-60°$ and $|b| \leq 1°$, a part of which is reproduced in Figure 6. Similarity among the three maps is quite impressive; implying that the three components are closely correlated at least spatially and more probably genetically.

In addition to the discrete sources, there exists a background component extending diffusely along the galactic plane in finite width. It may correspond to the ELD HII region proposed by Mezger (1978). The contribution of the discrete sources is, however, dominant and their distribution is very complex; therefore separation of the extended component is not so straightforward. Actually, there is some discrepancy in the observed widths (FWHM) of Maihara et al. (1979); $b_{1/2} \sim 3°$, and Nishimura et al. (1980); $b_{1/2} \sim 1.5°$, probably due to differences in the angular resolutions.

A survey in submillimeter region was conducted by Owens et al. (1979). The observed width of $b_{1/2}=4 \sim 5°$ is considerably broader than that observed in the shorter range. Colder dust might be distributed more diffusely, but extremely large amounts of low temperature dust should be assumed to broaden the distribution to such a width, surmounting the emission of the high temperature dust. A further observation is important.

As for the dust temperature, 50 K and 18 K are estimated for the galactic center and the general diffuse components respectively (Nishimura et al. 1980). A complete set of the bi-spectral observation of the galactic plane has just been presented by Boissé et al. (1980), giving a little higher temperature, $25 \sim 30$K, for the ridge component.

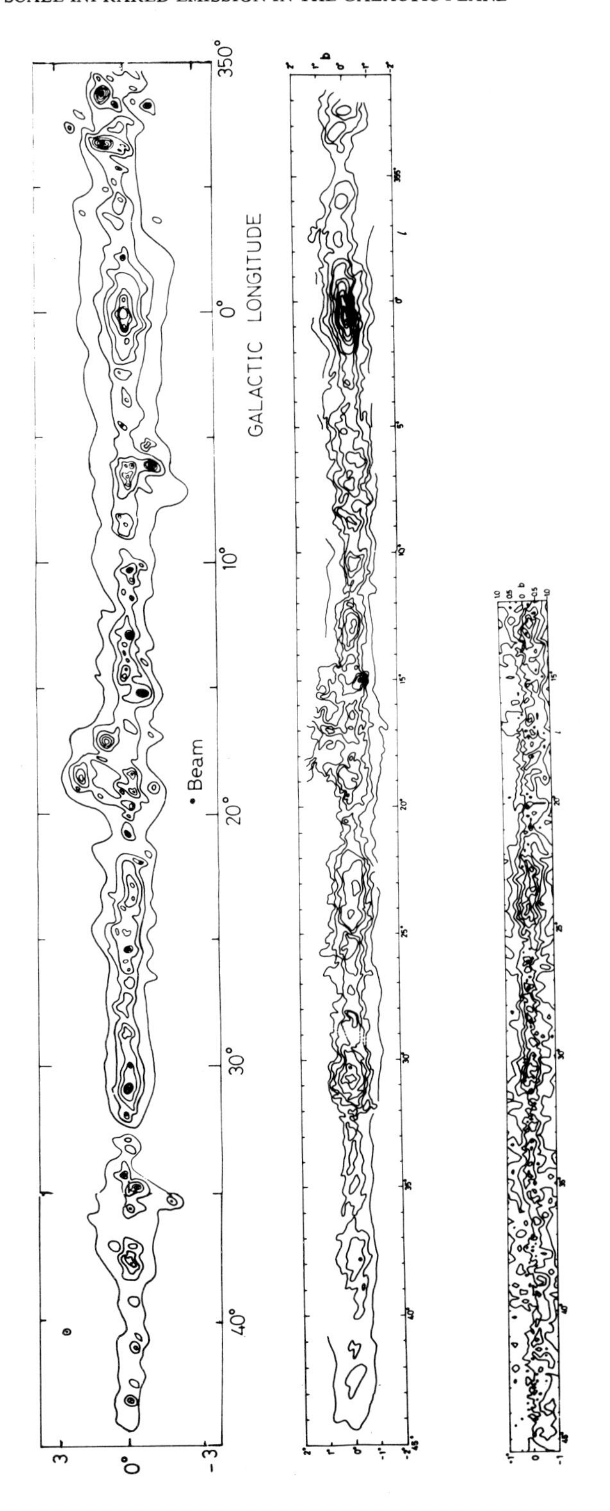

Fig. 6. Comparison of the intensity maps of the 5 GHz radio continuum (Altenhoff et al. 1970), the FIR emission (Nishimura et al. 1980), and CO emission (Cohen et al. 1980).

IV. A BRIEF SKETCH OF THE INNER GALAXY

As has been mentioned so far, the infrared data for the inner region of the Galaxy have been accumulated in considerable amount and variety. Combining these data together with those in radio and other spectral ranges, we are now able to draw a comprehensive picture of the inner structure of the Galaxy. Here we try to sketch it briefly and rather qualitatively.

1) The Galactic Central Region

So far as the central bulge is concerned, it is of medium size ($R \sim 1.5$ kpc) and normal luminosity ($3 \cdot 10^{10}$ $L_\odot$); almost compatible to M31 (Maihara et al. 1978). In contrast with M31, however, the nuclear region is still very active and abundant in gas and dust, forming a giant complex of HII regions and molecular clouds. The far infrared luminosity from the whole region ($R < 300$ pc) amounts to of the order of 10^9 $L_\odot$, from which the total mass of dust is estimated to be $10^5 M_\odot$, if the dust temperature is adopted as 40 K (Nishimura et al. 1980, Maihara et al. 1980). The dust mass thus obtained is consistent with the interstellar extinction of Av=10-15 mag, shared by the Galactic nuclear region (Oda et al. 1979). The concentration of dust in the nuclear region is also well understood from the existence of large complex of CO clouds ($M_{H2}=3 \sim 7 \cdot 10^8$ $M_\odot$, Bania, 1977). Because the far infrared emission has a broader distribution than the radio continuum, some unknown sources must exist to heat the dust; the contribution of starlight from old stars is supposed to be comparable to that from HII regions or young stars (Mezger and Pauls 1978).

2) The 5 kpc Complex

One of the most interesting discoveries in the past few years is that of the so called 5 kpc ring. The near infrared observations extended to the southern sky, however, suggest that it is more like an arm than a ring structure. In any case, it is undoubtedly an extremely active region filled with almost every constituent: young and old stars, dust, gases in neutral and ionized as well as molecular states, and possibly cosmic ray and γ-ray sources.

As for the stellar component, it is suggested from its extremely thin distribution in z-plane ($z_{1/2} \sim 140$ pc) that it consists of predominantly very young objects, the age of which is estimated to be of the order of 10^7 years (Hayakawa et al. 1980) or 10^8 years (Boissé et al. 1980).

The possible candidates for such young objects are OB stars or M-supergiants. The former possibility is, however, excluded simply because too high a stellar density would have to be assumed as compared with that expected from the radio data (see Table 2).

If it were M-supergiants, the relative stellar density required is still a little too high compared with that expected from the general population of stars in the solar vicinity (Allen 1973). The

TABLE 2. STELLAR POPULATION IN THE 5 kpc COMPLEX

	Energy Release Rates			
	Near IR (2.4 μm) $5 \sim 6 \times 10^{35}$ W/kpc^3			UV (radio continuum) $1 \sim 2 \times 10^{35}$ W/kpc^3
	M-type Giants	M-type Supergiants	O-type Supergiants	O-type Supergiants
5 kpc (observed)[a]	5×10^{-4}	3×10^{-5}	1×10^{-3}	8×10^{-6}
10 kpc (observed)[b]	10^{-5}	10^{-7}	10^{-7}	10^{-7}
5 kpc (predicted)[c]	10^{-4}	10^{-6}	10^{-6}	10^{-6}
5 kpc excess[d]	5	30	10^3	8

[a]Stellar number density (no/pc^3) at 5 kpc as deduced from energy releases.
[b]Stellar number density (no/pc^3) at 10 kpc taken from Allen (1973).
[c]Observed 10 kpc stellar density multiplied × 10, the mass ratio between 5 and 10 kpc.
[d]Ratio of observed to predicted 5 kpc stellar densities.

excess becomes moderate only when M-giants are assumed to be the main constituent. In this case, however, the z-distribution would become considerably broader ($z_{1/2} \sim 400$-500pc) in contrast with the observed value of 50-140 pc (Hayakawa et al. 1980b, Okuda et al. 1979).

In this regard, it is interesting here to remember that the AFGL sources and OH/IR sources are also confined in the galactic plane in a similar way to the near infrared sources. The AFGL sources are frequently identified as carbon or late M stars with thick dust shells (Allen et al. 1977), while the OH/IR sources are identified as Mira variables with thick circumstellar dust (Bowers 1978). Generally, these stars are believed to be at a rather advanced stage of stellar evolution, and therefore their scale heights should be relatively large, somewhat contradictory to the observed results.

These facts suggest that the stellar population in the region is somewhat different from that in the solar neighborhood. It might be rich in evolutionarily young but apparently late type stars, such as massive stars which evolved quickly with large mass loss and hence are surrounded by a thick dust shell.

On the other hand, the far infrared and radio observations have revealed the fact that the region is full of young objects, i.e. HII regions containing many OB stars, as well as dense molecular clouds in which star formation is actively in progress. This favors the argument that the near IR emission should be attributed mostly to M-supergiants, but in that case their population would become too large compared with the other stellar constituents.

The far infrared, radio continuum and CO emissions are correlated spatially but in a very complicated way. More detailed and extensive observations with higher spatial and spectral resolution are essential for fully understanding their genetic relation. Spectroscopic observations are especially important for studies of physical states or dynamics in the region, as well as for distance determinations of the sources, which are crucial to delineate the galactic structure.

As has been discussed, the inner region of our Galaxy is one of the most intriguing regions to be explored. Infrared techniques are undoubtedly becoming powerful and indispensable tools for its complete understanding together with radio techniques.

The author is very grateful to Drs. T. Nishimura, R. Cohen, J. Serra, S. D. Price, T. Matsumoto, and K. Noguchi for informing him of their unpublished data.

REFERENCES

Allen, C.W., 1973, Astrophysical Quantities, 3rd ed., 247.
Allen, D.A., Hyland, A.R., Longmore, A.J., Caswell, J.L., Goss, W.M. and Haynes, R.F., 1977, Astrophys. J. 217, 109.
Altenhoff, W.J., Downes, D., Goad, L., Maxwell, A. and Rinehart, R., 1970, Astron. Astrophys. Suppl. 1, 319.
Bania, T.M., 1977, Astrophys.J. 216,381.
Bennett, K., Bignami, G.F., Buccheri, R., Hermsen, W., Kanbach, G. Lebrun, F., Mayer-Hasselwander, H.A., Piccinotti, G., Scarsi, L., Soroka, F., Swanenburg, B.N., and Wills, R.D., Proc. of the 12th ESLAB Symposium, Frascati, ESA SP-124, 83, 1977.
Boissé, P., Gispert, R., Corn, N., Wijnbergen, J., Serra, G., Ryter, C. and Puget, J.L., 1980, preprint.
Bowers, P.F., 1978, Astron. Astrophys. 64, 307.
Burton, W.B., 1976, Ann. Rev. Astr. Astrophys. 14, 275.
Cohen, R., Dame, T., Tomasevich, G. and Thaddeus, P., 1980, private communication.
Georgelin, Y.M., and Georgelin, Y.P., 1976, Astron. Astrophys. 49, 57.
Hauser, M.G., Silverberg, R.F., Gezari, D.Y., Kelsall, T., Mather, J., Steier, M., and Cheung, L., 1980 (private communication).
Hart, L. and Pedlar, A., 1976, Monthly Notices Roy. Astron. Soc. 176, 547.
Hayakawa, S., Ito, K., Matsumoto, T., Ono, T., and Uyama, K., 1976, Nature 261, 29.
Hayakawa, S., Ito, K., Matsumoto, T., Murakami, H. and Uyama, K., 1978, Publ. Astron. Soc. Japan, 30, 369.
Hayakawa, S., Matsumoto, T., Murakami, H., Uyama, K., Yamagami, T. and Thomas, J.A., 1979, Nature, 279, 510.
Hayakawa, S., Matsumoto, T., Noguchi, K., and Uyama, K., 1980a, private communication.
Hayakawa, S., Matsumoto, T., Murakami, H. and Uyama, K., Yamagami, T. and Thomas, J.A., 1980b, preprint.

Hoffmann, W.F., Frederick, C.L., and Emery, R.J., 1971, Astrophys. J. Letters 170, L89.
Hofmann, W., Lemke, D., and Thum, C., 1977, Astron. Astrophys. 57, 111.
Hofmann, W., Lemke, D., and Frey, A., 1978, Astron. Astrophys. 70, 427.
Houck, J.R., Soifer, B.T., Pipher, J.L., and Harwit, M., 1971, Astrophys. J. 169, L31.
Ito, K., Matsumoto, T. and Uyama, K., 1977, Nature 265, 517.
Kawara, K., Kobayashi, Y., Kozasa, T., Sato, S. and Okuda, H., 1980, private communication.
Lockman, F.J.,1976, Astrophys. J., 209, 429.
Low, F.J., Kurtz, R.F. Poteet, W.M. and Nishimura, T., 1977, Astrophys. J. Letters 214, L115.
Maihara, T., Oda, N., Sugiyama, T. and Okuda, H., 1978, Publ. Astron. Soc. Japan 30, 1.
Maihara, T., Oda, N. and Okuda, H., 1979, Astrophys. J. Letters, 227, L129.
Maihara, T. Shibai, H., Oda, N. and Okuda, H., 1980, private communication.
Mezger, P.G., 1978, Astron. Astrophys. 70, 565.
Mezger, P.G. and Pauls, T., 1979, IAU Symposium No. 84, p.357.
Nishimura, T., Low, F.J. and Kurtz, R.F., 1980, preprint.
Oda, N., Maihara, T., Sugiyama, T. and Okuda, H., 1979, Astron. Astrophys. 72, 309.
Okuda, H., Maihara, T., Oda, N. and Sugiyama, T., 1977, Nature 265, 515.
Okuda, H., Maihara, T., Oda, N. and Sugiyama, T., 1979 IAU Symposium No.84, p.377.
Owens, D.K., Muehlner, D.J. and Weiss, R., 1979, Astrophys. J. 231, 702.
Pipher, J.L., 1973, IAU Symposium No. 52, p. 559.
Price, S.D., and Walker, R.G., 1976, Airforce Geophysics Laboratory Technical Report No. 760208.
Price, S.D., 1979, Bull. Am. Astron. Soc. 11, 706.
Price, S.D., 1980 (preprint).
Rouan, D., Léna, P.J., Puget, J.L., de Boer, K.S., and Wijnbergen, J.J., 1977, Astrophys. J. Letters 213, L35.
Serra, G., Puget, J.L., Ryter, C.E., and Wijnbergen, J.J., 1978, Astrophys. J. Letters 222, L21.
Serra, G., Boissé, P., Gispert, R., Wijnbergen, J.J., Ryter, C., and Puget, J.L., 1979, Astron. Astrophys. 76, 259.
Sugiyama, T., Maihara, T., and Okuda, H., 1973, Nature Phys. Sci. 246, 57.
Viallefond, F., Léna, P., de Muizon, M., Nicollier, C., Rouan, D., and Wijnbergen, J.J., 1980, Astron. Astrophys. 83, 22.

DISCUSSION FOLLOWING PAPER PRESENTED BY H. OKUDA

T. JONES: What exactly is the cause of the bumps in the 2.4 μm flux in the plane?

OKUDA: I think they are partly due to the anisotropic distribution of sources and partly to variations of interstellar extinction; there is some anti-correlation between peaks in the 2.4 μm and peaks in the mm-wave CO intensity which suggest that molecular clouds may be causing part of the extinction.

T. JONES: If these bumps are due to excess star counts, what is the population type?

OKUDA: We are not yet certain, but I would believe they are either M giants or young M supergiants.

T. JONES: In your histogram [Figure 5] you indicated you needed hundreds of stars per square degree; this would represent a very large number of supergiants.

OKUDA: I agree that not all can be supergiants, but I think that the contribution of supergiant stars is significant.

PUGET: The component of the 2.4 μm distribution which is associated with a young population, namely the one which is narrower in the latitude direction and which peaks at 30° longitude, can be compared with the far infrared luminosity. In making this comparison we find that the 2.4 μm emission is significantly stronger than is expected based on a standard Population I initial mass function. In other words the Population I excess is partly unexplained if you make the assumption that the stellar population is the same throughout the Galaxy.

AARONSON: It will be very interesting to compare the total 2-μm luminosity of our Galaxy with that of other spirals. When will an accurate estimate of this luminosity be available, or do you have such an estimate now?

OKUDA: The faintest regions of the Galaxy are difficult to map from a balloon at 2.4 μm because of atmospheric OH air glow. We will have to do the experiment by rocket or satellite.

BAUD: CO surveys show that the CO midplane lies below the IAU-plane in the first quadrant. Do the FIR and MIR surveys show the same phenomenon?

OKUDA: Yes, they do.

INTERPRETATION OF THE LARGE-SCALE EMISSION FROM THE GALACTIC PLANE

S. Drapatz
Max-Planck-Institut für Physik und Astrophysik
Institut für Extraterrestrische Physik
Garching, W.- Germany

I. INTRODUCTION

To obtain information on the large-scale structure of our Galaxy, one has to investigate the radiation in the short-wavelength and long-wavelength portion of the electromagnetic spectrum where attenuation throughout the whole galactic disk is very low: gamma-radiation on the one-hand side and infrared and radio radiation on the other side. But since most of the different modes of radiation are generated by interaction of two or more basic galactic constituents one derives only indirect information on a specific component. To come as close to a unique solution as possible the results of as many different spectral regions as possible should be combined. From this point of view it is very encouraging that infrared astronomy has been entering the field and will continue to contribute in the years to come. While gamma-radiation presents mainly information about the cosmic rays and the interstellar gas and radio astronomy about early-type stars and the interstellar gas, infrared astronomy has opened a new way to investigate a major fraction of the stellar population and the inter-

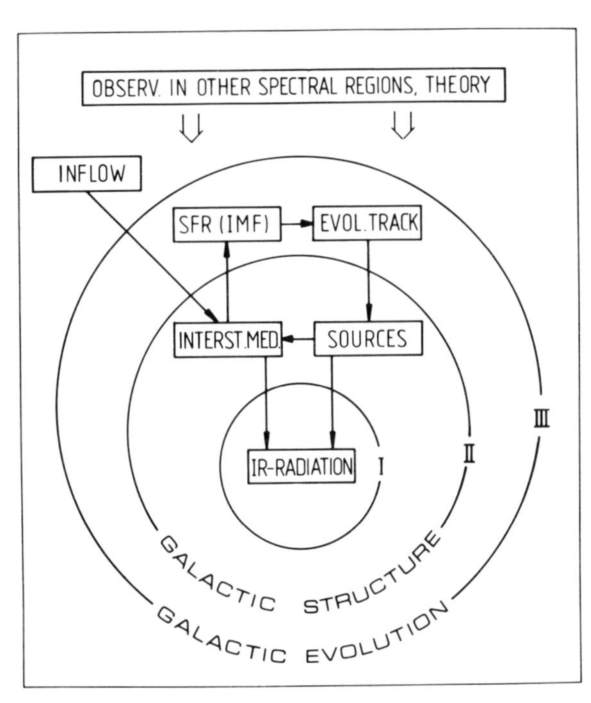

Figure 1: Concept of Interpretation of the Large-Scale IR-Emission

C. G. Wynn-Williams and D. P. Cruikshank (eds.), Infrared Astronomy, 261–274.

stellar dust. The information is contained in the diffuse galactic emission, which is observed at low galactic latitudes with a field of view large enough to discriminate against point sources, i.e. the emission is averaged over typical dimensions of some 100 pc.

A summary of the corresponding observations has been given in the previous talk (Okuda). In the following the interpretation of these data will be dealt with (Fig. 1). Firstly, the present state of the Galaxy will be described by modelling the distribution of stars, dust and radiation field on the basis of these observational results including data from other wavelength regions (Section II). In a next step first attempts will be reported to draw conclusions about the interaction of ISM and stars inferring basic parameters like star formation rate, initial mass function and chemical abundance (Section III). The galactic center will not be discussed, since it is the subject of another talk.

II. MODELLING OF THE LARGE-SCALE EMISSION

Simple models are used to investigate whether the experimental results fit into our overall picture of the Galaxy. The emission of the sources (mainly stars) is modified by the interstellar medium (treatment of radiative transfer) before it can be observed at the earth. In the models quoted it is generally assumed that the dust properties and the gas/dust ratio are the same throughout the Galaxy, so that these quantities can be scaled by the gas densities. Of course, the relative amount of dust could be smaller (e.g. thermal evaporation and sputtering of grains in hot regions) as well as larger (heavy element gradients). If the fraction of heavy elements locked up in grains and molecules is constant throughout the Galaxy, then the dust density scaled with the CO density is unaffected by heavy element gradients. While it is generally adopted that abundance gradients exist in the Galaxy (e.g. Janes 1979 and ref. therein), their precise nature remains to be determined (existence of more noticeable gradients in CNO elements than in heavier elements, local inhomogeneities, etc.).

1) Near-Infrared (NIR) and Middle-Infrared (MIR) Diffuse Emission

The interpretation of the NIR-radiation starts with the assumption that it is due to integrated light of late-type stars attenuated by interstellar extinction along the line-of-sight. For the solar neighborhood different authors deduce the 2.4 μm emissivity from star counts per spectral type and luminosity class (McCuskey 1969, Allen 1973) to be around $2 \cdot 10^{24}$W pc^{-3} μm^{-1} (Maihara et al. 1978, Hayakawa et al. 1978, Serra et al. 1980) in good agreement with the observed values. On the basis that in spiral galaxies there exists a homogeneous disk of stellar population with the same color as in the solar vicinity (Schweizer 1976) these data can be extrapolated using a conservative mass model of the Galaxy (e.g. Schmidt 1965). The extinction of interstellar dust is taken to be correlated with the gas density (e.g. Gordon and Burton

1976). The data at $l^{II} < 10^o$ and $l^{II} > 60^o$ can be explained in terms of a simple model, but one obtains a strong excess (factor 3) of the observed brightness over the predicted one in the region $10^o < l^{II} < 60^o$ (Hayakawa et al. 1977). Its association with spiral arms and local active regions (correlation with observations in other spectral regions, see Okuda-paper) implies young objects. For a determination of the corresponding emissivity several authors have discussed models where a Population I component is superimposed upon an 'old disk' component. Although the observations reveal that NIR sources are associated with spiral arms the axisymmetric approximation is assumed for simplicity in these models and no peculiar features are included in the discussion.

The Kyoto group (Maihara et al. 1978) has obtained numerical values for the excess emissivity so as to fit the observations for $l^{II} \sim 0° - 30°$. The distribution of dust is taken as comprising two components: a uniform component (A_V = 1.5 mag/kpc) and an additional concentration associated with the 5 kpc-ring (maximum A_V = 2.2 mag/kpc), where $A(2.4\ \mu m)/A_V = 0.08$. An exponential type dust density distribution with a scale height of 150 pc is used.

The Nagoya group has revised a previous model (Hayakawa et al. 1977) by improving on the treatment of extinction (Hayakawa et al. 1980). Including the clumpiness effect the optical depth is scaled with the column density of HI (scale height 120 pc) and CO (scale height 50 pc) as $A_V = 4 \cdot 10^{-22}\ N(HI) + 1.05 \cdot 10^{-17}\ N(CO)$ mag cm^2. It is shown that this optical depth is consistent with the observed longitude and latitude profiles of the 2.4 μm relative to the 3.4 μm observations (using model No. 15 of van de Hulst 1957). The excess luminosity is modelled by two rings (at R = 4.6 and 7.2 kpc) where the source function decreases exponentially with R and z. Since there exists no dip or flattening at the equator in the latitudinal brightness distribution, the emitting sources have to be highly concentrated to the galactic plane to compensate the absorption feature by dust grains associated with molecular clouds. According to the model, the scale height of the sources (z_o = 50 pc) is the same as for molecular clouds, i.e. they also seem to be extreme Population I objects. The volume emissivity averaged over 1 kpc in the radial direction of the 5 kpc-arm is $5 \cdot 10^{25}$ W pc^{-3} μm^{-1}.

The French group (Serra et al. 1980) uses three rings (one at 5 kpc and two additional ones). The radial and z-distribution are given by Gaussian profiles and calculations are done for different scale heights z_o. Finally z_o = 200 pc is selected, where the maximum emissivity at 5 kpc agrees with the value of the Kyoto group ($4 \cdot 10^{25}$W pc^{-3} μm^{-1}). This scale height implies objects of older Population I. The volume emissivity derived by all three groups is in good agreement; the corresponding surface flux density of the Nagoya group is smaller (factor ∿ 2) than for the other groups, since all models use nearly the same value for the optical depth in the galactic plane but different scale heights.

The mass of stars producing the infrared excess represents a minor fraction of the total mass since no mass excess at 5 kpc is observed. Evidence quoted above suggests that the sources are (extreme) Population I objects. Two candidate sources have been considered: (proto)stars embedded in circumstellar dust and late-type (super)giants.

The first alternative has been discussed by Hayakawa et al. (1977). They compare the 5 kpc region with a 'typical' dust cloud, the Ophiuchus dark cloud (located at a distance of $\sim$ 160 pc from the Sun). Triggered by a cloud-cloud collision the formation of a star cluster occurs in this region (Wilking et al. 1979). From its luminosity function (Vrba et al. 1975) an emissivity is derived which agrees with the emissivity in the 5 kpc region, if a reasonable filling factor of 10^{-2} for dust clouds is assumed. But the interpretation would require that practically all dust clouds are heated by newly born stars associated with them. Furthermore, a fraction of the stars in the Ophiuchus cloud seems to be either non-main sequence or background stars (Fazio et al. 1976). So, while dust-embedded stars seem not to be the dominant source of the NIR excess, they probably are responsible for a larger fraction of the MIR emission, since corresponding observations show a color temperature of about 500 K for the excess emission (Price 1980), a value characteristic of sources with circumstellar shells. The MIR emission is well correlated with the intensity distribution obtained in the 2.7 GHz Survey (Altenhoff et al. 1970). Rowan-Robinson (1979) finds the same correlation for the distribution of hot-centered molecular clouds. The cloud spectra are composed of a far-infrared peak (color temperature 50 K) and a weaker mid-infrared emission (color temperature 500 K). For the few nearby clouds, where far-infrared resolution is sufficient, one finds that the radiation comes from dust illuminated by sources embedded in the cloud, rather than by the stars responsible for the H II region.

The second alternative (red giants) has been put forward by Puget et al. (1979) and has been further discussed in connection with star formation rates (see Section III). Hayakawa et al. (1980), who find that the sources belong to extreme Population I, consider late-type supergiants. Assuming that the 5 kpc region has nearly the same spectrum of radiation as the Per OB 1 association, its surface brightness in blue light is obtained (21.3 mag arcsec^{-2}), which is in approximate agreement with corresponding values of other galaxies (M31, M51, M101). At 5 kpc the surface flux density at 2.4 μm corresponds to a surface density of $5 \cdot 10^{-4}$pc^{-2} M2 supergiants, whereas the Lyc-photon density (Mezger 1978) can be provided by $\sim 2 \cdot 10^{-3}$ pc^{-2} B supergiants. The ratio of the star densities of about 4 agrees with the corresponding ratio in Per OB1, where most of the early-type stars are of early B-type. But the Perseus cluster, which is located outside the solar circle (distance $\sim$ 2 kpc from the Sun), may have relatively more massive stars than the clusters in the 5 kpc region (Burki 1977). This will be discussed in Section III.

In summary the NIR excess seems to be due to late-type giants/ supergiants, whereas some MIR-excess should be provided by circumstellar shells around young and/or evolved stars. From a theoretical point of view the two classes of sources are not easy to separate (Yorke and Shustov 1980). It is possible for cool late-type stars with high rates of mass loss to have infrared spectra similar to protostars with accretion envelopes, provided the same type of dust is present in each case. Information pertaining to the intrinsic spectra of the central source is not available in the resulting infrared continuum. Nevertheless, the product of luminosity and lifetime for protostars ($\sim 10^9$ $L_\odot$yr) seems to be lower than for giants ($\sim 10^{10}$ $L_\odot$yr), so that an unrealistically high number of protostars would be needed to explain the NIR brightness.

2) Far-Infrared (FIR) Diffuse Emission

The UV and visible radiation of (mainly) stellar objects is partly absorbed by interstellar grains and reradiated at FIR-wavelengths, where the dust is nearly transparent. For the prediction of the diffuse emission simple models have been used (Stein 1967, Fazio and Stecker 1976) on the basis of a uniform stellar radiation field throughout the Galaxy. The fact that the FIR radiation observed was more than one order of magnitude higher than predicted generated the interest in somewhat more refined models.

In a first step the total (wavelength-integrated) diffuse emission in the galactic equator has been derived as a function of longitude and compared to observational results. The basic idea in such a model (Drapatz 1979) is that the total FIR emissivity ε (which equals its total short-wavelength absorptivity) at location R in the galactic plane may be related to its value ε^0 in the solar neighborhood by $\varepsilon(R) = \varepsilon^0\ S(R)\ D(R)$. Here S represents the stellar distribution in other regions of the Galaxy and D the influence of the dust distribution, which consists of two opposing effects: a larger amount of dust reduces the short wavelength radiation field by extinction but increases the emitting mass. The FIR intensity is then obtained by integrating the emissivity along the line-of-sight.

The quantity ε^0 is derived using a compilation of observed optical properties of the interstellar medium and a mean interstellar radiation field in the solar vicinity, which is formally approximated by a sum of three dilute black-body spectra, each characterizing one group of stars (O/B, A/F, late-type stars). This procedure is justified by more sophisticated derivations of the interstellar radiation field (Henry 1977, Mattila 1980) and by good agreement with observations. One obtains $\varepsilon^0 = 1 \cdot 10^{25}$ Wpc^{-3} with roughly an equal contribution of all three groups of stars. The distribution of young stars throughout the Galaxy is represented by a simple spiral model based on the work by Georgelin and Georgelin (1976). The radial distribution in the arms follows the relative distribution of giant H II regions (Mezger 1970); the arm/interarm contrast is obtained from the ratio of Lyc-photons of giant and small H II regions (Smith et al. 1978). The stars in the arm region obey the initial luminosity function, the stars in the interarm region the observed luminosity function (McCuskey 1966). The older stars (A-K) are not affected by the spiral structure and their scaling factors S are obtained from Schmidt's model (Schmidt 1965). The dust scaling factor D has been derived under certain assumptions for a homogenous medium with dust particles having either a completely forward or a totally isotropic scattering pattern and for a medium where dense clouds are superimposed upon a homogenous background. In all three cases the mean dust density $< n >$ averaged over cloud and intercloud regions is of moderate influence ($D \sim \sqrt{< n >}$). The calculated diffuse emission shows a maximum intensity around $l^{II} = 30^o$ ($I = 1.1 \cdot 10^{-8}$ W $cm^{-2}sr^{-1}$) and a slow decrease around $l^{II} = 45^o$ and is in quite good agreement with the observations. The total FIR luminosity of the Galaxy ($4 < R < 12$ kpc) is $\gtrsim 5 \cdot 10^9$ $L_\odot$.

Ryter and Puget (1977) have correlated the FIR emission of massive molecular clouds associated with well-known H II regions with their gas column densities to derive a luminosity normalized per hydrogen atom $L_H \sim 2 \cdot 10^{-30}$ W (H atom)$^{-1}$. It is argued that this quantity should prevail in most of the interstellar medium leading to an intensity $I \sim 3 \cdot 10^{-8}$ W $cm^{-2}sr^{-1}$ at $l^{II} \sim 30^o$, which is high. Here one should keep in mind that most of the H_2 resides in molecular clouds, where no massive stars are formed at present. A relation between FIR intensity and gas density can have a meaning on the basis of the good correlation of the longitude profiles of CO and FIR emission and the suggestion that the star formation rate varies with a certain power of the overall mean gas density (Schmidt 1959).

The French group has then built a model (Boissé et al. 1980) by discussing the efficiency of absorption of starlight by dust as a function of the mass of the star. One fraction of the stellar radiation is absorbed by the molecular cloud in which the star is formed, another fraction is absorbed by dust in the diffuse interstellar medium and a third one is absorbed by all other molecular clouds. An evaluation of the efficiency factors as a function of the mass of main sequence stars yields $f \sim 0.14(M/M_\odot)^{0.4}$ Averaging over the stellar population in the solar neighborhood gives $f \sim 0.25$, whereas $f \sim 0.4$ in the 5-kpc region. The corresponding efficiency factors in the model of Drapatz (1979) would have values 0.21 and 0.44 respectively.

An interpretation of the observed FIR intensity is given by Mezger (1978). He finds that the main portion ($\sim$ 90%) of the early-type stars are located outside radio H II regions. The corresponding Lyc-photons either ionize an extended low-density (ELD) medium or are absorbed by the dust or escape from the Galaxy. The first fraction is determined from measurements of the diffuse thermal radio emission, the other two contributions are estimated. Of the number of Lyc-photons absorbed by the gas in the H II regions a fraction 2/3 is degraded into $L\alpha$-photons. Comparing this $L\alpha$ intensity $I(L\alpha)$ with the observed intensity $I(FIR)$ by use of the relation $I(FIR) = IRE \cdot I(L\alpha)$ yields IRE = 8.

This value can be met, if either all non-ionizing photons from cluster stars get absorbed by grains inside the ELD H II regions and the neutral clouds or if only half of these photons is absorbed and absorption of photons of the general stellar radiation field (stars of later type) account for the other half.

All three models agree that all star classes contribute to some extent to the FIR emission, early-type stars being especially important in the 5 kpc arm. Since there exists a good correlation of the longitude profile of the FIR emission with the corresponding H 166α-observations (typical of diffuse radio H II regions with large spatial extent, e.g. Lockman 1976) as well as with CO-observations (molecular clouds, e.g. Solomon and Sanders 1979) different dust containing regions are assumed to be dominant in the transformation of star light into FIR radiation: Mezger stresses the ELD H II regions, the French group emphasizes parent clouds of newly born stars and Drapatz the ionized, atomic, and molecular component. When young massive stars form, they are located in or close to the edge of their parent cloud, which absorbes their radiation in this initial phase. Rowan-Robinson (1979) finds from observations that hot-centered clouds contribute roughly $\sim 8 \cdot 10^8 L_\odot$ to the total FIR luminosity of the Galaxy. The massive stars then form H II regions, which by the champagne effect (Tenorio Tagle et al. 1979) evolve into ELD H II regions and finally form associations. In the galactic plane (scale height $\sim$ 50 pc) the absorption due to dust in molecular clouds and in ELD H II regions is comparable (absorption-length $\sim$ 1 kpc), whereas absorption by dust located in diffuse H I and H II regions dominates in directions out of the plane. The short wavelength radiation penetrates into the molecular clouds heating up the grains in the outer shells (Flannery et al. 1980). These warm regions comprise some 10% of the volume and mass of the cloud fairly independent of grain parameters and cloud sizes.

So far the 'total' FIR radiation has been discussed, although observations do not cover the whole wavelength region (especially not the 30 - 60 μm range). Furthermore, the measurements have been carried through in certain bands, so that the general shape of the whole spectrum and the integrated total radiation has to be attained on the basis of theoretical models. They have to include the microscopic properties of the dust (chemical nature, size distribution). The spectrum for spherical dust particles is given by

$$I(\lambda) = \iint B\left[\lambda, T(a, U(\lambda, 1))\right] Q(a, \lambda)\pi a^2 \, n(a, 1) \, da \, dl \qquad (1)$$

where B is the Planck-function (grain temperature T), U the stellar radiation field, Q the emission efficiency ($Q = \sigma/\pi a^2$), n the spatial and size distribution of grains and the integration is over all grain radii a and the line-of-sight 1. Drapatz and Michel (1976) have carried through a specific model calculation. They first derive and discuss a certain grain model (grain-size distribution a^{-4}, chemical composition taken as silicates perturbed by impurities and radiation defects generating absorption maxima due to local lattice vibrations in the infrared and due to color centers in the UV). The high specific heat of the material reduces the appearance of temperature spikes of small grains (Drapatz and Michel 1977, Aannestad & Kenyon 1979). On the basis of this grain model the diffuse galactic emission is obtained, where the stellar radiation field is dealt with like in the work by Drapatz (1979) already quoted, but with a more qualitative treatment of radiative transfer. At $l^{II} = 30^o$ the maximum emission occurs around $\lambda = 80$ μm decreasing to half that value at 150 μm. The wavelength of the maximum intensity is determined by the absorption efficiency of grains $Q \sim \lambda^{-2}$ as $\lambda(max) \sim 0.2/T$ cm. If the interstellar dust has a different composition the particle temperature will be higher for a larger ratio of short wavelength absorption to infrared emission and for smaller particles. So the color temperature of the diffuse emission will only indirectly be related to the dust temperature distribution. Also, more refined models have to include the small-scale distribution of the radiation sources as well as that of the interstellar matter, i.e. the spectrum of the radiation will show a spatial distribution. Some general remarks related to these problems will be given in the following.

The distribution of the dust temperature as induced by the random distribution of stars has been computed by Rouan (1980). The distribution is quite narrow (typically $\Delta T = 2K$ at half maximum) with a sharp cut-off on the low temperature side, reflecting the presence of a uniform starlight background. This is due to the fact that the absorption length in the interstellar medium is much larger than the mean distance of stars. Most of the mass of the dust has therefore the temperature T_m of the maximum of the distribution and emits at $\gtrsim 100$ μm. The temperature distribution P(T) has an extended wing on the high temperature side, however, due to the more scarcely distributed hot stars. It converges towards the nearest star approximation $P(T) \sim T^{-10}$ (for $Q \sim \lambda^{-2}$). Therefore, the small fraction of grains that is hotter (say $1.5 \cdot T_m$) will emit a fraction of the total energy ($\gtrsim 10\%$) at shorter wavelengths and determine the spectrum in that range ($\lambda \lesssim 80$ μm).

In the FIR/submm region the spectrum will be influenced by an additional component originating from the inner parts of giant molecular clouds, where the gas is heated to 10 K by cosmic rays, H_2 formation, gravitational collapse or magnetic ion-neutral slip heating (Goldsmith and Langer 1978). The dust is not in thermal equilibrium with the gas for small densities ($n < 10^4$ cm^{-3}); still the temperatures will be similar in the interiors of dense clouds (Leung 1975). The corresponding diffuse emission should be comparable in intensity with the emission in the long wavelength wing of the warm component (diffuse medium plus outer layers of molecular clouds). If the dust properties are similar, the optical depths through both regions along the line-of-sight will be comparable (τ(350 μm $\sim 10^{-2}$ at $l^{II} = 30°$). Recent observational results (Owens et al. 1979) are compatible with this interpretation.

III. IMPLICATIONS FOR GALACTIC EVOLUTION

One fundamental feature of galactic evolution is the conversion of interstellar matter into stars (formation of stars from the ISM) and the ejection of matter back into the ISM (during stellar evolution and after nuclear processing). The first process is described by the Star Formation Rate (SFR)

$$SFR = \dot{\xi}(M,R,t) = d\dot{N}(M,R,t)/d\ln M \qquad (2)$$

which is the number of stars arriving (per unit area of the galactic plane and per unit time) at the Main Sequence with a mass distribution given by the Initial Mass Function ξ. The description of the second process follows from that equation by taking into account the evolution and lifetime $\tau(M)$ of the stars formed. The observed surface flux densities F_i are connected with the above quantities as

$$F_i = \iint k(R,t) \cdot \dot{\xi}_o(M) \cdot L(M) \cdot f_i(M) \, d\ln M \, dt \qquad (3)$$

Observations present values of F_i for the NIR (subscript 1), FIR, (subscript 2) and Lyc-radiation (subscript 3), the last one being deduced from radio continuum measurements. The integral includes stars of all ages, which have been formed during time t_a, the age of the stellar population. The upper time limit of the integration is t_a for $\tau(M) > t_a$ and $\tau(M)$ for $\tau(M) < t_a$. Integration limits for post main sequence stars have to be considered separately. The assumption is made that the SFR at other locations in the Galaxy and at other times equals its present value in the solar neighbourhood ξ_o multiplied by a dimensionless scaling factor $k(R,t)$. The quantities L (stellar luminosity), τ, and f depend on the star class. Serra et al. (1980) have defined the efficiency factors as follows: f_1 = fraction of the total energy of a star radiated in the NIR region, f_2 = fraction of the total energy of a star absorbed by dust and reemitted in the FIR-region (see Section II.2) and f_3 = fraction of the total energy of a star emitted by Lyc-radiation. Mean values $< f_i >$ can be defined as the ratio F_i/F_{tot} with $F_{tot} = F(f=1)$. Although equation (3) includes all stars, different F_i are dominated by different stellar populations: F_3 is due to high mass stars, while F_2 includes high and intermediate mass stars ($M \gtrsim 1 M_\odot$). F_1 is determined by evolved stars of intermediate mass ($1 \lesssim M \lesssim 5\ M_\odot$). The low mass stars ($M < 1\ M_\odot$) are not directly accessible to observations and are subject to present theoretical investigations (e.g. Zinnecker and Drapatz 1979, Zinnecker 1980). While stars of high and intermediate mass are formed in open clusters and OB associations (Mezger and Smith 1977) this seems not to be true for low mass stars. First quantitative arguments for the necessity for two modes of star formation are given by Talbot (1980). For the high mass mode Lequeux (1979) has recalculated the IMF for the solar vicinity following the method of Salpeter (1955) but using more recent catalogues. This IMF will be used in the following to discuss the results of Section II on the basis of equ. (3).

1) Old Disk Population

Calculating F_1 and F_2 from equ. (3) for t_a = 10 Gyr and a constant SFR(i.e. k=1) reproduces quite well the observed values of NIR (Serra et al. 1980) and FIR surface flux density (Boissé et al. 1980) in the solar neighbourhood. Since F_1 is dominated by older stars and F_2 by all the stars it implies that k cannot have changed much in time, if the IMF was the same during the lifetime of the Galaxy. This is in agreement with the work of Miller and Scalo (1979), who derive a constant SFR in the solar vicinity from a critical reanalysis of observational material and the condition that the IMF should be continuous. A physical interpretation of this constant SFR is given in the model of Kaufman (1979), where after a disk age t $\sim 10^9$ yr the decaying exponential birthrate for star formation by high mass stars (triggered by expanding H II regions and supernova shock waves) is dominated by a constant star formation rate triggered by a galactic spiral density wave. However, Cassé et al. (1979) show that gas flows, such as differential infall or radial flows, have to be included to reproduce the observed gas distribution in the Galaxy. These gas flows are also needed to explain the abundance of elements (difference in metallicity between young and old stars in the solar vicinity, heavy element gradients throughout the Galaxy). Corresponding models have been suggested (Tinsley and Larson 1978, Chiosi 1980). Also, a closed system seems unrealistic for a constant SFR, if the frequently used law should prevail that the SFR depends on a power m of the gas density. While Lockman (1979) states that this law is unlikely to be correct over scales of a few kpc, it can be used for parametrization purposes (Guibert et al. 1978).

For decreasing galactocentric distance the fraction of F_1 which is due to the old disk population (wide scale height) follows Schmidt's mass law, so that the same IMF holds with k proportional to the local mass density in stars: k (5 kpc, t) $\sim$ 5 and t_a = 10 Gyr, as before.

2) Population I Excess

The present integrated (over all masses) SFR in the arm region is obtained from the fact that the rate of formation of early-type stars is proportional to F_2. The result is $\sim 20 M_\odot (pc^2 Gyr)^{-1}$ since k(5 kpc,t) > 5 in good agreement with the values obtained from F_3 by Mezger (1978).

There have been undertaken first attempts to investigate the SFR as a function of time. Serra et al. (1980) solve equations (3) for a given region in the Galaxy (R = 5 kpc) by maintaining ξ_o and representing k as a time step function (3 steps with $t_a \sim 10^7$, 10^8, 10^9 yr). These values take into account that the excess population is young (small scale height, no mass excess above Schmidt mass distribution). Since the F_i are of comparable size the system would have a trivial solution with a constant SFR, if the $< f_i >$ are also of comparable size. Not only are the $< f_i >$ different, but they also change in opposite directions when changing the age of the stellar population ($\langle f_1 \rangle$ increases, $\langle f_{2,3} \rangle$ decrease with the age of the population). As the authors correctly state the derivation of the $\langle f_i \rangle$ depends crucially on the stellar parameters L, τ (mainly their product). While those parameters seem to be fairly well known for the main sequence stars a problem arises from the uncertainties in the evolutionary tracks of post main sequence stars which are severely affected by mass loss and the chemical abundance of elements. With comparison to recent theoretical investigations the authors seem to have overestimated the low mass contributions (see Schönberner 1979) and underestimated the high mass contribution (see Weaver et al. 1978). Still, accepting their parameters $\langle f_i \rangle$ at face value a relative increase of $\langle f_1 \rangle$ is necessary which could be obtained either by increasing the SFR by a factor of 5 - 10 about 10^9 years ago (to increase the number of red giants) or by steepening the IMF (changing ξ) or by decreasing the upper mass limit M_u. It should be kept in mind that the variation in heavy element abundances could also mimic a different M_u. For higher abundances, stars of a given mass have lower T_{eff} and therefore fewer Lyc-photons, as if they were less massive stars. But even taking that effect into account Panagia (1980) finds a deficiency of massive stars. Evidence for a lower rate of formation of massive stars with decreasing galactocentric distance is also obtained from other observations: from UBV photometry of very young clusters (Burki 1977) and from the determination of the IRE of large H II region associated with molecular cloud complexes (Boissé et al. 1980). Also Hayakawa et al. (see Section II.1) state that a larger number of B stars (rather than a few O stars) is compatible with their NIR observation. From a theoretical point of view the limiting mass of stars with $L \sim M^3$ is proportional to $Z^{-1/2}$ (Shields and Tinsley 1976) in agreement with the above results for higher abundances of heavy elements Z inside the solar circle. Each piece of evidence stated may not be fully convincing for itself, but taken together they seem to indicate a change of ξ or of M_u with galactocentric distance.

In conclusion, the investigation of the large-scale infrared emission of the Galaxy starts to deliver important input data for the development of models of galactic structure and evolution. The increasing accuracy, resolution, and completeness of observations will allow for more sophisticated models and better understanding of the parameters involved.

I want to thank my colleagues from the Max-Planck-Institut, especially H. Zinnecker, for interesting discussions and some colleagues from other institutes for communication of not yet published results.

REFERENCES

Aannestad, P.A., Kenyon, S.J., 1979, Astrophys. J. 230, 771
Allen, C.W., 1973, Astrophysical Quantities, The Athlone Press, London
Altenhoff, W.J., Downes, D., Goad, L., Maxwell, A., Rinehart, R., 1970 Astron. Astrophys. Suppl. 1, 319
Boissé, P., Gispert, R., Coron, N., Wijnbergen, J., Serra, G., Ryter C., Puget, J.L., 1980, (preprint).
Burki, G., 1977, Astron. Astrophys. 57, 135
Cassé, M., Kunth, D., Scalo, J.M., 1979, Astron. Astrophys. 76, 346
Chiosi, C., 1980, Astron. Astrophys. 83, 206
Drapatz, S., Michel, K.W., 1976, Mitt. Astr. Ges. 40, 187
Drapatz, S., Michel, K.W., 1977, Astron. Astrophys. 56, 353
Drapatz, S., 1979, Astron. Astrophys. 75, 26
Fazio, G.G., Stecker, F.W., 1976, Astrophys. J. 207, L49
Fazio, G.G., Wright, E.L., Zeilik, M., II, Low, F.J., 1976, Astrophys. J. 206, L165
Flannery, B.P., Roberge, W., Rybicki, G.B., 1980, Astrophys. J. 236, 598
Georgelin, Y.M., Georgelin, Y.P., 1976, Astron. Astrophys. 49, 57
Goldsmith, P.F., Langer, W.D., 1978, Astrophys. J. 222, 881
Gordon, M.A., Burton, W.B., 1976, Astrophys. J. 208, 346
Guibert, J., Lequeux, J., Viallefond, F., 1978, Astron. Astrophys. 68, 1
Hayakawa, S., Ito, K., Matsumoto, T., Uyama, K., 1977, Astron. Astrophys. 58, 325
Hayakawa, S., Ito, K., Matsumoto, T., Murakami, H., Uyama, K., 1978, Publ. Astron. Soc. Japan 30, 369
Hayakawa, S., Matsumoto, T., Murakami, H., Uyama, K., Thomas, J.A., Yamagami, T., 1980, (preprint).
Henry, R.C., 1977, Astrophys. J. Suppl. 33, 451
Janes, K.A., 1979, Astrophys. J. Suppl. 39, 135
Kaufman, M., 1979, Astrophys. J. 232, 707
Lequeux, J., 1979, Astron. Astrophys. 71, 1
Leung, Ch., M., 1975, Astrophys. J. 199, 340
Lockman, F.J., 1976, Astrophys. J. 209, 429
Lockman, F.J., 1979, Astrophys. J. 232, 761
Maihara, T., Oda, N., Sugiyama, T., Okuda, H., 1978, Publ. Astron. Soc. Japan 30, 1
Mattila, K., 1980, Astron. Astrophys. 82, 373
McCuskey, S.W., 1966, Vistas in Astronomy 7, 141
McCuskey, S.W., 1969, Astron. J. 74, 807
Mezger, P.G., 1970, IAU-Symp. 38, 107
Mezger, P.G., Smith, L.F., 1977, IAU-Symp. 75, 133
Mezger, P.G., 1978, Astron. Astrophys. 70, 565
Miller, G.E., Scalo, J.M., 1979, Astrophys. J. Suppl. 41, 513
Owens, D.K., Mühlner, D.J., Weiss, R., 1979, Astrophys. J. 231, 702
Panagia, N., 1980, in P.A. Shaver (ed.), Radio Recombination Lines, D. Reidel Publ. Co., Dordrecht, Boston, London, p. 99
Price, S.D., 1980, Bull. AAS, 11, 706

Puget, J.L., Serra, G., Ryter, C., 1979, IAU-Symp. 84, 105
Rouan, D., 1980, (preprint)
Rowan-Robinson, M., 1979, Astrophys. J. 234, 111
Ryter, C.E., Puget, J.L., 1977, Astrophys. J., 215, 775
Salpeter, E.E., 1955, Astrophys. J. 121, 161
Schmidt, M., 1959, Astrophys. J. 129, 243
Schmidt, M., 1965, Stars and Stellar Systems, Vol. V: Galactic Structure, Univ. of Chicago Press, p. 513.
Schönberner, D., 1979, Astron. Astrophys. 79, 108
Schweizer, F., 1976, Astrophys. J. Suppl. 31, 313
Serra, G., Puget, J.L. Ryter, C.E., 1980, Astron. Astrophys. 84, 220
Shields, G.A., Tinsley, B.M., 1976, Astrophys. J. 203, 66
Smith, L.F., Biermann, P., Mezger, P.G., 1978, Astron. Astrophys. 66, 65
Solomon, P.M., Sanders, D.B., 1979, in: Giant Molecular Clouds in the Galaxy Pergamon Press, Oxford
Stein, W.A., 1967, Interstellar Grains NASA Doc 67-60065
Talbot, R.J., 1980, Astrophys. J. 235, 821
Tenorio-Tagle, G., Yorke, H.W., Bodenheimer, P., 1979, Astron. Astrophys. 80, 110
Tinsley, B.M., Larson, R.B., 1978, Astrophys. J. 221, 554
v.d. Hulst, H.C., 1957, in: Light Scattering by Small Particles Wiley, Chapman and Hall
Vrba, F.J., Strom, K.M., Strom, S.E., Grasdalen, G.L., 1975, Astrophys. J. 197, 77
Weaver, T.A., Zimmerman, G.B., Woosley, S.E., 1978, Astrophys. J. 225, 1021
Wilking B.A., Lebofsky, M.J., Rieke G.H. and Kemp J.C. 1979, Astron. J. 84, 199
Yorke, H.W., Shustov, B.M., 1980, Astron. Astrophys. (to be published)
Zinnecker, H., Drapatz, S., 1979, Mitt. Astr. Ges. 45, 22
Zinnecker, H., 1980, (Ph.D.Thesis) Techn. Universität München

DISCUSSION FOLLOWING PAPER DELIVERED BY S. DRAPATZ

BLITZ: Is it possible to model the existing infrared data to determine what the radial exponential scale length of the Galaxy is, or are there too many free parameters in the models?

DRAPATZ: Most of the work to date has been concerned with the spiral or ring component of the disk rather than the exponential component. However, from near infrared data Hayakawa et al. estimate the scale length in the solar vicinity to be 1 kpc and the spheroidal component in the inner region of the Galaxy to be somewhat steeper than the Schmidt Law mass distribution. I find that far infrared data are compatible with a distribution of early-type stars following giant radio H II regions and of other stars following the Schmidt Law.

BLITZ: Are the near infrared data good enough to attempt to model the distribution of disk stars in the outer Galaxy to see if the warp in the gas is mimicked by a warp in the stellar disk? The maximum warp occurs toward ℓ = 60°, and knowing whether the warp in the stars follows that of the gas has important implications for understanding the origin of the warp.

DRAPATZ: At present the sensitivity of surveys towards the outer region of the Galaxy is insufficient to show structure beyond 1-2 kpc. We cannot yet say if there are warps in the outermost regions of the Galaxy. However, for the inner regions the near infrared ridge shows the same deviations from the galactic plane as other Population I objects.

HABING: Can you characterize the difference in population at R = 4 kpc and R = 5 kpc?

DRAPATZ: Inside of the ring, say at 4 kpc, we can fit the data with a population of stars similar to that in the solar neighborhood, but with a deficiency of early-type stars. In the ring itself at 5 kpc, there is an extra population of younger age, possibly due to a higher star formation rate in the past.

T. JONES: How do you reconcile the use of a few luminous Population I stars to explain the IR excess above the old disk population with the fact that at 28° longitude the actual source counts are 2 to 3 times above the disk at adjacent longitudes?

DRAPATZ: Even if the 5 kpc ring did not exist, there would still be a factor of five more stars there than in the solar neighborhood, due to the Schmidt Law. Part of the problem you raise may also be due to anisotropies in the stellar distribution and to holes in the extinction.

AARONSON: The near-IR data for other spirals indicate that in general, old disk M giants dominate the light at 2 μm. Although there may be regions of our Galaxy dominated by supergiants, such regions are probably not typical of the Galaxy as a whole.

DRAPATZ: Though I agree that these are basically giants, I think that currently there are uncertainties in the evolutionary tracks of giants and supergiants, partly due to the previous omission of effects such as mass loss, which may lead to order-of magnitude changes in the luminosities of these stars.

KRISCIUNAS: Miller and Scalo (Ap.J. Suppl. 41, 513-547, 1979) note that a function whose slope smoothly steepens with increasing mass is more valid than the power law in star formation and the stellar mass function. They favor a half-Gaussian distribution in log $[M/M_\odot]$. This is important for models of star formation.

DRAPATZ: The interpretation of the infrared data given deals with stars of mass $\geq 1\ M_\odot$. In that mass range the power law and half-Gaussian distribution do not differ sufficiently to influence the results. For the shape of the complete initial mass function, however, different functions must be considered, also to prevent a divergence for low masses.

INFRARED STUDIES OF STAR-FORMING REGIONS - SUMMARY

B. Zuckerman
University of Maryland
College Park, Maryland 20740

When Gareth Wynn-Williams first asked me to give this summary talk he specifically requested that I be "provocative." Having already been to Hawaii three times I realized how much easier it is to be laid back rather than provocative here - indeed, I but rarely rise above the ground energy state. So, the prospect of giving this talk has so terrified me that I have been sitting in the front row taking notes for two days now - something I have never done at any meeting. I even sought out the advice of an expert summarizer and first row sitter, my colleague Virginia Trimble, who advised me not to worry, but to simply "tell them what they should have said but didn't."

Well, fortunately, I need not have worried so much since there have been a remarkable number of exciting new results on star formation presented at this meeting. Indeed in order to remember one as interesting I must go back over five years to the meeting that Peter Mezger organized at an Austrian ski resort in January 1975. Many of my friends and colleagues who are here today were also at that meeting. I can still remember, as if it were only yesterday, sitting at a table with Eric Becklin and Nick Scoville and watching, with amazement, the amount of German beer they could consume at a single sitting.

Since it is impossible to review everything of the last two days in fifteen minutes my remarks will be limited to three areas: (1) the largest scale - galactic structure; (2) the intermediate scale - giant molecular clouds; and (3) the small scale - protostars and protoplanets.

In his beautiful paper, Reinhard Genzel much too modestly called the measurement of the distance to Orion by measurement of the proper motions in the H_2O masers a "minor sidelight." His group has also measured the distance to W51 in the same way. This represents a potential breakthrough in our ability to determine the spiral structure of the Milky Way.

We live in a spiral galaxy, but what does it look like? The HI and CO pictures are still quite controversial. Does the CO show spiral

C. G. Wynn-Williams and D. P. Cruikshank (eds.), Infrared Astronomy, 275–279.

structure? Maps of extragalactic CO, although helpful, can never entirely solve this problem for our own galaxy. We must find a method for determining distances in the Milky Way that is independent of kinematics. Optical wavelengths are out because of extinction due to the interstellar dust. Radio astronomers have no "standard candles" but the H_2O maser technique seems promising if we can measure the proper motions of many maser spot groups throughout the Milky Way. Improvements in VLBI sensitivities and the passage of time are both required, so this is a difficult experiment.

I have long believed that measuring the galactic distance scale is the birthright of infrared astronomers (perhaps my colleague at Maryland, Frank Kerr, has helped to prejudice me here). They must search for standard infrared candles in the galactic plane at large distances from the Earth. Bob McLaren mentioned to me a few weeks ago that he is planning to study Cepheids in the infrared. This welcome step should help to improve the extragalactic distance scale but won't do much for galactic astronomy. Perhaps IRAS or some other upcoming satellite will detect many infrared sources that ground based astronomers, by measurement of the Brackett lines, can establish to be main-sequence early B and late O-type stars located at large distances from the Earth.

The challenge to infrared astronomers is to find a standard candle before the radio astronomers are able to use the H_2O masers to map out the arms of the Milky Way.

Ultimately, distances will be measured by interplanetary interferometry. Even here infrared astronomers have a fundamental advantage. A recently proposed giant space interferometer (Buyakas et al., 1979) consisting of individual telescopes a few kilometers in size separated by 10 Astronomical Units can, in principle, achieve an angular resolution of 10^{-10} arc seconds at 1-mm wavelength. In practice, scattering by interstellar and interplanetary plasma irregularities may prevent us from reaching this limit. Since this effect varies as λ^2 the best angular resolution will be achieved in the infrared. (Again, optical observations will be limited by interstellar dust.)

The second topic that I would like to discuss are the giant molecular clouds (GMC's). Radio astronomers have discovered GMC's and have shown that there is a problem in understanding their kinematics and long term stability. We have been searching for years for a solution without any resounding successes by studying things such as "clumping." I believe that we can study clumping until the cows come home without really solving the problem. I think that the solution to the mystery probably lies in infrared observations and their interpretation but infrared astronomers must search harder and over larger areas. Perhaps IRAS will do some of the work, for example at 20 μm, as mentioned by Neal Evans in his talk.

To indicate more precisely what I have in mind, I would like to mention two models for GMC's that have recently been proposed. I emphasize that it is entirely possible that neither of these models supplies the correct explanation for the supersonic velocities observed in GMC's but they do illustrate the kinds of infrared observations that might be usefully attempted.

The first model is due to Norman and Silk (1980) who suggest that molecular clouds contain numerous hidden, embedded T Tauri-like stars which possess very energetic winds (of the type that I discuss below in conjunction with Table 1) that stir up the clouds. There are at least two potential problems with this model - the T Tauri stars may not exist in the requisite numbers and, even if they do, they may not possess energetic winds. A second model for supplying large amounts of energy of bulk motion to the GMC's is due to Bash et al. (1980), who propose that as GMC's orbit in the galaxy they are struck by numerous smaller molecular clouds. This model also has some potential problems which I will not discuss here.

At any rate, models such as these can be checked by a variety of infrared observations. The embedded stars will appear as continuum sources. The energetic winds and cloud-cloud collisions will produce shock waves which will heat the gas. This should result in far infrared atomic and molecular line emission of the type observed near HII regions by the Cornell and Berkeley groups. Hill and Hollenbach (1978) have considered the propagation of a 10 km/s shock in a dense cloud. When T has fallen to 100 K behind the shock, far infrared lines from species such as OI, CII, CI, CO, OH may be observable with sensitivities attainable during the 1980's.

It is conceivable that bulk motions in GMC's may be coupled to and controlled by embedded magnetic fields. If so, it may be very difficult to detect the fields at either radio or infrared wavelengths. Polarization measurements in the far infrared may be of some help.

In the past two days we have heard a fantastic array of new material related to the formation of stars, especially in the Kleinmann-Low/Becklin-Neugebauer region in Orion. I long for the simple, "good old days" at the Austrian ski resort where I could relax and drink beer and pretend that I understood Orion.

Probably the most exciting current problem is understanding the physical mechanism responsible for the high-velocity, high-energy phenomena seen by infrared and radio spectroscopists in Orion and other regions of star formation. In the most extreme regions the energy in bulk motions is a few percent of the energy in radiation.

The underlying physical mechanism responsible for this mass motion is still obscure. Indeed the only mention of physical mechanisms at this meeting was the brief exchange between Drs. Beckwith and Beckman regarding radiation pressure driven winds. Dr. Beckman informs me that he has a paper in press in MNRAS on this topic. But radiation driven winds are very controversial as the explanation of the pre-main sequence phenomena listed in Table 1 and other mechanisms that rely on conversion of rotational, pulsational, magnetic, or gravitational energy into kinetic energy of mass outflow deserve consideration. SS433 may indicate that such conversions are possible. Presently unknown factors, such as whether the underlying protostar is single or multiple, may enter into the solution.

At any rate, there now exists a tremendous opportunity for infrared theoreticians and observers to work together to elucidate the fundamental underlying mechanism(s). This certainly does not mean that we should stop gathering new data on the high energy protostellar phenomena but only that the time may already be ripe for some clever intuitive calculations and inspired guesses as to fundamental mechanisms.

To help, I would like to present a Table, a version of which I first showed at a review talk on Orion at the Mexico City AAS meeting in January 1979. A similar table also appears in a paper submitted to the Ap. J. by the Kuipers and me. But first it is necessary to define a quantity I call "Over Pressure" (O.P.): O.P. $\equiv (\dot{M}V_\infty)/(L/c)$. For mass loss driven by radiation pressure, $\dot{M}V_\infty = \tau L/c$. Here $\dot{M}$ is the rate of mass lost, V_∞ is the terminal velocity of the flow, L is the underlying source luminosity, and τ is the effective optical depth for coupling this luminosity to the expanding gas (and dust) cloud. If most of the underlying luminosity escapes without being either scattered or absorbed by the gas and dust then $\tau \ll 1$. If the radiation suffers multiple absorptions and scatterings before it escapes, then $\tau > 1$.

TABLE 1.

OVER PRESSURES

OBJECT	$L/L_\odot$	A_V(Mag)	O.P.
Main Sequence O-Type Star	10^5	0	$\sim$1
Infrared Giants (IRC+10216)	10^4	$\sim$15	$\sim$1
Planetary Nebulae	10^4	Moderate	$\sim$1
T Tauri	20	Moderate	
Optical lines			$\sim$20(?)
2 μm H_2 Emission			25
Herbig Haro Objects	10-1000	$\gtrsim$20(?)	
Shock Models } Bullet Models			10^2-10^3
2 μm H_2 Emission			$\gtrsim 10^2$(?)
Orion, DR21, etc.	10^5	20-2000(??)	
2 μm H_2 Emission			10^2-10^3

As an example, consider a planetary nebula for which the following parameters probably apply: $\dot{M} \sim 10^{-5}\ M_{\odot}$/year, $V_{\infty} \sim 20$ km/s, $L \sim 10^4\ L_{\odot}$. Then, as indicated in Table 1, O.P. ~ 1. Other assumed values for $\dot{M}$ and V_{∞} may be found in Kuiper et al. (1980). The basic problem is to explain the large over pressures indicated for the pre-main sequence objects.

The over pressure given for T Tauri based on optical data is due to the classical interpretation of Kuhi. This interpretation, strong outflow, has been questioned by Ulrich and Knapp (1979) hence the "?" in the O.P. column. The other question marks are mainly due to the still uncertain observational situation in the infrared.

A correlation between A_V and O.P. which <u>would</u> be expected for a radiation driven wind but <u>would not</u> necessarily be expected for other driving mechanisms mentioned above is <u>not</u> evident in Table 1. Future observations should clarify this point. Measurement of the total far infrared luminosity of low mass star forming regions, such as those described by Dr. Nordh, is very important to establish L for T Tauri stars and the exciting stars of HH objects.

Interferometric measurements such as those discussed by Drs. Howell and Sibille may tell us whether the stars that excite the winds are single or multiple or whether they are surrounded by massive disks lying close to the stars. Infrared interferometers may also help us to learn more about protoplanets (and hence planets themselves). Since there is more mass and luminosity in the protoplanetary nebulae than in the planets themselves infrared observers have an advantage over astrometric types. A particularly interesting possibility would be to see if preplanetary disks are found in multiple star systems as well as around single stars.

It is hard to think of an area in astronomy that is as rich in promise in the coming decades as is the infrared.

REFERENCES

Bash, F., Hausman, M., and Papaloizou, J.: 1980 (preprint).
Buyakas, V.I., Danilov, Yu.I., Dolgopolov, G.A., Feoktistov, K.P., Gorshkov, L.A., Gvamichava, A.S., Kardashev, N.S., Klimashin, V.V., Komarov, V.I., Melnikov, N.P., Narimanov, G.S., Prilutsky, O.F., Pshennikov, A.S., Rodin, V.G., Rudakov, V.A., Sagdeev, R.Z., Savin, A.I., Semenov, Yu.P., Shklovskii, I.S., Sokolov, A.G., Tsarevsky, G.S., Usyukin, V.I., and Zakson, M.B.: 1979, Acta Astronautica 6, p. 175.
Hill, J.K., and Hollenbach, D.J.: 1978, Astrophys. J. 225, p. 390.
Kuiper, T.B.H., Zuckerman, B., and Kuiper, E.N.R.: 1980, Astrophys. J. (submitted).
Norman, C., and Silk, J.: 1980, Astrophys. J. 238, p. 158.
Ulrich, R., and Knapp, G.R.: 1979, Astrophys. J. 230, p. L99.

THE GALACTIC CENTER

Ian Gatley
United Kingdom Infrared Telescope

E. E. Becklin
University of Hawaii

Recent infrared and radio observations of the Galactic Center are reviewed. For the region between 1 and 100 pc most of the observed phenomena can be explained by a large density of late-type stars, a ring of molecular material, and a number of regions of active star formation. The central parsec (Sgr A) appears to be a unique region of activity in the Galaxy; this result is based on recent high angular resolution data at 30 to 100 μm and high resolution spectral line observations at 12.8 μm. The observations are discussed in terms of the mass, density structure, and luminosity of the region; the ultimate source of the activity is discussed.

I. INTRODUCTION

The subject of the Galactic Center has recently been reviewed in a major work by Oort (1977). More recent reviews have been given by Wollman (1978) and Lacy (1980). In this paper we will concentrate chiefly on the discussion and interpretation of data obtained since Oort's review, and will further restrict our topic to two particular scale sizes. Firstly we will inventory the types of sources found within the central few hundred parsecs of the Galactic Center. Our purpose is to provide a frame of reference for the interpretation of observations of galactic nuclei such as those discussed at this meeting by Rieke; the Galaxy provides our only realistic chance to resolve spatially the various components within this region. Secondly we will discuss the nature of the central parsec of the Galaxy; again the proximity of the source provides an unparalleled opportunity for detailed study. We will see that this central region is unique in the Galaxy.

The numbers given in this review are order-of-magnitude only. At a distance of 10 kpc our two scale sizes correspond roughly to the central 1° and central ½' of the Galaxy respectively.

C. G. Wynn-Williams and D. P. Cruikshank (eds.), Infrared Astronomy, 281–295.

II. THE REGION 1 < R < 200 PARSECS

a) Near Infrared Observations

The near infrared radiation from the Galactic Center comes predominantly from late type stars (Becklin and Neugebauer 1968). Figure 1 shows the 2.2 μm emission of the central 1°.1 presented in the form of a photograph. As in the larger scale 2 μm maps presented at this meeting by Okuda the emission is extended along the Galactic plane. The peak in the 2 μm surface brightness is the dynamical center of the Galaxy (Becklin and Neugebauer 1968; Sanders and Lowinger 1972). The extinction over much of the region corresponds to $A_V \sim 30$ mag (Becklin and Neugebauer 1968; Becklin et al. 1978b). The presence of patches of rather different extinction has been demonstrated (Rieke, Telesco, and Harper 1978; Becklin and Neugebauer 1978; Lebofsky 1979);a good example is the dark patch in Figure 1 directly below the center, which coincides in position with a molecular cloud (Solomon et al. 1972; Hildebrand et al. 1978). For our purposes the distortion to our interpretation caused by variable extinction is slight.

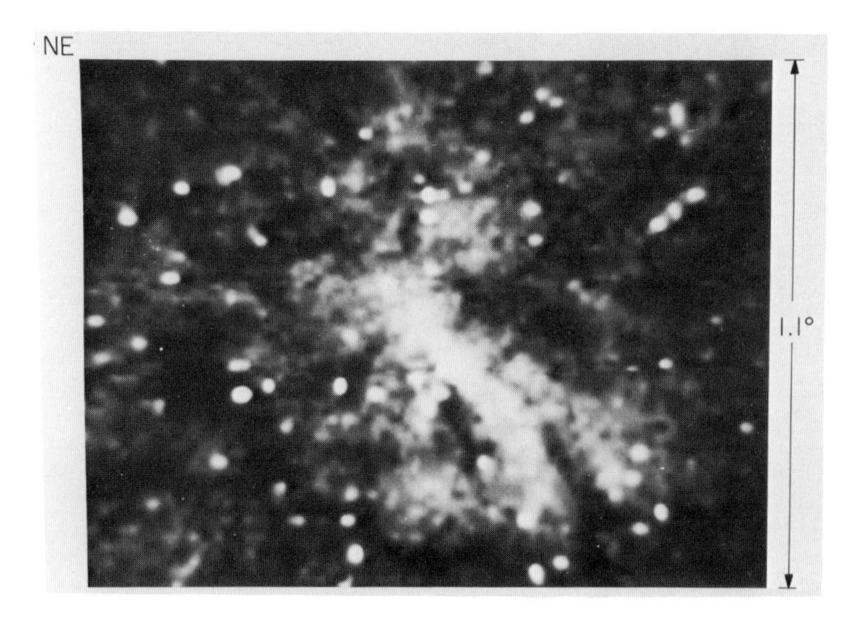

Figure 1. A 2.2 μm map of the central 1° of the Galactic Center in the form of a black and white photograph. The angular resolution is 1'.2 (Becklin, Neugebauer, and Early 1974).

b) Radio Continuum Observations

Figure 2 is a composite map of the radio continuum emission from the central 1° of the Galaxy (Downes et al. 1979); the solid contours

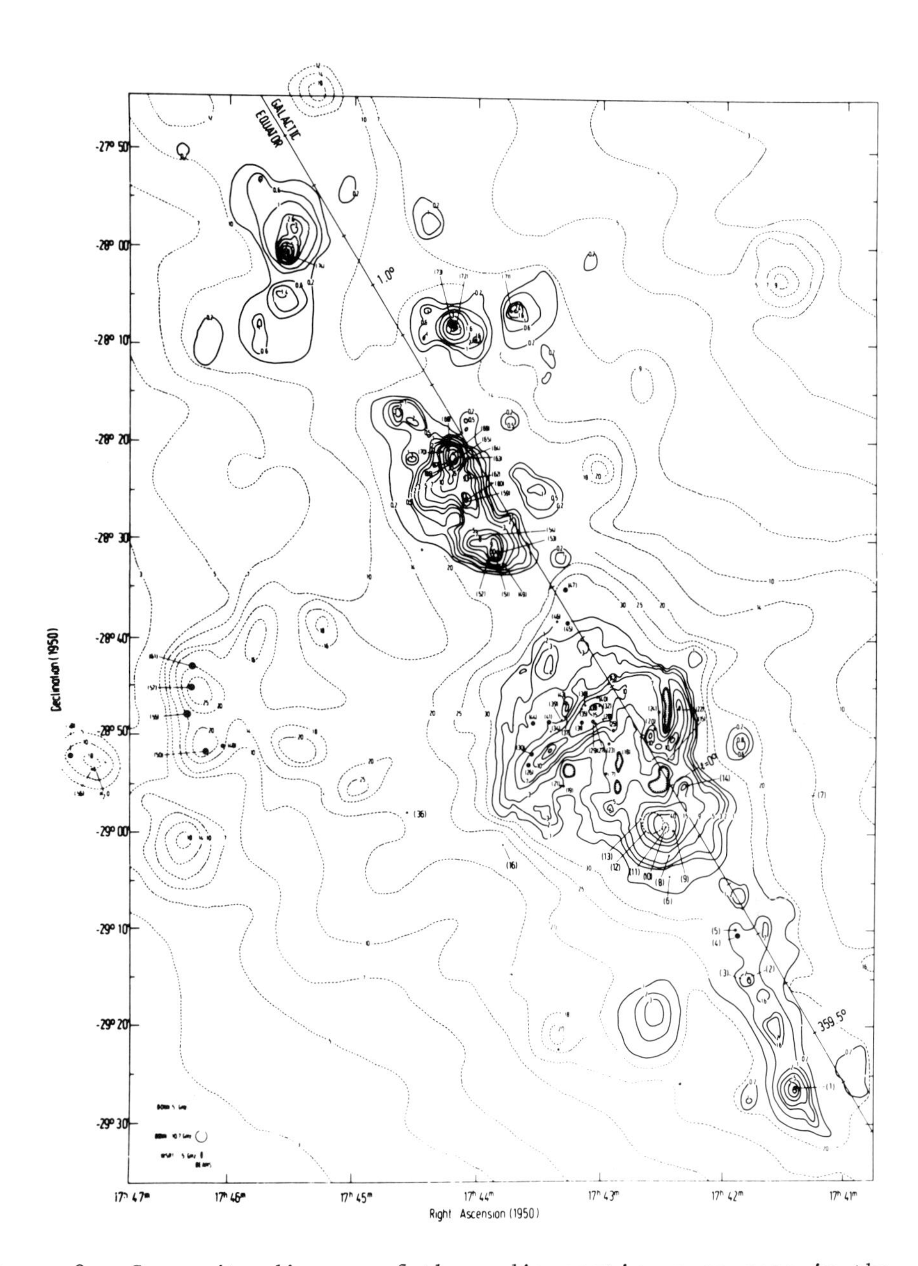

Figure 2. Composite diagram of the radio continuum sources in the Galactic Center region (Downes et al. 1979).

show the 10.7 GHz flux. The radio emission is also extended chiefly along the Galactic plane. Recombination line measurements (Pauls et al. 1976; Gardner and Whiteoak 1977) show that much of the radio continuum emission is free-free radiation from HII regions; the origin and nature of the extended non-thermal emission will not be discussed here.

The HII region Sgr A (West) (Ekers et al. 1975; Pauls et al. 1976) coincides with the peak of the 2 μm emission, and is located in the Galactic nucleus.

c) Molecular Line Observations

Many authors (Scoville 1972; Kaifu, Kato, and Iguchi 1972; Scoville and Solomon 1973; Bania 1977; Cohen 1977) have interpreted longitude-velocity plots of molecular emission from the central 2° of the Galaxy in terms of an expanding ring of material (see Oort 1977, §5.5). The observations suggest that the ring has a current radius R ∿ 200 pc, an age of several million years, and a total mass of $\sim 10^7$ $M_\odot$ (e.g., Gusten and Downes 1980). If the presence of such an expanding ring is a persistent or recurrent event in the Galactic Center, an average mass loss rate from the inner Galaxy of 1 $M_\odot$ yr^{-1} is required to sustain it.

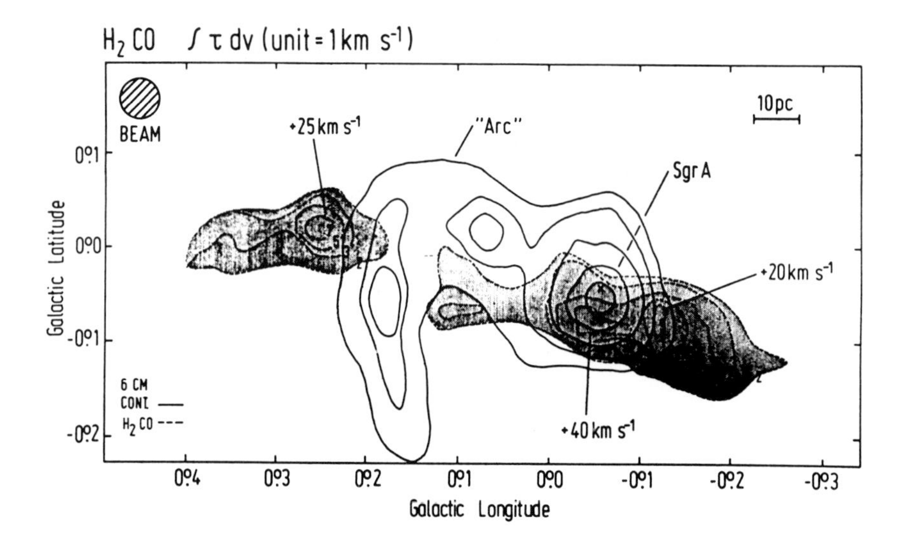

Figure 3. Maps of the integrated H_2CO optical depth of the dense molecular clouds in the vicinity of Sgr A (Gusten and Downes 1980).

Many discrete molecular clouds are also seen in the Galactic plane; Sgr B2 is a famous example (e.g., Oort 1977). The line of sight location of these clouds is a problem; in particular, molecular clouds are seen near the position of Sgr A. Figure 3 (Gusten and Downes 1980) shows the integrated H_2CO optical depth map superposed on the radio continuum contours (cf. Figure 2). In the past such data have sometimes been interpreted as showing that Sgr A is located in a molecular cloud (e.g., Figure 1, Wollman 1978). <u>This is unlikely to be true</u>. Recent data suggest a model in which no molecular clouds lie within 50 pc of the Galactic center (Gusten and Downes 1980, Figure 4). Dynamical arguments based on tidal disruption considerations give a similar limit if the clouds are to have significant lifetimes.

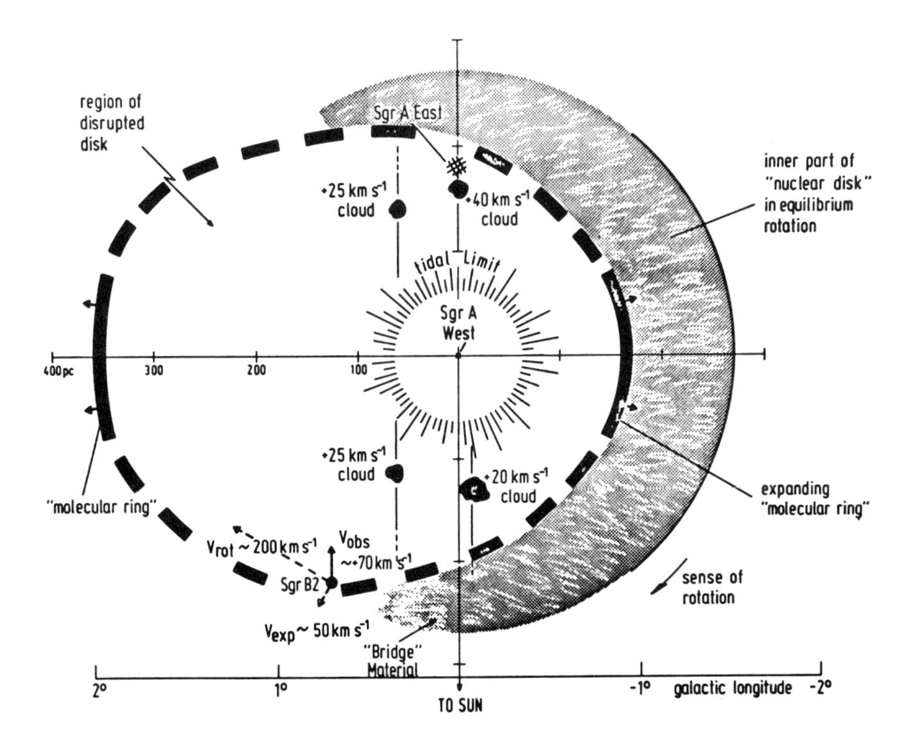

Figure 4. Possible locations of molecular clouds in the Galactic Center region, as viewed from above the galactic plane (Gusten and Downes 1980).

d) Far Infrared Observations

In reviews at this meeting by Okuda and Drapatz we saw that far infrared emission at wavelengths around 100 μm is common throughout the Galactic plane. This radiation is thermal emission from dust heated by starlight. As we saw, much effort has been invested in defining the dominant luminosity source in heating the dust. Provided that the absorption optical depth is not less than unity, little power will escape conversion into the far infrared regardless of its source. It is most important to appreciate that the dust density necessary to satisfy this condition is low, and does not imply, for example, densities typical of molecular clouds.

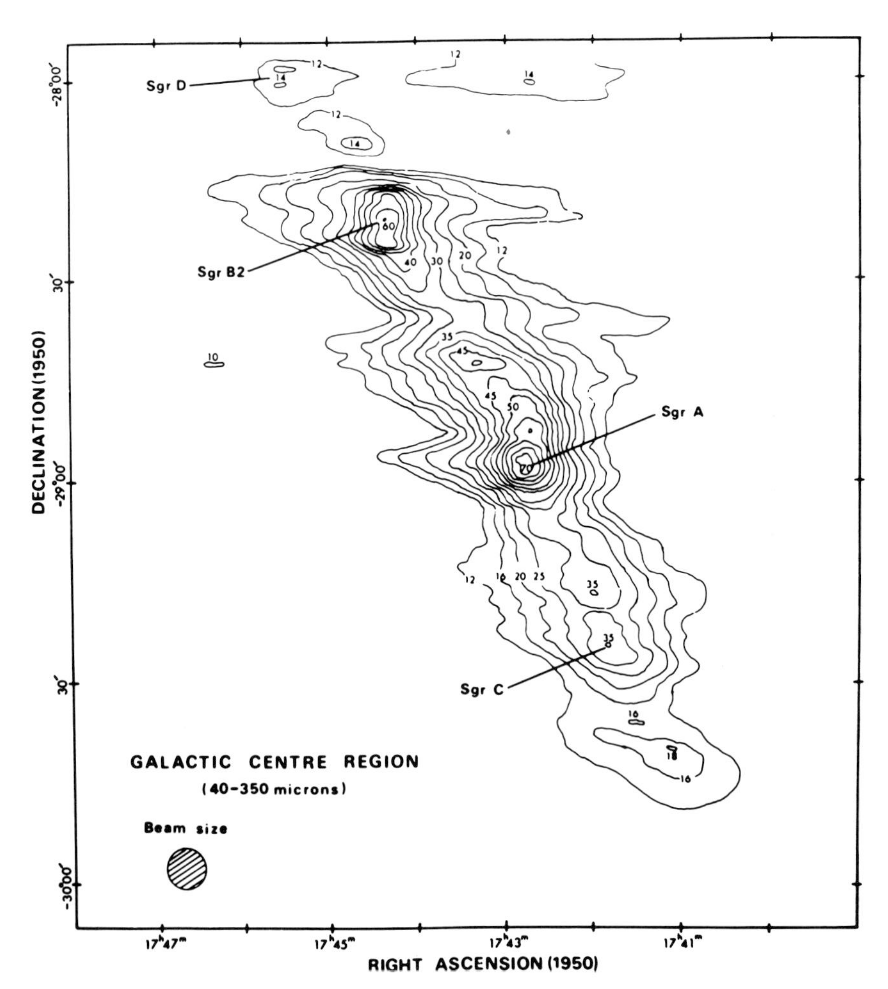

Figure 5. Far-infrared map of the Galactic Center region. Contour unit: 0.95×10^{-10} W m^{-2} (Alvarez et al. 1974).

Figure 5 (Alvarez et al. 1975) shows the far infrared emission from the central 1° of the Galaxy. Again the emission is elongated along the Galactic plane. There is a correlation between far infrared and radio continuum surface brightness (cf. Figure 2). Clearly this is because the luminosity available to heat the dust is locally increased at the HII regions. If the dust density also rises locally, no more luminosity will be available to heat it.

e) Star Formation and Molecular Clouds

There is a correlation between the occurrence of molecular clouds and HII regions in the region $10 < R < 200$ pc. For example, Sgr B2 is one of the densest molecular clouds in the Galaxy (Scoville, Solomon, and Penzias 1975) and contains seven HII regions each comparable to the Orion Nebula (Martin and Downes 1972). There is very strong evidence that star formation is occurring at a high rate (Mezger and Smith 1977) in the molecular clouds near the Galactic Center (Gatley et al. 1978; Gusten and Downes 1980).

III. THE CENTRAL PARSEC

The position of the Galactic Center is the peak of the 2 μm distribution (Figure 1) and thus the peak of the density of the late type stars. It coincides with the HII region Sgr A (Figure 2) and a cluster of compact 10 μm sources (Rieke and Low 1973). The supernova remnant Sgr A (East) (Pauls et al. 1976) is not within the central parsec, and will not be discussed. There is an ultracompact non-thermal source (Balick and Brown 1974; Lo et al. 1977) which probably is within the central parsec.

The late type stars within the central parsec provide $L \gtrsim 10^6\ L_\odot$ and their mass is deduced to be $\gtrsim 3 \times 10^6\ M_\odot$ (Becklin and Neugebauer 1968). The HII region requires $L > 10^6\ L_\odot$ in the ultraviolet to maintain the ionization.

The concentration of observable phenomena toward the Galactic Center makes confusion a problem; only angular resolutions of 30" or better can lead directly to understanding of the central parsec. Recent improvements in technology have now allowed study at this level of detail. These observations clearly show that conditions within the central parsec are unique.

a) The NeII Line Observations

The properties of the ionized plasma in Sgr A can be deduced from observations of the NeII line at 12.8 μm (Aitken et al. 1974; Willner 1978; Lacy et al. 1979; Lacy et al. 1980). These observations are of critical importance because they have both high angular resolution and high spectral resolution, which gives velocity information.

Figure 6 shows the central NeII line velocity (Lacy et al. 1980) at the positions of the 10 μm continuum features (Becklin et al. 1978a). The distribution of integrated NeII line intensity is similar in appearance to the 10 μm continuum map of the hot dust.

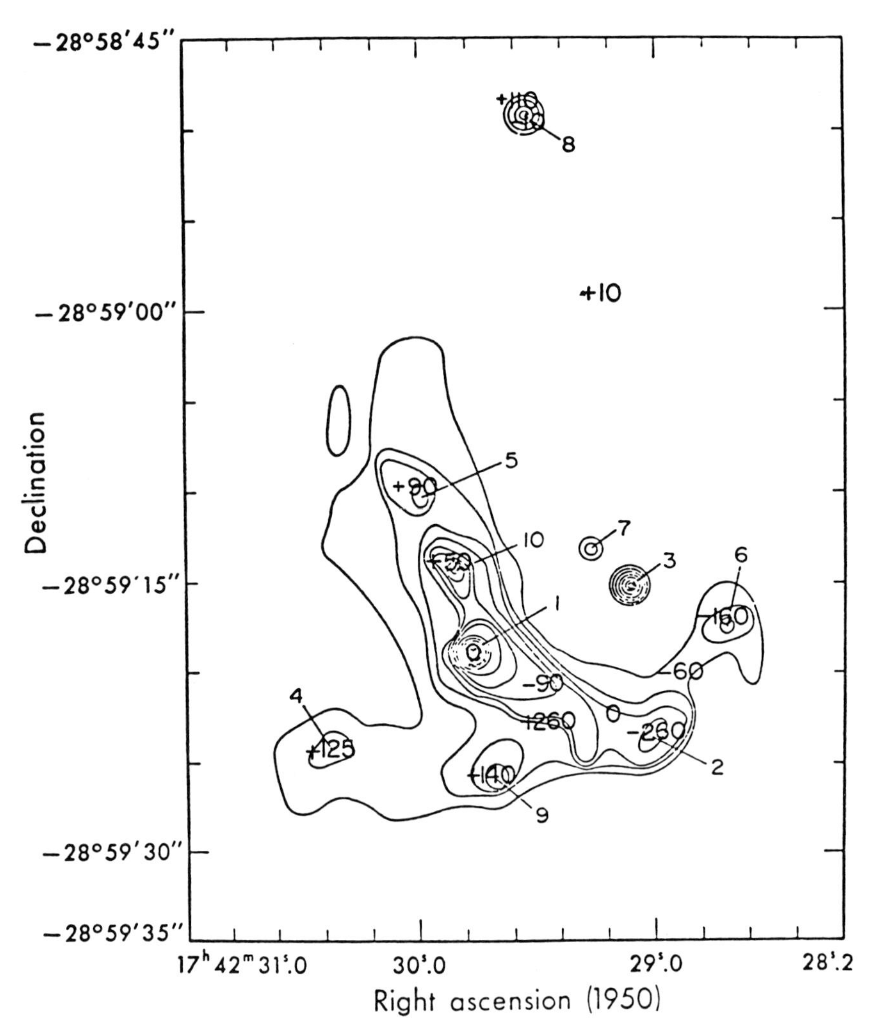

Figure 6. Velocities of the ionized gas clouds superimposed on the 10 μm continuum map (Lacy et al. 1980).

The major results of the NeII line observations can be summarized as follows:

1) The plasma is very clumpy.

2) The velocities of these plasma clumps are high (200 km/sec, Figure 6).

3) The velocity dispersion in the individual plasma clumps is high (50 km/sec).

4) The plasma excitation is low, implying an effective temperature for the exciting source (if radiative) of $T_{ex} < 35{,}000$ K. (This result follows from observations of fine structure lines other than NeII).

5) A typical plasma clump has

Mass: $M \sim 1\ M_\odot$
Luminosity: $L \sim 10^5\ L_\odot$
Internal velocity dispersion: $\Delta v \sim 50$ km/sec
Size: $D \sim 0.1 - 0.5$ pc
Lifetime: $T \sim 10^4$ years

6) The individual plasma clouds

(a) Show no intrinsic silicate absorption (Becklin et al. 1978a);
(b) Show no unidentified spectral features common in the 3-8 μm spectra of galactic HII regions (Willner et al. 1979; Gatley et al., unpublished);
(c) Are collectively responsible for the bulk of the free-free emission from Sgr A (West);
(d) Fill only a few percent of the total volume of Sgr A (West).

7) The most likely source of the gas in the plasma clumps is mass loss from the late type stars. There must also be a gas sink, in order to maintain the very low gas density throughout most of the volume of the central parsec.

In summary, then, the NeII observations clearly show that Sgr A is not a normal galactic HII region.

b) The Far Infrared Observations

In normal galactic HII regions the far infrared measurements will give the total luminosity, as we saw earlier in §1(d). Indeed low resolution ($\geq$ 1') observations of Sgr A (e.g., Gatley et al. 1977) have been interpreted in just this way.

High resolution ($\leq$ 30") far infrared observations (Harvey, Campbell, and Hoffman 1976; Rieke, Telesco, and Harper 1978) show a double lobed structure. Figure 7 (Gatley, Becklin, and Werner 1981) shows new 30 and 100 μm maps of Sgr A. The maps reveal the striking result that the 100 μm surface brightness <u>does not peak at the position</u>

of the center of the Galaxy[1]; the 30 μm surface brightness does peak at the center. Independent of any model, these observations require that the dust absorption optical depth at uv and visual wavelengths be much less than unity within the central parsec of the Galaxy.

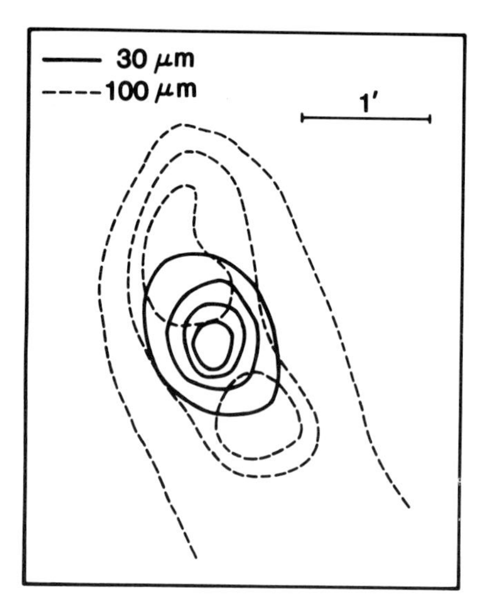

Figure 7. 30 μm and 100 μm maps of the Galactic Center obtained with coincident 30" beams (Gatley, Becklin, and Werner 1981).

The observed double-lobed structure can be generated by a central source of luminosity $L \sim 1 - 3 \times 10^7 L_\odot$. Such a central source is consistent with the observed fact that the far infrared color temperature is highest at the position of Sgr A. This luminosity source can ionize the plasma clumps, suggesting that the individual clumps may not contain luminosity sources. The double-lobed appearance of the far infrared surface brightness occurs naturally if the dust density is very low in the central parsec, and rises beyond 1 pc with the dust somewhat confined in the plane of the Galaxy.

c) The Excitation of the Plasma Clumps

Observations of infrared fine structure lines (Lacy et al. 1980) appear to rule out shock excitation rather conclusively.

---

[1] The original paper by Harvey, Campbell, and Hoffmann (1976) contains a positional error in declination; all the contours in their Figure 1 should be moved 15 arcsec to the north. This erratum was kindly communicated to us by Dr. Harvey.

Three sources of radiative ionization seem worthy of consideration, namely:

1) Star Formation

The absence of molecular clouds within 50 pc of Sgr A suggests that conditions appropriate for star formation do not currently exist at the Galactic Center. This is not an insuperable objection, as is shown by comparison with the source 30 Doradus in the Large Magellanic Cloud (Werner et al. 1978).

2) Planetary Nebulae

The plasma clumps have been compared to planetary nebula. (Becklin and Neugebauer 1975; Alloin, Cruz-Gonzáles, and Peimbert 1976; Rieke, Telesco, and Harper 1978). There are problems with the production rate, mass, excitation, and luminosity if the sources are planetary nebulae, but the problems of understanding stellar evolution in the environment of the Galactic Center remain to be addressed (e.g., Edwards 1980).

3) Exotic Phenomena

Many of the phenomena observed throughout the central few hundred parsecs can be explained via a single hypothesis, namely a central object currently radiating $\sim 10^7$ $L_\odot$ largely in the soft ultraviolet, and capable of periodic outbursts. This exotic object can be responsible for:

(a) The excitation of the plasma clumps;

(b) The far infrared luminosity;

(c) Removing gas to maintain the low densities either through accretion or expulsion;

(d) The expansion of the molecular ring at $R \sim 200$ pc through an outburst some 10^7 years ago;

(e) The high rate of star formation, which can be triggered by the passage of the expanding molecular ring;

(f) The occurrence of the ultracompact non-thermal source;

(g) A link between the Galactic nucleus and active nuclei observed in other galaxies.

REFERENCES

Aitken, D. K., Jones, B., and Penman, J. M.: 1974, Monthly Notices Roy. Astron. Soc. 169, 35P.
Alloin, D., Cruz-Gonzáles, C., and Peimbert, M.: 1976, Astrophys. J. 205, 74.
Alvarez, J. A., Furniss, I., Jennings, R. E., King, K. J., and Moorwood, A. F. M.: 1974, in A. F. M. Moorwood (ed.), "HII Regions and the Galactic Centre,"Proceedings of the Eighth ESLAB Symposium, ESTEC, Noordwijk, Netherlands, p. 69.
Balick, B., and Brown, R. L.: 1974, Astrophys. J. 194, 265.
Bania, T. M.: 1977, Astrophys. J. 216, 381.
Becklin, E. E., and Neugebauer, G.: 1968, Astrophys. J. 151, 145.
Becklin, E. E., and Neugebauer, G.: 1978, Publ. Astron. Soc. Pacific 90, 657.
Becklin, E. E., and Neugebauer, G.: 1975, Astrophys. J. Letters 200, L71.
Becklin, E. E., Matthews, K., Neugebauer, G., and Willner, S. P.: 1978a, Astrophys. J. 219, 121.
Becklin, E. E., Matthews, K., Neugebauer, G., and Willner, S. P.: 1978b, Astrophys. J. 220, 831.
Becklin, E. E., Neugebauer, G., and Early, D.: 1974, in A. F. M. Moorwood (ed.), "HII Regions and the Galactic Centre," Proceedings of the Eighth ESLAB Symposium, ESTEC, Noordwijk, Netherlands, p. 227.
Cohen, R. J.: 1977, Monthly Notices Roy. Astron. Soc. 178, 547.
Downes, D., Goss, W. M., Schwarz, U. J., Wouterloot, J. G. A.: 1979, Astron. Astrophys. Suppl. Ser. 35, 1.
Edwards, A. C.,: 1980, Monthly Notices Roy. Astron. Soc. 190, 757.
Ekers, R. D., Goss, W. M., Schwarz, U. J., Downes, D., and Rogstad, D.: 1975, Astron. Astrophys. 43, 159.
Gatley, I., Becklin, E. E., Werner, M. W., and Wynn-Williams, C. G.: 1977, Astrophys. J. 216, 277.
Gatley, I., Becklin, E. E., Werner, M. W., and Harper, D. A.: 1978, Astrophys. J. 220, 822.
Gatley, I., Becklin, E. E., Werner, M. W.: 1981 (in preparation).
Gardner, F. F., and Whiteoak, J. B.: 1977, Proc. Astron. Soc. Australia 3, 150.
Gusten, R., and Downes, D.: 1980, Astron. Astrophys. (submitted).
Harvey, P. M., Campbell, M. F., and Hoffmann, W. F.: 1976, Astrophys. J. Letters 205, L69.
Hildebrand, R. H., Whitcomb, S. E., Winston, R., Stiening, R. F., Harper, D. A., and Moseley, S. H.: 1978, Astrophys. J. Letters 219, L101.
Kaifu, J., Kato, T., and Iguchi, T.: 1972, Nature Phys. Sci. 238, 105.
Lacy, J. H., Townes, C. H., Geballe, T. R., and Hollenbach, D. J.: 1980, Astrophys. J. 241 (in press).
Lacy, J. H., Baas, F., Townes, C. H., and Geballe, T. R.: 1979, Astrophys. J. Letters 227, L17.

Lacy, J. H.: 1980, in P. A. Wayman (ed.), "Highlights of Astronomy 5", Dordrecht: Reidel, p. 163.
Lebofsky, M. J.: 1979, Astron. J. 84, 324.
Lo, K. Y., Cohen, M. H., Schilizzi, R. T., and Ross, H. N.: 1977, Astrophys. J. 218, 668.
Martin, A. H. M., and Downes, D.: 1972, Astrophys. Letters 11, 219.
Mezger, P. G., and Smith, L. F.: 1977, in T. de Jong and A. Maeder (eds.), "Star Formation," IAU Symp. No. 75, Dordrecht: Reidel. p. 133.
Oort, J. H.: 1977, Ann. Rev. Astron. Astrophys. 15, 295.
Pauls, T., Downes, D., Mezger, P. G., and Churchwell, E.: 1976, Astr. Astrophys. 46, 407.
Rieke, G. H., and Low, F. J.: 1973, Astrophys. J. 184, 415.
Rieke, G. H., Telesco, C. M., and Harper, D. A.: 1978, Astrophys. J. 220, 556.
Sanders, R. H., and Lowinger, T.: 1972, Astron. J. 77, 292.
Scoville, N. Z.: 1972, Astrophys. J. Letters 175, L127.
Scoville, N. Z., and Solomon, P. M.: 1973, Astrophys. J. 180, 55.
Scoville, N. Z., Solomon, P. M., and Penzias, A. A.: 1975, Astrophys. J. 201, 352.
Solomon, P. M., Scoville, N. Z., Jefferts, K. B., Penzias, A. A., and Wilson, R. W.: 1972, Astrophys. J. 178, 125.
Werner, M. W., Becklin, E. E., Gatley, I., Ellis, M. J., Hyland, A. R., and Robinson, G., and Thomas, J. A.: 1978, Monthly Notices Roy. Astron. Soc. 184, 365.
Willner, S. P.: 1978, Astrophys. J. 219, 870.
Willner, S. P., Russell, R. W., Puetter, R. C., Soifer, B. T., and Harvey, P. M.: 1979, Astrophys. J. Letters 229, L65.
Wollman, E. R.: 1978, in W. B. Burton (ed.) "The Large-Scale Characteristics of the Galaxy," IAU Symp. No. 84, Dordrecht: Reidel, p. 367.

DISCUSSION FOLLOWING PAPER DELIVERED BY I. GATLEY

BLITZ: A recent article by Burton and Liszt uses elliptical streamlines to successfully model the tilted distribution of molecular gas seen toward the Galactic Center. It therefore appears unnecessary to require an explanation for the radial motions of this molecular gas (i.e. expansion) in the Galactic Center. That is, such motions may not exist for the overwhelming majority of molecular gas.

GATLEY: That is certainly a possibility, and the interpretations of the dynamical data are by no means unequivocal. In particular, Oort (1977) in his review lists various phenomena at larger radii which indicate that an explosion has taken place.

BLITZ: Oort pointed out a discrepancy of a factor of 10 (greater, in fact, according to more recent observations) between the extinction derived toward the Galactic Center from the infrared observations, and those implied by CO observations. Do you have any new thoughts on this problem?

GATLEY: While the extinction through a typical molecular cloud is certainly much larger than 27 magnitudes, the data indicate that these clouds are displaced from the peak of Sgr A, and that there is no such cloud in the line of sight to the center.

HYLAND: What kind of exotic object do you postulate as the power source for the Galactic Center?

GATLEY: The neon velocities can be interpreted to give a limit on the mass of a central object. Two possible distributions of matter have been suggested by the Berkeley Group. One is a uniform distribution of stars following the contours of the 2 μm profile, giving some 10^7 $M_\odot$ of stars. A second model comprises, in addition to the stars, a compact 3×10^6 $M_\odot$ object at the center. According to those authors, a black hole accreting 10^{-5} $M_\odot$ yr^{-1} will emit very close to the luminosity and the exciting spectrum we require to explain both the far infrared luminosity and the ionization we see in the plasma clouds.

HOLLENBACH: A fundamental problem with having a single ionization source in the Galactic Center is that it should be brighter at 2 μm than any observed source. This is true for an accretion disk around a massive black hole as well as a single blackbody source with $T < 35\,000$ K.

GATLEY: If you tilt your accretion disk so that its axis is along the axis of the Galaxy, does this reduce its 2 μm flux density to the level of IRS 16, the source which coincides with the VLBI radio source?

HOLLENBACH: We need more than a factor of 10, and we have doubts about the validity of tilting the disk in the direction you suggest, because the neon data are indicative of rotation in quite a different direction to the rotation axis of the Galaxy.

GATLEY: I would answer that by saying that firstly the neon data are based on only about 14 points. Secondly, the plasma clouds are young objects so that the velocities of the gas reflect the motions of the objects producing the gas, not the overall infall motions towards the center.

T. L. WILSON: Determining the lifetimes of the clouds by dividing their diameters by their expansion velocities is open to question. Such an argument applied to Orion gives a dynamical age much less than the age of the stars.

SCOVILLE: Also, if each H II region has a lifetime of about 10^4 years and 1 $M_\odot$ of gas, I do not see how you would produce enough material out of them to give the 1 $M_\odot$ yr^{-1} needed for the expanding molecular ring.

GATLEY: I agree with you. Either there has to be some sweeping-up of mass, or else we have to let things accumulate for a length of time. It is interesting that if we take the expected mass loss rate for late-type stars you accumulate more gas and dust than you see over very large-scale sizes out to 100 pc, in only about 10^8 years.

SCOVILLE: Why is this material not ionized?

GATLEY: Perhaps the source of excitation is located in the very nucleus. Also in the plane itself, where the material will accumulate, one will achieve optical depth unity.

INFRARED STUDIES OF THE STELLAR CONTENT IN EXTRAGALACTIC SYSTEMS

Marc Aaronson
Steward Observatory
University of Arizona

ABSTRACT

Normal galaxies emit most of their radiation longward of one micron, and many problems related to our understanding of galaxy formation and evolution can be fruitfully addressed with measurements at near-infrared wavelengths. Such problems include the make-up of the red stellar population, the star formation rate, the initial mass function, metallicity effects, and mass-to-light ratio. How these various quantities depend on morphological type, on total mass (or absolute magnitude), on radial position, and on environment is also of great interest. In this review recent infrared observations of extragalactic stars, star clusters, and galaxies having important bearing on these questions are discussed. Particular emphasis is placed on new evidence for the presence of a finite intermediate age population in early-type systems. This evidence comes from observations of intermediate age stars in many Magellanic Cloud globular clusters, observations of such stars in at least one nearby dwarf spheroidal (Fornax), the difficulties of fitting theoretical isochrone models to the red V-K colors of E and S0 galaxies, and the differences in the infrared color-magnitude relations for the Virgo and Coma clusters.

"It is not very bright to measure a blue magnitude for a red object."
-- Vera Rubin

I. INTRODUCTION

Knowledge about the stellar content in other galaxies is essential for understanding the processes underlying galaxy formation, structure, and evolution. Because most galaxies emit the bulk of their radiation in the near infrared, measurements at these wavelengths provide a uniquely important view of the universe. The development of highly

C. G. Wynn-Williams and D. P. Cruikshank (eds.), Infrared Astronomy, 297–316.

sensitive InSb detector systems within the last five years has enabled a number of detailed infrared stellar-content studies to be undertaken, and as we shall see, unexpected and intriguing results have often emerged.

This article will mostly be concerned with observations conducted in the one to three micron region of objects whose luminosities are contributed mainly by direct stellar radiation. The emphasis will be placed on newer results, which in many cases have yet to appear in print. For longer wavelength studies, and a more comprehensive listing of the published literature, the reader is referred to the recent review article of Rieke and Lebofsky (1979). Note that the topic of active star formation in spiral galaxy nuclei will scarcely be touched upon here, as the subject is discussed in another article in this volume by G. Rieke. Also, I have not attempted to cover the entire field of stellar populations, but rather, to indicate those areas where the infrared can make and has made a particularly worthwhile and interesting impact.

The plan is to first discuss IR observations of individual stars and star clusters and what they have revealed about the stellar content in nearby galaxies (including our own). Next, attempts to synthesize the light of E and S0 galaxies, systems which are thought to have ceased all star formation long ago, will be summarized. Emphasis will be placed on metallicity effects and the nature of the coolest stellar component. Finally, studies involving spiral galaxies will be examined.

Certain photometric observations have proven especially useful in analyzing composite red stellar content. These include broad band measurements at J (1.2 microns), H (1.6 microns), K (2.2 microns), and L (3.5 microns); and narrow band measurements of 2.3 micron CO and 1.9 micron H_2O indices. The CO feature is a sensitive measure of luminosity in late-type stars, and is well-suited for determining the giant-to-dwarf ratio; while H_2O absorption provides information about the very coolest stars present (see Baldwin, Frogel and Persson 1973). Further uses for the IR data will be developed as we proceed.

II. STARS

IIa. Relevant Galactic Work

In order to construct stellar synthesis models of the integrated light from extra-galactic systems, it is desirable to have a library of calibrating stars having as wide a range as possible in temperature, luminosity, and metallicity [M/H]. Calibrations of the various IR colors and band strengths of concern here as a function of spectral type and luminosity are given by Johnson (1966a); Lee (1970); Persson, Aaronson and Frogel (1977); Frogel et al. (1978); Aaronson, Frogel and Persson (1978b); and Ridgway et al. (1980). The main drawback of these compilations is that they are primarily applicable to stars believed to be near or perhaps a bit below solar [M/H] values.

Some progress in establishing an empirical library having wider metallicity coverage is being made through the study of giant stars in galactic clusters (Cohen, Frogel and Persson 1978; Pilachowski 1978; Frogel, Persson and Cohen 1979; Persson et al. 1980c). These authors measure JHK magnitudes and narrow-band CO and H_2O indices for cluster giants reaching typically 3 mag below the giant branch tips. An accurate bolometric magnitude (M_{bol}) and effective temperature (T_{eff}) can be calculated from the JHK data, leading to reliable placement of the cluster stars on the physical HR diagram for comparison with theory. The model atmosphere calculations in Cohen et al. (1978) suggest that both J-K and V-K colors can be used to derive proper effective temperatures for metal-poor giants, independent of gravity or metallicity. The models, however, do not include molecules, which may partially explain their discrepancy with recent empirical evidence suggesting that a metallicity dependence in the (V-K, T_{eff}) relation does exist for the coolest stars (Mould and Aaronson 1980; and see below).

An important result from this series of papers, and a point I will come back to several times again, concerns the good agreement between the empirical giant branches and the theoretical models of Rood (1972) and Sweigart and Gross (1978). In particular, it appears that in the HR diagram, asymptotic giant branch (AGB) stars in metal poor galactic globulars do not rise above the first giant branch tip, a limit believed to be imposed from mass loss.

One other interesting result has developed from the cluster studies. At the metal poor and rich ends, both the CO strength and the effective temperature of the giant branch correlate well with other [M/H] estimators. However, there is some breakdown of this correlation with the intermediate-metallicity clusters. Pilachowski (1978) has pointed out that for these clusters the CO index correlates not with [M/H] but with horizontal branch morphology. The infrared observations thus lend support to the idea that variation in the [CNO/Fe] ratio relates to the well-known second parameter problem in galactic globulars.

IIb. Stars in Nearby Galaxies

The galaxies in the Local Group are sufficiently close to be easily resolved into stars. The Magellanic Clouds provide a particularly happy hunting ground for infrared studies of the cool stellar content, as there is available a large sample of stars with little reddening all at the same distance. Glass (1979) has published JHKL photometry of late-type supergiants in both the LMC and SMC. He finds that the bolometric magnitudes of these stars decrease with advancing spectral type, in apparent contrast to such stars in our own galaxies. Glass also finds that for the brightest supergiants, types M0-M1, $|M_{bol}$ (LMC) $|>|$ M_{bol} (SMC) $|>|$ M_{bol} (Milky Way)$|$. However, this last result may be subject to some serious selection effects.

Humphreys and Warner (1978) have detected several luminous blue variables in both M31 and M33 at K. These stars appear similar to Eta Carinae-type and S-Doradus type variables in the Milky Way and LMC.

Mould and Aaronson (1979, 1980) have spectroscopically identified numerous carbon and M stars at the tip of the giant branch in a number of red globulars in the Magellanic Clouds. JHK photometry for many of these stars has been obtained by Frogel, Persson and Cohen (1980a) and Mould and Aaronson (1980); the IR data is essential for accurately calculating the bolometric luminosities of these very cool giants.

The M_{bol}'s of the Cloud cluster carbon stars are found to range between -4.5 and -5.5 mag, which places them 1-2 magnitudes above the giant branch tip of galactic globulars. Mould and Aaronson (1979, 1980) have argued that these are AGB stars which can only be produced by clusters of intermediate age (e.g. 1 - 8 billion years old). These authors have developed a relation between the age of a cluster and the luminosity at the top of the AGB. Physically, such a relation exists because more massive (i.e. younger) stars have larger envelopes and can attain higher luminosities before attrition of the envelope due to mass loss. From the IR data, Mould and Aaronson have derived independent age estimates for the clusters in their sample, and find strong hints of a cluster age-metallicity correlation, which is expected if chemical enrichment has occurred over the cluster formation period. In the Clouds the presence of both young globulars and old ones resembling those in the Galaxy was known for years. The identification of intermediate age clusters suggests that the birth of Cloud globulars has been a continuing process, in marked contrast to the situation in the Milky Way.

The interpretation that carbon stars imply (relative) youth in metal-poor populations provides a natural explanation for the remarkable observations of Blanco, Blanco and McCarthy (1978). These authors find that in the central bulge of the galaxy, M stars outnumber carbon stars 300 to 1; while in the field of the LMC carbon stars outnumber M stars 2 to 1; and in the SMC they outnumber M stars 50 to 1. Thus, the predominance of oxygen-rich stars in the Milky Way is consistent with the presence of an old enriched stellar content; while the Clouds would appear to have a large intermediate-age, metal-poor population, with the absence of M stars in the SMC reflecting a lower [M/H] than is found in the LMC. This view agrees well with Butcher's (1977) finding from an analysis of the LMC luminosity function that a large LMC intermediate-age field population is indeed present.

There are several further aspects of the Cloud cluster IR data worth mentioning. First, the JHK colors of the SMC cluster carbon stars are in the mean bluer than those in the LMC, which supports the view that the SMC has lower mean metallicity. Second, the non-carbon cluster stars scatter nicely about the mean galactic field (J-H, H-K) two-color relation, and yet depart significantly from the field (J-K, V-K) relation. The departure is in the sense expected if V-K is metallicity dependent, and yet surprisingly large. The upshot is that values of T_{eff} calculated from J-K are significantly less (typically 200K) than values calculated from V-K. Since J-K appears less sensitive to changes in [M/H] than V-K, a (J-K, T_{eff}) relation seems preferable for use with cool, metal-poor stars. Using a T_{eff} calibration based on model atmospheres, Frogel _et al_. (1980a) have argued that the location in the HR diagram of M-type cluster stars was inconsistent with theoretical isochrones if the stars are of intermediate age. This problem is resolved when a (J-K, T_{eff}) scale tied to the stellar diameter work of Ridgway _et al_. (1980) is used instead.

Aaronson and Mould (1980) have observed a number of the very red stars found by Demers and Kunkel (1980) in the Fornax dwarf spheroidal galaxy, both spectroscopically and in the infrared. These stars appear very similar in nature to the Cloud cluster carbon stars. Most significantly, the M_{bol}'s calculated from the JHK photometry place these stars $\sim$ 2 magnitudes above the first giant branch tips of galactic clusters. In analogy with the Clouds, it would appear that Fornax also contains an intermediate-age population. By comparing the fractional carbon star light in Fornax and the Cloud globulars, Aaronson and Mould estimate that 20% of Fornax is of intermediate age. These authors have suggested a correlation between age spread in dwarf spheroidals and their total masses, with Fornax, the most massive dwarf spheroidal, retaining longest the gas necessary for continued star formation. Small numbers of carbon stars have now also been identified in Sculptor, the next most massive spheroidal known. IR photometry of those stars is needed to confirm their similarity with the Fornax and Cloud carbon stars.

In the next few years we can expect to see an increasing number of papers related to IR measurements of extragalactic stars, which leads me to an area where the infrared could prove particularly valuable -- the determination of the distance scale. The two most powerful stellar distance indicators may be Cepheids, and the brightest M supergiants. The latter _appear_ to attain a remarkably constant maximum visual luminosity in galaxies with widely varying absolute magnitude (Humphreys 1980). However, two familiar problems plague the use of these indicators -- reddening and metallicity. For instance, some of the galactic calibrating Cephids have E(B-V) as large as 1.5 mag! Much of the well-known dispute between Sandage and Tammann, and de Vaucouleurs rests with how one models the correction for galactic absorption. Reddening in other galaxies may also be important. Work by Humphreys (1980) suggests that the brightest supergiants are reddened in their parent galaxies by typically 1/2 magnitude in A_V. This effect has not been previously accounted for, but that it exists should not be too surprising when one considers that the brightest supergiants are often found near dusty, active regions of star formation. An example of a possible metallicity problem is the different (B-V, period) relations for galactic and LMC Cepheids, a result attributed to differences in [M/H] and requiring a somewhat uncertain correction (e.g. Martin, Warren and Feast 1979). Furthermore, very little is understood about how changes in [M/H] might affect the calibration of the M supergiants.

Infrared observations of Cepheids and M supergiants have the potential for completely circumventing the problem with reddening, and perhaps alleviating any problems with metallicity as well. With present IR equipment, extra-galactic Cepheids are not measurable to very great distances, perhaps just barely to M31. The M supergiants are another story. These are well within the capability of detection in galaxies as far away as M101, at about 6.5 Mpc. Crowding problems are not as severe as one might think if the chopping is done against a reasonably uniform background since the objects are going to be much redder than anything else around. How rewarding IR studies of these distance indicators will actually be is unclear, and will depend on such factors as whether or not the M supergiants evolve in constant M_{bol}, as opposed to constant M_V. Nevertheless, the potential rewards are sufficiently great that I believe the effort involved is worth making.

III. STAR CLUSTERS

Infrared photometry of star clusters provides a powerful diagnostic tool in constructing stellar synthesis models for the integrated light of galaxies. Because clusters are coeval and of uniform metallicity, they enable a straightforward empirical check to be made against theoretical models of varying [M/H] and age. IR measurements have now been conducted for clusters in the Galaxy (Aaronson et al. 1978a, hereafter ACMM), in the Magellanic Clouds (Persson et al. 1980a), in Fornax (Aaronson and Mould 1980), and in M31 (Frogel, Persson and Cohen 1980b). Each of these studies will be briefly discussed.

ACMM measured JHK magnitudes and CO and H_2O indices for about 40 galactic globulars. The V-K, J-K, and CO indices were found to correlate well with [M/H]. This result was not unexpected, since the integrated colors should become redder with increasing [M/H] owing to greater line blanketing, shifting of the horizontal branch, and cooling of the giant branch. Because V-K colors are particularly sensitive to the last effect it was hoped that they would provide a more accurate measure of [M/H] than U-V. In fact only a marginally better correlation was seen for V-K than for U-V, but the errors in V-K are considerably larger owing to both instrumental and reddening uncertainties.

Persson et al. (1980a) have obtained IR data for some 84 clusters in the LMC and SMC. Included in this sample are open clusters, blue populous clusters, and red (including many intermediate age) globulars. An important result of this work is the degree to which carbon stars dominate the integrated light in some of the intermediate age clusters. NGC 2209 provides a particularly striking case. This cluster has a well determined age from the main sequence turnoff of 0.8 billion years (Gascoigne et al. 1976), and a comparatively high value of -0.5 for [M/H] (Gustafsson, Bell and Hejlesen 1976). Its U-V color is $\sim$ 0.9 (Bernard 1975), which is about 0.3 mag bluer than what one might expect for a galactic globular of comparable metallicity. However, the V-K color of this object is about 3.6 $\pm$ 0.2 (the large error being due to a significant mismatch in the optical and IR aperture sizes). This V-K color is a magnitude redder than comparable galactic globulars, and is in fact as red as the reddest early-type galaxies. It turns out the extraordinary colors of this object may be due to the presence of a single carbon star in the beam!

The IR Cloud cluster data will provide a much needed empirical constraint on future synthesis models of active star-forming galaxies. The only published models of this kind with IR colors are those of Struck-Marcell and Tinsley (1978). These models do not include objects with anything like the colors of NGC 2209, nor do they successfully fit the cloud open cluster data, suggesting a revision is required in the supergiant evolutionary tracks.

To determine whether Fornax might also contain intermediate age globulars, Aaronson and Mould (1980) measured JHK colors for four of this galaxy's clusters. One of these objects, Hodge 2, has quite red H-K colors and probably contains carbon stars.

Frogel _et al_. (1980b) have obtained infrared data for some 40 globular clusters in M31. The various IR colors are found to be tightly correlated with optical colors and line strengths. The spectral flux from 0.3 to 2 microns thus appears uniquely defined by a one parameter relation with [M/H]. Frogel _et al_. derive metallicities for the M31 from V-K colors; these authors agree with the earlier conclusion of van den Bergh (1969) that M31 globulars are weighted toward larger [M/H] values, but do not find that the upper bound of [M/H] in M31 is any greater. It should be noted that the _absolute_ [M/H] calibrations presented by ACMM and Frogel _et al_. (1980b) are called into question by recent evidence from Pilachowski, Canterna and Wallerstein (1980) and Cohen (1980) that the metal rich globulars 47 Tuc and M71, thought to have [Fe/H] $\sim$ -0.4, are actually metal poor, with [Fe/H] $\sim$ -1.1. However, the _relative_ ranking in [M/H] from IR colors should still be basically correct.

IV. GALAXIES

In this section we will see how the advent of modern infrared measurements has affected present day understanding of the composite light from galaxies.

IVa. Early-Type Galaxies

An extensive survey of early-type (E and S0) infrared galaxy colors was undertaken by Frogel _et al_. (1978) and Aaronson _et al_. (1978b). JHK colors were measured for some 50 nearby ellipticals, along with CO and H_2O band strengths. (Additional JHK data is also in Persson, Frogel and Aaronson 1979). These data placed some stringent contraints on the stellar synthesis models of Tinsley and Gunn (1976) and O'Connell (1976), favoring those models having giant-star dominated populations with small mass-to-light ratios ($M/L \sim 5$). Models having a slope in the initial mass function (IMF) of $x \lesssim 1$ (following Tinsley's 1972 notation) agreed best with the data. In addition, the strength of the H_2O index suggested the presence of a stellar component at least as late as M5. An important implication of all this was that a large evolutionary correction ($\Delta q_0 > 1$) was required in cosmological tests of the deceleration parameter (Tinsley 1972).

A further finding of Frogel et al. (1978) was that the V-K and J-K colors tended to redden with decreasing aperture size and increasing luminosity, trends that also occur with optical colors and line indices. Infrared color changes with radial position have also been reported by Strom et al. (1976, 1978). Strom et al. (1978) argued that the qualitative similarity in the ratio of optical to IR color changes seen between and within galaxies supported a similar causal origin for the color-aperture and color-magnitude effects. On the other hand, Frogel et al. (1978) reported marginal evidence for a difference in the ratio of color changes among and within galaxies, the effect existing largely in the J-K color. Curiously, such a result was not obtained by Aaronson (1977) for early-type spirals, which are largely bulge dominated. I believe the effect found by Frogel et al. arises in part from instrumental difficulties associated with measurements in the J band; in any event, the question of relative color changes is worth further pursuit.

It is probably safe to say that the color variations discussed above are today almost universally accepted as arising from changes in metallicity. In fact Faber (1973) has argued that for E galaxies the color-luminosity effect can be ascribed to a one parameter relation with [M/H], although the existence of a second parameter, perhaps related to axial ratio, has recently been proposed by Terlevich et al. 1980.

Tinsley (1978) and ACMM attempted to calibrate the integrated optical and IR colors of early-type galaxies as a function of [M/H]. Both the details of these models and the results differed in a number of important respects. In particular, Tinsley found that bright ellipticals were somewhat metal poor ($[M/H] \sim -0.3$) and that $x < 1$; whereas ACMM concluded the E galaxies were metal rich ($0 \lesssim [M/H] \lesssim 0.3$) and that x could be larger than 1 and perhaps as large as 2. The reason for these differences has sometimes been misunderstood. The principle cause is traced back to the fact that the giant branch used by ACMM is based entirely on the Ciardullo and Demarque (1977) theoretical isochrones, with AGB stars not rising in M_{bol} above the first giant branch tip, in agreement with the empirical evidence seen in galactic globulars; whereas Tinsley's giant branch is based on a semiempirical luminosity function for old disk stars and contains objects with M_{bol}'s up to -5.5 mag , that is, two magnitudes above the first giant branch tip.

A revision of the ACMM models has been calculated by Frogel et al. (1980b), incorporating the new temperature scale of Ridgway et al. (1980) and a new globular [M/H] scale at the metal rich end (see § III above).

These models are somewhat redder at a given [M/H] than the ACMM models. The models which best fit the early-type galaxies have solar metallicity and x = 1.35 (the Salpeter value).

ACMM and Frogel et al. (1980b) show that bright early-type galaxies have redder broad band colors and stronger narrow band indices than either galactic or M31 globulars. In a (U-V, V-K) color diagram the galaxy sequence is considerably displaced from the cluster sequence -- at fixed U-V galaxies are about 0.3 mag redder in V-K, and no combination of clusters can reproduce the galaxy colors.

Both the ACMM and Frogel et al. models adequately fit the cluster data, and the galaxy infrared color-color and color-band strength relations. However, there is a serious problem with both sets of models in that they do not at all match the UVK colors of E galaxies. (This problem is present in Tinsley's 1978 models, but apparently to a much lesser extent.)

Something appears to be missing in both the ACMM and Frogel et al. models which is present in early-type galaxies. The precise nature of this missing component rests with whether the problem lies in U-V or V-K; that is, whether a hot or cool luminous component needs accounting for. This is in fact not an obvious choice. A hot ultraviolet component has been detected in a number of bulge-dominated galaxies with IUE, but its size is not nearly enough to account for the discrepancy here (cf. Wu et al. 1980). Frogel et al. (1980b) concluded the problem exists with V-K, and I concur with this point of view. Before considering the nature of the missing cool component, further evidence for its existence will be discussed.

Aaronson, Persson and Frogel (1980) have carefully examined the U-K and V-K color-magnitude (CM) relations in the Virgo and Coma clusters. The original motivation for this study was the hope that the CM method of determining relative cluster distance moduli might be improved, since the size of the effect is doubled in going from a U-V to a U-K color. Whether this hope can be realized rests with the answer to two questions. First, for a given cluster sample is the scatter in the U-K CM relation smaller than in the U-V CM relation? Second, is the U-K CM relation universal? This latter point is a necessary requirement if we are to interpret the apparent magnitude difference between two clusters at some fixed color as being in reality an accurate reflection of relative distance modulus.

It turns out the answer to the first question is that the scatter in magnitude is the same when U-K is used instead of U-V. Thus no advantage is gained by going to the IR. This result provides strong support for the conclusion reached by Visvanathan and Sandage (1977) that the scatter in the CM effect is dominated not by observational error, but by true cosmic variation.

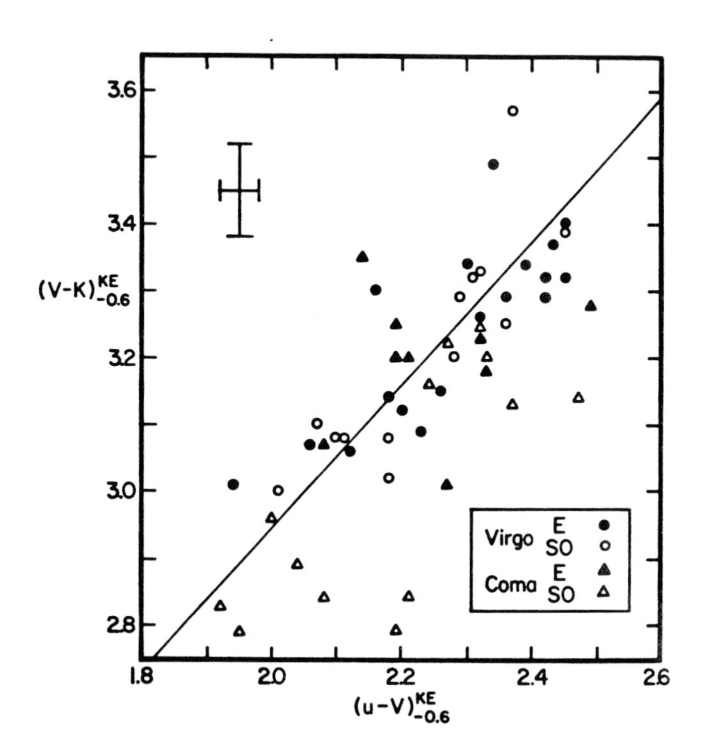

Figure 1. A (V-K, u-V) two-color diagram for early-type galaxies in the Virgo and Coma clusters. The data have been corrected for reddening, redshift, and aperture effect. Most Coma objects, especially the SO's, lie to the right of the least-squares line, which has been fit to the Virgo data only.

The answer to the second question is unexpected -- the (U-K, V) CM relation does not appear to have a universal form. To see this, first consider Figure 1, which shows a plot of U-V against V-K for galaxies in the Virgo and Coma clusters. (The notation in the Figure follows that of Visvanathan and Sandage 1977.) It is clearly seen in Figure 1 that the distribution of points in the two clusters does not overlap -- the majority of Coma points lie to the right of the least squares line fit to just the Virgo data. This shift could result from a color change in either U-V and/or V-K, but the following argument suggests that the shift is in fact primarily in V-K: the relative distance moduli between Coma and Virgo predicted from the ratio of redshifts is 4.17 mag (see Aaronson et al. 1980). Now the infall velocity toward Virgo is at present a controversial subject, but current estimates range between 200 - 500 km sec^{-1}. This leads to a predicted relative magnitude shift at fixed color of Δm = 3.50 - 3.85 mag. From the U-K CM relation

Aaronson et al. find a difference of $\Delta m = 3.00 \pm 0.2$ mag; the infall velocity implied from this magnitude difference seems unquestionally too large. (The author does not understand the value of 3.66 ± 0.35 mag quoted by Tammann, Sandage and Yahil 1979, which is based on these same data.) However, when separate solutions are made for U-V and V-K colors, the results found are $\Delta m = 3.49 \pm 0.2$ mag and 2.56 ± 0.3 mag, respectively. In other words, the value of Δm calculated from U-V colors is close to the expected range, while the value found from V-K colors is considerably too small. At fixed U-V color (and absolute magnitude), galaxies in Coma appear to be 10% bluer in V-K than those in Virgo, and it would seem that a second parameter is required to fully describe the CM effect. (Whether this parameter is related to the one proposed by Terlevich et al. 1980 and mentioned above is presently unclear.)

What might the nature of this second parameter be? First, note that in Figure 1 the color-shift effect is especially pronounced for the S0 galaxies. Second, when the UVK colors of nearby field galaxies are plotted, they fall along a relation similar to Virgo galaxies, with the brighter field S0's appearing to have somewhat redder V-K colors than those in Virgo. Thus, Coma galaxies (especially S0's) tend to be missing a cool stellar component found in Virgo and field galaxies and perhaps enhanced in field S0's, assuming dust effects can be neglected. Is this cool component related to the missing cool component discussed earlier in regard to the isochrone models? I suspect the answer is yes.

Let us turn to the question of what these cool stars might be. One possibility is red dwarfs, but this would appear to be ruled out by the observed CO indices. A second possibility relates to the idea of a very metal rich population missing from the isochrone models and the Coma galaxies. Perhaps galaxies in the dense regions of Coma were stripped of their gas before being able to form such a population.

A third possibility is that both the isochrone models and the Coma galaxies are missing giant stars which populate the HR diagram 1 to 2 magnitudes above the first giant branch tip. Recall that it is the presence of such stars in Tinsley's (1978) models which yield better agreement with the UVK colors. By analogy with the Magellanic Cloud work discussed earlier, we are led to the implication that early-type galaxies contain a finite population of intermediate age stars. In this regard, it is interesting that evidence of a significant intermediate age population has been presented for M32 by O'Connell (1980) and (as discussed earlier) for Fornax by Aaronson and Mould (1980). Stripping of Coma cluster galaxies at an early time could again account for the differences seen there.

Conclusive evidence for an intermediate age population in early-type galaxies would have a profound impact on the conventional wisdom related to these systems, so it is important to consider whether stars 2 magnitudes above the first giant branch tip necessarily implies youth.

The answer may depend on whether the mass loss rate is independent of [M/H]. The absence of high luminosity AGB stars from galactic globulars certainly argues that the mass loss rate does not change much from an [M/H] of -2 to -1 (although Frogel et al. 1980a do show that the reddest variables in 47 Tuc rise $\sim$ 0.4 mag above the first giant branch tip, so perhaps a small [M/H] dependence is present). If we are to avoid the notion of intermediate age stars, it would seem that the mass loss rate in one solar mass objects with [M/H] $\sim$ 0 must be significantly less than in metal-poor stars.

A number of arguments suggest that the intermediate age stars being considered here are not likely to be carbon stars, as in the Clouds. First, there is the circumstantial evidence provided by the absence of these stars in the Milky Way's bulge (Blanco et al. 1978). Second, the observed CO and H_2O indices appear incompatible with the presence of carbon stars (Aaronson et al. 1978b), as do K-L measurements (Aaronson 1978b; Frogel, unpublished data). Finally, FTS spectra discussed below show no evidence for the 1.8 micron Ballick-Ramsay bands, a strong carbon star discriminant.

A key debate occupying much of the astronomical literature in recent times has concerned the evolutionary status of S0 galaxies. One point of view (e.g. Sandage and Visvanathan 1978) is that environmental effects do not play a significant role in determining the formation and evolution of S0's. The opposite viewpoint (see Strom and Strom 1978) is that some (if not most) S0's are, in effect, stripped spirals. To support the former view, Sandage and Visvanathan cite the similar UBV color distribution for E and S0 galaxies both inside and outside clusters. However, we have seen evidence here that the UVK color distribution does depend on cluster environment. It is clearly of importance to confirm this effect with further observations of galaxies in clusters of varying richness and density.

Before leaving this thorny subject let me note that considerable progress in understanding the stellar content of other early-type systems may come about through several studies currently underway of stars in the Baade's window region. Coordinated optical and IR measurements will hopefully enable the luminosity and metallicity distribution of these stars to be determined, providing our best opportunity of studying a galaxy's bulge component in detail.

Finally, I should mention that several groups are actively pursuing near-infrared measurements of high-z ellipticals. Theoretical predictions based on single-burst models suggest that such measurements may have considerable advantage over optical data in constructing the classic Hubble diagram, in that evolutionary effects should be less important (Spinrad and Bruzual 1980). This will clearly not be the case if residual star formation in ellipticals has continued for a significant length of time, as suggested above. It is thus of great interest that Lebofsky (1980) may be finding significant H-K color evolution at redshifts too small to show such evolution in the single burst models.

IVb. Spiral Galaxies

In a seminal study, Johnson (1966b) found that the red stellar component in galaxies having a wide range in type was rather similar, that most of the longer wavelength radiation came from K and M giants, and that some very cool stars are present. More recently, Aaronson (1977, 1978a) has measured integrated JHK colors and CO and H_2O indices for a large number (∿ 90) nearby bright spiral galaxies. To first order, these data confirm Johnson's (1966b) conclusion that the underlying red population in spirals and ellipticals is the same: the JHK colors show little dependence on morphology; only for type Im is a significant "blueing" of the colors seen. Further, the narrow-band indices hardly vary through types E to Sbc. However, for types Sc and later there is evidence the CO index decreases, going from a mean value of 0.16 mag for type Sbc to 0.11 mag for type Im. This result seems to contradict star formation models of the type calculated by Huchra (1977) and Struck-Marcell and Tinsley (1978), which suggest an increasing dominance of supergiants in late-type spirals which would drive the CO index up. I believe the observed downturn in CO index is real and reflects the lower metallicity late-type spirals are thought to have, but the sample is small and additional systems should be measured to confirm the effect.

The UVK colors of spirals exhibit a rough segregation with type that is analogous to the trend seen with UBV colors. This spiral sequence exhibits a clear separation from the elliptical galaxy-globular cluster sequence in that, for the spirals, V-K changes as a much slower function of U-V. This effect is simply accounted for by the difference between a metallicity change, which immediately affects the V-K color, and a population change produced by adding young blue stars to old red ones, which has little effect on the V-K color.

The V-K and J-K colors of spirals become redder with decreasing aperture size, as in E and S0 galaxies. It is interesting to compare the relative size of the U-V to V-K color change as a function of type. For the Sab galaxies studied by Aaronson the changes in the two colors are comparable, both becoming redder by about 0.13 mag in the interval -0.3 to -1.3 in log A/D(0) (following the notation of de Vaucouleurs, de Vaucouleurs, and Corwin 1976). For Sb and Sbc galaxies the color change over this same interval is 0.32 mag and 0.36 mag for U-V, but only 0.13 mag and 0.22 mag for V-K, respectively. The color changes seen in V-K thus appear to lag behind those found in U-V. A similar sort of effect is seen when optical and infrared growth curves are compared -- for a given spiral type the IR growth curve is shallower than the optical one. Both of these results can be understood as the composite effect of superimposing a young population with a blue radial gradient on an old red population which is perhaps more spherically symmetric. For spiral galaxies of type Sa - Sb the IR colors appear in fact to be largely bulge dominated, and they very much support the notion that the stellar content in the bulges of E, S0, and spiral galaxies is the same. In particular, if the V-K color-aperture relation is interpreted as indicating a metallicity change, then it seems that in the mean the bulges of early-type spirals have the same composition gradients as are found in E and S0's, which suggests that bulge dominated galaxies have similar star formation and chemical enrichment histories _in their inner regions_. The last qualification is important, as Strom and Strom (1978) offer evidence that the halo color gradients found in E and S0 galaxies differ.

Aaronson (1977) also looked for a V-K color-magnitude effect in his spiral galaxy data, but found none. However, the sample was rather ill-suited for this purpose owing to the rather narrow magnitude coverage. Visvanathan and Griersmith (1977) have identified a significant U-V color magnitude relation for Sa spirals in Virgo and 11 other groups, and infrared observations of this sample would be of interest.

While a number of synthesis models for spiral galaxies have been published, two considerations make detailed comparison between the models and the IR data difficult. First, M giant stars are often either ignored completely or treated in a very superficial fashion as, for instance, by lumping all stars in a single M5 bin. Second, the IR data has been obtained with aperture sizes typically 4 - 5 times larger than the optical data on which most models have been based. Nevertheless, by analogy with early-type galaxies, it would seem that the spiral galaxy IR data can only support giant dominated models with small mass-to-light ratios. Dwarf enhanced models such as those calculated by Williams (1976) having $M/L \sim 50$ seem definitely ruled out.

Recently, Wynn-Williams _et al_. (1979) and Rieke _et al_. (1980) have considered models for the infrared emission of active spiral galaxy nuclei. The latter authors have argued that bursts of star formation having a low-mass cut-off in the IMF can account for most of the observed properties in M82 and NGC 253. A related result is that of Rieke and Lebofsky (1978), who find that significant 10 micron emission from the nuclei of spiral galaxies is quite common. Their measurements imply that such nuclei have a typical M/L_{bol} of about 0.2, a value that again suggests recent intense star formation. The cause behind such bursts and why they occur in some galaxies and not others is I believe a major unsolved mystery.

Infrared models for star formation on a more global scale have been constructed by Struck-Marcell and Tinsley (1978). These authors point out that UBV colors do not suitably distinguish age effects and cannot, for instance, discriminate between old galaxies with a recent star formation burst and truly young objects. They find that V-K colors provide about three-fold better age discrimination than U-V. With the hope of identifying possible young galaxies, Aaronson and Huchra (1980) have obtained IR data for a number of blue Markarians. The (reddish) V-K colors for these objects seem to favor the star-burst rather than young galaxy picture, assuming the Struck-Marcell and Tinsley models are even approximately correct. It is interesting, though, to consider the bluest object measured by Aaronson and Huchra - MK116, one of the so-called isolated extra-galactic H II regions of Sargent and Searle (1970).

This object has a V-K color of 0.5 $\pm$ 0.3, implying that some 50% of the stars formed in a burst about 3 x 10^7 years ago (see Struck-Marcel and Tinsley 1978). MK116 is probably the best case for a young galaxy, and a more accurate V-K color would be useful to pin down whether there are any old stars in it at all.

A minor controversy has arisen related to the stellar content in normal spiral galaxy nuclei. Using various optical line indices, McClure, Cowley and Crampton (1980) have argued that old horizontal branch stars rather than young main sequence stars are major contributors in the ultraviolet. However, the opposite conclusion is reached by Boroson (1980). The relevant IR observations suggest that in fact both mechanisms for producing UV light occur, but perhaps not in the same object. (Note that the overlap of galaxies in these two studies is rather small.) On the one hand, the late-type galaxies studied by McClure *et al*. show the strongest evidence for a decrease in [M/H], consistent with the downturn in CO index reported earlier. However, Rieke and Lebofsky's (1978) 10 micron measurements indicate considerable recent star formation in many spiral nuclei. Also, the two most extreme cases of young star contamination in Boroson (1980) - NGC 2681 and 5194 - do have strong CO indices, supporting the young star view. Nuclear (as opposed to large aperture) CO measurements for the objects in the Boroson and McClure *et al*. samples would be of interest in further pinning down the youth versus metallicity question.

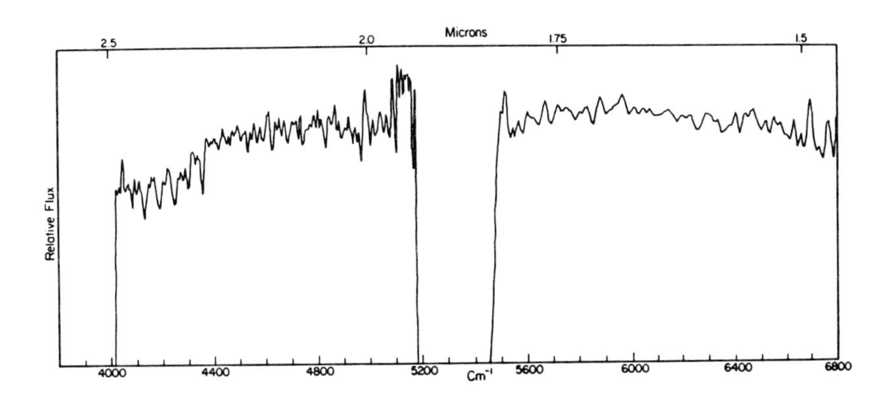

Figure 2. An infrared spectrum of the galaxy M81. The resolution is $\sim$ 8 cm^{-1} in the K-band region, and $\sim$ 16 cm^{-1} in the H-band region. Between 5000 and 5600 cm^{-1}, below 4100 cm^{-1}, and above 6600 cm^{-1} the spectrum is severely degraded by atmospheric extinction. The first five bands of first-overtone ^{12}CO are clearly visible. The broad continuum peak at 6000 cm^{-1} can be identified with the maximum in the continua of late K and M stars arising from the minimum in the H^- opacity.

Small aperture IR measurements of the semi-stellar nucleus in M31 have been obtained by Persson et al. (1980b) to test the proposal of Faber and French (1980) that the ratio of dwarf to giant light increases in going from the bulge to the nucleus of M31. The IR data does not support this contention. In particular, the CO index measured in a 2.5" aperture is 0.16 mag, identical to that found by Aaronson (1977) in a 107" aperture. Aaronson (unpublished data) has also measured the CO index of M31 in a 213" and 410" aperture, with the result being 0.16 mag in both cases. The CO index in this galaxy is thus constant over a factor of 164 in radius. Gradients in optical line indices attributed to [M/H] variations do occur in M31 (e.g. Cohen 1979), so it would seem that the CO index becomes saturated with [M/H] at the metal rich end.

In the next few years we can expect to see an increase in infrared spectroscopic studies of normal galaxies. The IR group at Arizona has been conducting such observations for the last few observing seasons. An example of an early effort is shown in Figure 2, which presents an FTS spectrum of M81 obtained by Aaronson and Boroson.

The first five first-overtone ^{12}CO bands are clearly visible in Figure 2, if little else, confirming that the index I have spoken of repeatedly is in fact really due to CO. The situation with regard to H_2O absorption is less clear. Although there is a hint of a downturn in the continuum above 4960 cm^{-1} and below 5700 cm^{-1}, the spectrum becomes so degraded by atmospheric extinction in the main region of interest that it cannot be regarded as either confirming or denying the presence of steam. As mentioned above, there is little hint of absorption due to the Ballick-Ramsay C_2 band at 1.8 micron, a strong signature of cool carbon stars. However, we can identify the broad continuum peak at about 6000 cm^{-1} as being due to the H^- opacity minimum which produces a maximum in the continua of cool K and M stars (cf. Catchpole and Glass 1974).

The spectrum in Figure 2 is too noisy to place a sound limit on the ^{12}C/^{13}C ratio, but this is a potentially very interesting number which spectra of only a little higher sensitivity can determine. In our own galaxy this ratio has been measured to vary from 90 in the sun to as low as 5 in some red giants (e.g. Dearborn, Eggleton and Schramm 1976). If mixing on the giant branch is responsible for the effect, we should expect to also find low values of the ratio in the giant-dominated light of other galaxies.

In closing, let me say that I have tried to demonstrate in this review how observations in the infrared have yielded important insight into many questions related to stellar populations that could not have been answered from optical data alone. With the development of large telescopes on Mauna Kea and Mt. Hopkins optimized for low background work, and continuing improvement in instrumental techniques (perhaps we will even see arrays someday), infrared studies in this fascinating area should continue to make a valuable contribution.

Preparation of this article was partially supported by the National Science Foundation.

REFERENCES:

Aaronson, M.: 1977, Ph.D. Thesis, Harvard University.
Aaronson, M.: 1978a, Ap. J. (Letters), 221, L103.
Aaronson, M.: 1978b, P.A.S.P., 90, 28.
Aaronson, M., Cohen, J. G., Mould, J. and Malkan, M.: 1978a, Ap. J., 223, 824 (ACMM).
Aaronson, M., Frogel, J. A. and Persson, S. E.: 1978b, Ap. J., 220, 442.
Aaronson, M. and Huchra, J.: 1980 (in preparation).
Aaronson, M. and Mould, J.: 1980, Ap. J. (in press).
Aaronson, M., Persson, S. E. and Frogel, J. A.: 1980 (in preparation).
Baldwin, J. R., Frogel, J. A. and Persson, S. E.: 1973, Ap. J., 184, 427.
Bernard, A.: 1975, Astr. Ap., 40, 199.
Boroson, T.: 1980, Ph.D. Thesis, University of Arizona.
Blanco, B. M., Blanco, V. M. and McCarthy, M. F.: 1978, Nature, 271, 638.
Butcher, H. R.: 1977, Ap. J., 216, 372.
Catchpole, R. M., and Glass, I. S.: 1974, M.N.R.A.S., 169, 69P.
Ciardullo, R.B., and Demarque, P.: 1977, Trans. Astr. Obs., Yale U., 35.
Cohen, J. G.: 1979, Ap. J., 228, 405.
Cohen, J. G.: 1980, private communication.
Cohen, J. G., Frogel, J. A., and Persson, S. E.: 1978, Ap. J., 222, 165.
de Vaucouleurs, G., de Vaucouleurs, A., and Corwin, H. G.: 1976, Second Reference Catalogue of Bright Galaxies (Austin: University of Texas Press).
Dearborn, D. S. P., Eggleton, P. P., and Schramm, D. N.: 1976, Ap. J., 203, 455.
Demers, S., and Kunkel, W. E.: 1979, P.A.S.P., 91, 761.
Faber, S. M.: 1973, Ap. J., 179, 731.
Faber, S. M., and French, H. B.: 1980, Ap. J., 235, 405.
Frogel, J. A., Persson, S. E., Aaronson, M. and Matthews, K.: 1978, Ap. J., 220, 75 (FPAM).
Frogel, J. A., Persson, S. E. and Cohen, J. G.: 1979, Ap. J., 227, 499.
Frogel, J. A., Persson, S. E. and Cohen, J. G.: 1980a, Ap. J., in press.
Frogel, J. A., Persson, S. E. and Cohen, J. G.: 1980b, Ap. J., in press.
Gascoigne, S. C. B., Norris, J., Bessell, M. S., Hyland, A. R. and Visvanathan, N.: 1976, Ap. J. (Letters), 209, L25.
Glass, I. S.: 1979, M.N.R.A.S., 186, 317.
Gustafsson, B., Bell, R. A. and Hejlesen, P. M.: 1977, Ap. J. (Letters), 216, L7.
Huchra, J. P.: 1977, Ap. J., 217, 928.
Humphreys, R. M.: 1980, Ap. J. (in press).
Humphreys, R. M. and Warner, J. W.: 1978, Ap. J. (Letters), 221, L73.
Johnson, H. L.: 1966a, Ann. Rev. Astr. Ap., 4, 193.
Johnson, H. L.: 1966b, Ap. J., 143, 187.
Lebofsky, M. J.: 1980, IAU Sym. No. 96, Infrared Astronomy (this volume).
Lee, T. A.: 1970, Ap. J., 162, 217.

Martin, W. L., Warren, P. R., and Feast, M. W.: 1979, M.N.R.A.S. 188, 139.
McClure, R. D., Cowley, A. P. and Crampton, D.: 1980, Ap. J. (in press).
Mould, J. and Aaronson, M.: 1979, Ap. J., 232, 421.
Mould, J. and Aaronson, M.: 1980, Ap. J. (in press).
O'Connell, R. W.: 1976, Ap. J., 206, 370.
O'Connell, R. W.: 1980, Ap. J., 236, 430.
Persson, S. E., Aaronson, M. and Frogel, J. A.: 1977, A. J., 82, 729.
Persson, S. E., Frogel, J. A. and Aaronson, M.: 1979, Ap. J. Suppl., 39, 61 (PFA).
Persson, S. E., Aaronson, M., Cohen, J. G., Frogel, J. A. and Matthews, K.: 1980a, (in preparation).
Persson, S. E., Cohen, J. G., Sellgren, K., Mould, J. and Frogel, J. A.: 1980b, Ap. J., (in press).
Persson, S. E., Frogel, J. A., Cohen, J. G., Aaronson, M. and Matthews, K.: 1980c, Ap. J., 235, 452.
Pilachowski, C. A.: 1978, Ap. J., 224, 412.
Pilachowski, C. A., Canterna, R. and Wallerstein, G.: 1980, Ap. J. (Letters), 235, L21.
Ridgway, S.T., Joyce, R.R., White, N.M. and Wing, R.F.: 1980, Ap. J., 235, 126.
Rieke, G. H. and Lebofsky, M. J.: 1978, Ap. J. (Letters), 220, L37.
Rieke, G. H. and Lebofsky, M. J.: 1979, Ann. Rev. Astr. Ap., 17, 477.
Rieke, G. H., Lebofsky, M. J., Thompson, R. I., Low, F. J. and Tokunaga, A. T.: 1980, Ap. J. (in press).
Rood, R. T.: 1972, Ap. J., 177, 681.
Sandage, A. and Visvanathan, N.: 1978, Ap. J., 225, 742.
Sargent, W. L. W. and Searle, L.: 1970, Ap. J. (Letters), 162, L155.
Spinrad, H., and Bruzual, G.: 1980, Paper presented at IAU Symp. No. 96.
Strom, S. E. and Strom, K. M.: 1978, IAU Symp. No. 84, The Large-Scale Characteristics of the Galaxy (Dordrecht: D. Reidel Publ. Co.).
Strom, S. E., Strom, K. M., Goad, J. W., Vrba, F. J., and Rice, W.: 1976, Ap. J., 204, 684.
Strom, K.M., Strom, S.E., Wells, D.C. and Romanishin, W.: 1978, Ap. J. 220, 62.
Struck-Marcell, C., and Tinsley, B. M.: 1978, Ap. J., 221, 562.
Sweigart, A. V., and Gross, P. G.: 1978, Ap. J. Suppl., 36, 405.
Tammann, G. A., Sandage, A. and Yahil, A.: 1979, Determination of Cosmological Parameters, Les Houches Summer School Lecture Notes, U. of Basel.
Terlevich, R., Davies, R. L., Faber, S. M., and Burstein, D.: 1980, (preprint).
Tinsley, B. M.: 1972, Ap. J., 178, 319.
Tinsley, B. M.: 1978, Ap. J., 222, 14.
Tinsley, B. M., and Gunn, J. E.: 1976, Ap. J., 203, 52.
van den Bergh, S.: 1969, Ap. J. Suppl., 19, 145.
Visvanathan, N. and Griersmith, D.: 1977, Astr. Ap., 59, 317.
Visvanathan, N. and Sandage, A.: 1977, Ap. J., 216, 214.
Williams, T. B.: 1976, Ap. J., 209, 716.
Wu, C.-C., Faber, S. M., Gallagher, J. S., Peck, M. and Tinsley, B. M.: 1980, Ap. J., 237, 290.
Wynn-Williams, C. G., Becklin, E. E., Matthews, K. and Neugebauer, G.: 1979, M.N.R.A.S., 189, 163.

DISCUSSION FOLLOWING PAPER DELIVERED BY M. AARONSON

WYNN-WILLIAMS: What are the prospects of obtaining a good observational discriminant between populations of late-type giants and late-type supergiants in galaxies?

AARONSON: The prospects are good if you believe evolutionary models. Models predict that the later the galaxy type the stronger the CO index should be, because of the larger number of supergiants. Observationally the opposite is true, which indicates to me that supergiants do not dominate the populations except in very active regions of star formation. The decrease in CO index in late-type spirals is probably a metallicity effect.

TOVMASSIAN: You said that red supergiants are found in galaxies with signs of activity of their nuclei (M82, etc.). Does it mean that red supergiants are young stars?

AARONSON: I'm sure almost everyone would agree that supergiants are young stars.

PERSSON: With regard to the question of whether the U-V, V-K problem in elliptical galaxies is due to U-V or V-K, could you comment on the large CO band strengths measured for faint elliptical galaxies whose broad band colors are the same as those of globular clusters at the same blue broad band color, and second, on the recent IUE observations of elliptical galaxies which show that they have UV excesses.

AARONSON: The CO indices of even the very faint dwarf ellipticals are much stronger than those of globulars of similar U-V color, indicating that even in what are thought to be very metal-poor objects there is a very red, possibly metal-rich population. The IUE observations have in fact revealed the presence of a very hot component in elliptical and early-type spirals. The models that have been produced for this component predict significant contributions shortward of 2200 Å, but very little contribution at U.

THOMPSON: In NGC 2209 you commented that U-V was too blue but V-K was too red, and suggested that this effect could be due to a single carbon star in the beam. I would point out that carbon stars are much too deficient in the ultraviolet to produce this effect.

AARONSON: This object has a well-determined metallicity, [M/H] ~ -0.5, and age from the main sequence turn off (~0.8 billion years). The blue colors, as compared with galactic globulars of comparable [M/H], are from the young upper main sequence stars.

BECKLIN: Does the assumption, which everyone makes, that the IMF is a single power law have any validity and does this assumption make a big effect on the model results?

AARONSON: Although people usually use it, the assumption is certainly not a good one for our Galaxy, for example. Recent work by Scalo indicates an initial mass function broken into four parts, with the exponent generally ranging between 0 and 2. In terms of the models, the exponent is most critical near the turn-off point, but values in the range 0 to 2 do not greatly affect the colors that you see. You only get significant contributions from dwarf stars by having values of the exponent much greater than 2.

THOMPSON: It is important to look for carbon stars in galaxies, and it might be worthwhile to set up a photometric narrow band at 1.77 μm to detect the very strong Ballick-Ramsay band of C_2.

AARONSON: The absence of the 1.77 μm Ballick-Ramsay C_2 band in the FTS spectrum of M81 [Figure 2] provides, I believe, strong evidence against the presence of carbon stars.

ZUCKERMAN: Observations of carbon stars in intermediate age extragalactic clusters may help to solve a venerable problem in stellar evolution—the lower limit to the main sequence mass of stars that eventually become carbon stars.

Another problem of more recent vintage concerns some of the extremely red AFGL objects. Many of these strong 10 and 20 μm sources are carbon stars. We don't really know if all of these extremely red stars are also very massive. (IRC +10216 probably contains a few solar masses in its expanding molecular envelope.) Observations of extragalactic carbon stars (that are contained in clusters) at 10 and 20 μm may help to clarify this question.

AARONSON: That is an interesting point. I believe that the luminosities of objects such as IRC +10216 are much greater than the bolometric luminosities of the carbon stars in the Clouds, but it would be very interesting to observe some of these Cloud carbon stars at 10 μm to see if they do have any big excesses.

RAPID STAR FORMATION IN GALACTIC NUCLEI

G. H. Rieke*
University of Arizona

ABSTRACT

A large percentage of spiral galaxies are forming stars at a very high rate in their nuclei. There are indications that the process of star formation is modified significantly in these regions, compared with the solar neighborhood. The star formation may be fueled by interstellar material that is captured by the nucleus. Further work is needed to explore these possibilities and to establish the evolutionary connections among galactic nuclei, various types of which can be distinguished by their radically different infrared properties.

One of the big surprises from the pioneering infrared astronomy of the '60's was the large infrared excesses of a number of galactic nuclei (see, e.g., Pacholczyk and Wisniewski 1967; Kleinmann and Low 1970). Entering the '70's, three basic questions had grown out of this discovery: 1.) how prevalent are high nuclear luminosities in galaxies; 2.) what is the radiation mechanism; and 3.) how is the luminosity produced. During this past decade, we have been able to answer all three of these questions, at least in a tentative way. The answers reaffirm the suggestions already made in the 60's that infrared observations would force major reassessments in our view of galactic nuclei.

The first systematic attempt to measure the incidence of infrared excesses found that $\gtrsim$ 40% of bright, nearby spiral galaxies have strong nuclear emission at 10 μm (Rieke and Lebofsky 1978). Far infrared observations by Telesco and Harper (1980), Becklin et al. (1980), Rickard, Harvey, and Thronson (1980), and Harper (private communication) show that most of the galaxies bright at 10 μm have even larger excesses near 100 μm. The infrared emission accounts for luminosities in the range of $\sim 10^9$ to $\sim 3 \times 10^{10}$ $L_\odot$, from regions

* Alfred P. Sloan Fellow

C. G. Wynn-Williams and D. P. Cruikshank (eds.), Infrared Astronomy, 317–328.

typically a few hundred parsecs in diameter. The high incidence of such large power outputs is a new and important input to theories of galactic evolution.

The emission mechanism for the infrared fluxes has been identified as thermal reradiation by dust (except for QSOs and some type 1 Seyfert galaxies, where the mechanism is still uncertain). This conclusion was first suggested by the general shape of the far infrared energy distribution and its resemblance to the spectra of Galactic thermal sources such as HII regions (see, e.g., Telesco and Harper 1980). A beautiful confirmation of this conclusion is the spectrum of M82 from 5 to 30 μm (Gillett et al. 1975; Willner et al. 1977; Houck, Forrest, and McCarthy 1980), which shows a wealth of absorption and emission features also detected from Galactic thermal sources. Where observations are available (e.g., Gillett et al. 1975; Lebofsky and Rieke 1979), similar features are found in the spectra of other galaxies.

The identification of the luminosity source as hot, young stars rests largely on indirect evidence. The infrared emission is extended on a scale that is difficult to reconcile with the hypothesis of a single, compact, active nuclear source (see e.g., Rieke and Low 1975; Becklin et al. 1980; Rieke et al. 1980). The sizes and other characteristics of the sources strongly suggest an analogy with the inner few hundred parsecs of our Galaxy, which is generally agreed to be an important site of star formation (Oort 1977). The far infrared spectra already mentioned, the strong silicate absorption features at 10 μm, the probable association of far infrared emission with nuclei containing dense molecular clouds detected through mm-wave CO emission (Rickard, Harvey, and Thronson 1980), and the powerful emission-line spectra of the brightest infrared-emitting nuclei (see, e.g., Beck et al. 1978; Beck et al. 1979; Simon, Simon, and Joyce 1979; Rieke et al. 1980) are all strong indicators of similarities between conditions in these galactic nuclei and those in clearly recognized regions of star formation.

Other questions are suggested by the answers obtained so far. Three which seem especially important are: 1.) how is the process of star formation affected by conditions in galactic nuclei; 2.) can the wide range of properties from one nucleus to another be related in an evolutionary sequence; and 3.) what mechanism initiates and sustains the star-forming activity.

To illustrate possibilities for studying the process of star formation in infrared-luminous galactic nuclei, I will describe and expand on a recent study of M82 (Rieke et al. 1980). Given the continuing controversy over the stellar content of a well-behaved nucleus like that of M31 (Faber and French 1980; Persson et al. 1980), it may seem surprising that much of use can be deduced about the content of M82. However, conditions in the latter galaxy are sufficiently extreme to compensate for our inability to observe some of the traditional indicators of stellar content.

With currently available observations, five basic constraints can be placed on models of the stellar population in M82. These constraints are: 1.) the bolometric luminosity; 2.) the luminosity from red giants and supergiants; 3.) the ionizing flux; 4.) the mass; and 5.) star formation and evolution should be along "plausible" lines. The luminosity can be estimated from far infrared measurements and the distance of M82. From the depths of the CO bands near 2.3 μm and the general spectral behavior to 5 μm, it appears that the flux in the K photometric band (2.2 μm) is virtually all from red giants and supergiants. To estimate the luminosity from these stars, this flux must be corrected for extinction, which can be estimated from the relative prominence of the nucleus of the galaxy at various near infrared wavelengths to be $A_V \gtrsim 20$ (Rieke et al. 1980). The ionizing flux can be determined from the Bα line strength (Willner et al. 1977; Simon, Simon, and Joyce 1979), corrected for extinction by comparing the relative Bγ strength with the predictions of case B recombination. The mass can be estimated from the rotation or velocity dispersion, although errors could arise from the heavy extinction or from non-circular motions. At present, the most reliable rotation curve is probably that in the [NeII] line at 12.8 μm (Beck et al. 1978), although estimates of the mass by other means are in close agreement. An even more convincing determination of the mass could be made by measurements of velocities in the stellar CO bands near 2.3 μm, since non-circular motions would be unlikely to affect these observations. Figure 1 shows the velocity dispersion in these bands in an 8" beam from a spectrum at 2.6 cm^{-1} resolution (M. J. Lebofsky, private communication). The spectrum itself is shown at slightly reduced resolution in Figure 2. The dispersion has been analyzed by a correlation technique similar to that discussed by Tonry and Davis (1979). The upper limit on the velocity dispersion of ∿ 140 km/sec places an upper limit on the mass that is within a factor of two of the estimate from the [NeII] line. Finally, "plausible" patterns of star formation and evolution can be imposed by introducing an initial mass function as close as possible to that in the solar neighborhood, letting stars be born at some appropriate rate, and then permitting the stars to evolve along theoretical-empirical tracks to provide a series of population models for comparison with the other constraints (Rieke et al. 1980).

Three main conclusions can already be drawn from this approach. The first is that rapid star formation over a period of 10^7 to 10^8 years can explain all of the characteristics of M82 except possibly for the ultra-compact radio source. The models that account for the optical-IR properties as expressed by the five basic constraints discussed above predict a supernova rate adequate to account also for the X-ray emission and, if proper account is taken of the influence of the dense interstellar medium, for the nonthermal radio emission. This success adds weight to previous arguments that rapid star formation can provide the large infrared luminosities of galactic nuclei. The second conclusion is that the required mass of recently formed stars must be comparable to the total mass of gas in the same region

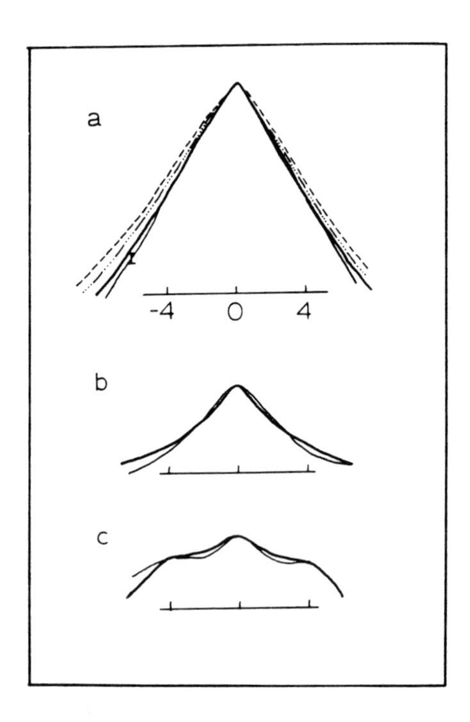

Figure 1. Velocity Dispersion for M82. Auto- and cross-correlations are plotted as a function of wavenumber (indicated in the horizontal scale under curves a) for the 2-0 CO bandhead (curves a), the 3-1 bandhead (b), and 4-2 bandhead (c). The heavy lines are the auto correlation of an artificial galaxy spectrum with zero velocity dispersion. The light, continuous lines are the cross-correlation of the spectrum of M82 with the artificial galaxy spectrum. The dashed line is the cross-correlation with the dispersion of M82 artificially broadened by 270 km/sec; the dash-dot line is the cross-correlation with an artificial broadening of 135 km/sec. A typical error bar is shown in a. All the spectra have an intrinsic resolution of 2.6 cm^{-1}, or 180 km/sec., and use a beam 8" in diameter. It can be seen that the dispersion of M82 is not detected, corresponding to an upper limit of $\sim$ 140 km/sec.

as estimated from the mm-wave CO observations. Thus, the star formation must be very efficient in terms of the proportion of the interstellar medium that is converted to stars. The third conclusion is that, unless <u>all</u> the methods of estimating rotational velocities are wrong by at least a factor of two, the relative number of solar mass stars formed must be far less than in the solar neighborhood.

This kind of study needs to be improved and expanded. For example, infrared observations can refine our estimates of the nuclear masses by

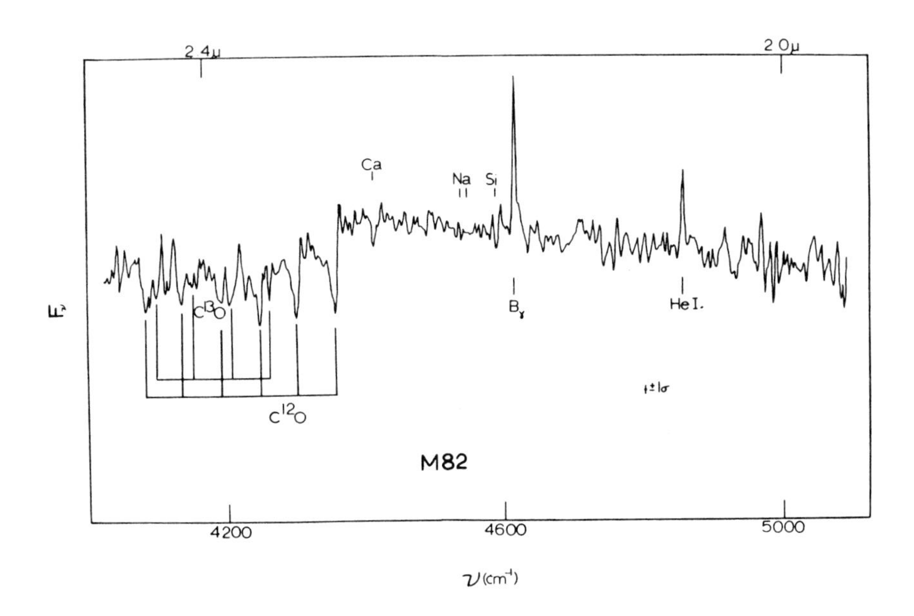

Figure 2. Spectrum of M82 near 2 μm. The beam size is 8", and the resolution is 4 cm^{-1}. The typical error bar applies to the spectral range 4250 cm^{-1} to 4800 cm^{-1}; outside this range, telluric absorptions decrease the signal-to-noise.

obtaining more detailed spectroscopy of the CO bands. New constraints on the stellar population can be obtained with high-quality infrared spectra, such as in Figure 2, combined with an adequate catalog of stellar spectra. Most importantly, these studies need to be extended to other galaxies, particularly those where the optical data are less compromised by extinction than in M82, in order to determine any biases resulting from some chance aspect of M82 itself.

The second area needing study is the evolutionary connection among different types of galaxy nuclei. Some examples are shown in Figure 3, which compares the nuclei of M31 (virtually no infrared excess), M81 (strong excess at 5, 10, and 20 μm; modest or no excess in the far infrared), M82 (typical of the nuclei whose luminosity is dominated by far infrared emission), and two Seyfert galaxies with strong thermal excesses. Many other distinctions can be made among these galaxies--e.g., M82 and NGC 1068 both contain dense interstellar clouds detected through CO emission, while such clouds are absent for M31 and M81 (Rickard, Harvey, and Thronson 1980). Comparison of the spectrum in Figure 2 with a composite spectrum of M81 and NGC 4736 shows that the stellar $C^{12}O$ absorption bands are stronger in M82 than in M81/NGC 4736. The $C^{13}O$ bands are much stronger in M82. The emission line spectrum is relatively weak in M81/NGC 4736.

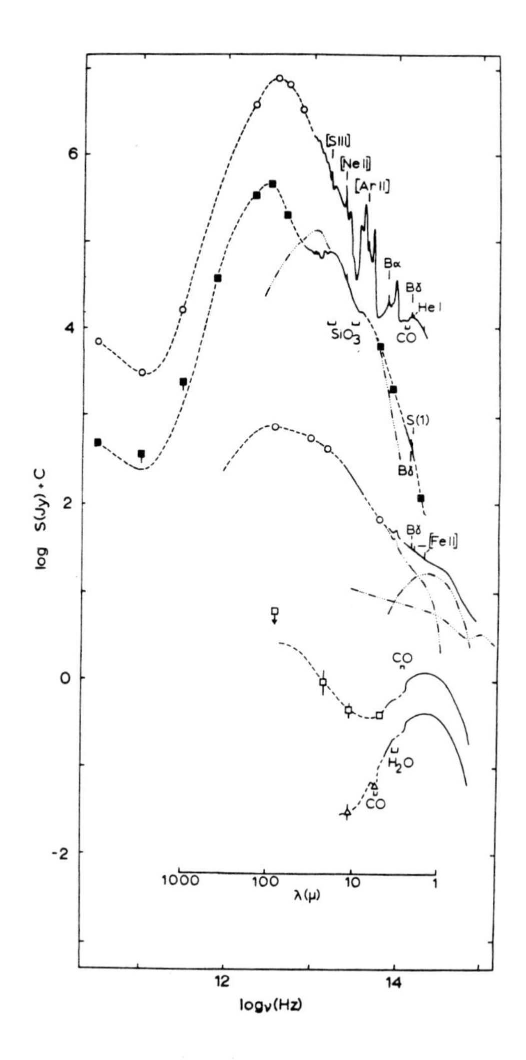

Figure 3. Spectral Energy Distribution of M82, NGC 1068, NGC 4151, M81, and M31 (reading top to bottom). The dash-dot curve for NGC 1068 represents the theoretical calculations of Jones et al. (1977): those for NGC 4151 show a possible division of its spectrum into thermal infrared, stellar, and nonthermal components. Various detected gaseous emission lines and stellar and interstellar absorption features are indicated.

Galaxies like M82 and NGC 253 have many of the attributes of type 2 Seyfert galaxies, including 1.) luminosities in the usual range; 2.) bright, compact nuclei that would appear as unresolved, and semi-stellar if the galaxies were a bit further away; 3.) radio and X-ray luminosities in the usual range; 4.) strong emission lines; and 5.) non-circular motions of ionized gas out of the galactic plane, so that the emission lines would have structure and possibly would appear broad if the galaxies were viewed face-on. Although there is no question that some Seyfert galaxies, such as NGC 4151, contain active nonthermal nuclear sources, it is likely that many type 2 Seyfert galaxies are very closely related to M82 and NGC 253, and that their nuclear activity is a result of star formation.

We have only begun to make observational distinctions based on the infrared properties of galactic nuclei. The three or four categories that can be identified now are presumably only samples of a whole continuum of properties. With more understanding of this continuum, we should be able to determine the lifetimes, frequencies of occurrence, and patterns of evolution for the episodes of rapid star formation.

Thirdly, we need to study mechanisms which could initiate and sustain the star-forming activity. For example, if the same body of gas is used for repeated episodes of star formation, we would expect enrichment of the heavy elements and isotopic anomalies (e.g., an increase in the C^{13}/C^{12} ratio) to be detectable. Most optical indications of abundance gradients in galaxies refer to an old stellar population and therefore to material that has not participated in the recent evolution of the nucleus. The present evidence from observation of fine structure lines of neon, sulphur, and argon is that the enrichment above solar-type abundances is small or non-existent (Gillett et al. 1975; Willner et al. 1977; Beck et al. 1978; Simon, Simon, and Joyce 1979). The relatively strong $C^{13}O$ absorption in the stars in the nucleus of M82 probably results at least in part from a population of luminous and cool stars. Thus, there is at present no evidence for extensive processing of the interstellar material in these galactic nuclei, although this conclusion is tentative until we improve our understanding of the extinction, stellar populations, and excitation states in these regions.

If the gas is relatively unprocessed, it may be that the star formation is fueled by material that only recently came into the nucleus. It would then be easy to understand why pairs of galaxies such as NGC 5194/5195 frequently show infrared excesses and other indications of star formation (Larson and Tinsley 1978), since tidal disturbances might inject gas into their nuclei. However, other galaxies bright in the far infrared--e.g., NGC 253, NGC 6946--are members of sparse clusters or are even isolated. The influence of the environment of the galaxy on its nuclear activity has so far not been studied systematically.

The problems I have posed here are not easy ones. However, we should be encouraged by the substantial progress during the past decade. With the sudden availability of three optimized infrared telescopes of 3 meters or larger aperture, the IRAS infrared survey, and the continuing development of infrared technology, the '80's promise exciting insights to rapid star formation in galactic nuclei.

This review was prepared with assistance from M. J. Lebofsky. The work was supported by the National Science Foundation.

REFERENCES

Beck, S. C., Lacy, J. H., and Geballe, T. R.: 1979, Ap. J., 231, p.28.
Beck, S. C., Lacy, J. H., Baas, F., and Townes, C. H.: 1978, Ap. J., 226, p. 545.
Becklin, E. E., Gatley, I., Matthews, K., Neugebauer, G., Sellgren, K., Werner, M. W., and Wynn-Williams, C. G.: 1980, Ap. J., 236, p 441.
Faber, S. M., and French, H. B.: 1980, Ap. J., 235, p. 405.
Gillett, F. C., Kleinmann, D. E., Wright, E. L., and Capps, R. W.: 1975, Ap. J. (Letters), 198, p. L65.
Houck, J. R., Forrest, W. J., and McCarthy, J. F.: 1980, preprint.
Jones, T. W., Leung, C. M., Gould, R. J., and Stein, W. A.: 1977, Ap. J., 212, p. 52.
Kleinmann, D. E., and Low, F. J.: 1970, Ap. J. (Letters), 159, p. L165.
Larson, R. B., and Tinsley, B. M.: 1978, Ap. J., 219, p. 46.
Lebofsky, M. J., and Rieke, G. H.: 1979, Ap. J., 229, p. 111.
Oort, J. H.: 1977, Ann. Rev. Ast. Astrophys., 15, p. 295.
Pacholczyk, A. G., and Wisniewski, W. Z.: 1967, Ap. J., 147, p. 394.
Persson, S. E., Cohen, J. G., Sellgren, K., Mould, J., and Frogel, J. A.: 1980, Ap. J. (in press).
Rickard, L. J., Harvey, P. M., and Thronson, H. A.: 1980, paper presented at IAU Symposium 96.
Rieke, G. H., and Lebofsky, M. J.: 1978, Ap. J. (Letters), 220, p. L37.
Rieke, G. H., and Low, F. J.: 1975, Ap. J., 197, p. 17.
Rieke, G. H., Lebofsky, M. J., Thompson, R. I., Low, F. J., and Tokunaga, A. T.: 1980, Ap. J., 238, p. 24.
Simon, M., Simon, T., and Joyce, R. R.: 1979, Ap. J., 227, p. 64.
Telesco, C. M., and Harper, D. A.: 1980, Ap. J. 235, p. 392.
Tonry, J., and Davis, M.: 1979, A. J., 84, p. 1511.
Willner, S. P., Soifer, B. T., Russell, R. W., Joyce, R. R., and Gillett, F. C.: 1977, Ap. J. (Letters), 217, p. L121.

DISCUSSION FOLLOWING PAPER DELIVERED BY G. H. RIEKE

T. L. WILSON: Arguments about mass of interstellar gas and isotope ratios based on radio observations of ^{12}CO and ^{13}CO should be treated with care, because of the low signal-to-noise ratio, the large optical depths in CO, excitation effects, and beam-filling factors.

RIEKE: Determining the mass by the use of CO measurements is only one of four or five methods used, and probably not the strongest of those.

WYNN-WILLIAMS: It seems to me that your derived mass in M82 is subject to serious systematic underestimation. M82 is an edge-on galaxy, so the velocity profiles you use are an integral along a line of sight. All the velocities along your line of sight will tend to be less than the velocity at the tangent point, which is the velocity you must use to get your mass. How do you make allowance for this effect, especially when your prime data, the neon velocities, are of such low signal-to-noise ratio?

RIEKE: The compactness of the 2 μm map shows that we are seeing predominantly the nucleus of the galaxy, so the effect you mention is not going to affect things there. There are also compensating effects which tend to lead to an overestimation of the mass on a theoretical basis, namely the fact that we take a spherical distribution of stars. There are systematic effects in each rotation curve, but the fact that they all agree in terms of upper limits or measured masses should seriously embarrass anyone who wants to say that the masses are substantially larger than those which lead to requiring the IMF to cut off towards 1 $M_\odot$ stars. To avoid this conclusion you would need peak velocities 2-3 times larger than the largest observed velocities.

WYNN-WILLIAMS: In our Galaxy discrete supernova remnants contribute only a small portion of the extended non-thermal radio radiation. As far as I know, efforts to link the strength of the extended Galactic non-thermal radio emission to the supernova rate, let alone the star formation rate, have been unsuccessful. In view of the uncertainty in our Galaxy, how can you use this as a constraint in M82?

RIEKE: In M82 we are dealing with recent rapid star formation, so we can attribute things rather directly to what has happened, without worrying about the pre-existing population of stars. The analysis is only qualitative in terms of the effects of supernovae, but if you take the radio output of Galactic supernova remnants as a function of age and apply that to the supernova rates predicted for M82 you find quite close agreement with observation for the radio output.

WYNN-WILLIAMS: But discrete supernova remnants in our Galaxy are seen only for a few times 10^4 years, whereas your starburst models last for 10^7 years.

RIEKE: The interstellar medium in M82 is sufficiently dense, and the photon field is sufficiently dense, that the effects of the energetic electrons are negligible after a few times 10^4 years.

BECKLIN: Your model of the star formation rates in M82 depends on the strength of the 2.2 μm continuum radiation from late type stars, yet in the figures you presented, the 2.2 μm distribution looked totally different from the 10 μm distribution. Would you like to comment?

RIEKE: The stars dominating the 2 μm emission are a few time 10^7 years old if you look at these models. In that time the stars will have rotated around the nucleus 3-10 times, so will have a more relaxed, symmetric distribution than the younger objects seen at 10 μm.

RICKARD: It should be noted that sometimes 10 μm is not infrared enough to pick out molecular clouds. NGC 5195 is a stronger 10 μm source than its companion, M51, yet it is weaker in the 2.6-mm CO line by more than a factor of 20. Clearly, if you select by 10 μm emission, you can still get a rather heterogeneous set of objects.

RIEKE: This is why I emphasize the importance of 100 μm observations. There are galaxies which have strong 10-20 μm excesses, yet have either no strong 100 μm flux density or no Bγ line flux. We now have about two dozen galaxies measured at 100 μm, and I hope we'll have two dozen dozen at least by the end of the decade.

WERNER: Many of the links between star formation and high luminosity which you mentioned are rather tenuous or circumstantial. I think the best indicator is the presence of ionized gas as revealed by radio continuum or Bγ emission. What evidence is there, based on energetic or statistical grounds, that the high far infrared luminosity in these galaxies is episodic rather than continuous?

RIEKE: There is a problem of fitting things together if they are continuous, in that the conversion of mass to luminosity is so high in some of these galaxies that the mass is used up very quickly. We do not yet know whether some galaxies have recurrent episodes and others none; there is a lot of work to do in terms of the evolution between one kind of galaxy and another.

PERSSON: Could you comment on the possibility of deriving abundances of sodium and calcium in the nucleus of M82 using the 2 μm spectrum?

RIEKE: It is an intriguing possibility, but so far people have not yet succeeded in using these lines to get abundances in Galactic stars, and it is certainly more difficult to determine them in galaxies like M82. As far as comparisons between galaxies go, this is already part of our long term spectroscopy program, but the strength of the lines is affected by things like surface gravity, temperature, luminosity, which all have to be corrected for.

PUGET: Can you say anything about the ratio of 5 GHz to far infrared flux densities as compared to Galactic complexes?

RIEKE: At 5 GHz the flux from M82 is almost completely dominated by the non-thermal component. From the Brackett lines we may deduce the radio thermal emission, and we find that, unless the H II regions are extraordinarily dense, we can account for virtually all of the measured 3-mm radio emission by free-free processes. This can be understood if Compton scattering of the energetic electrons steepens the non-thermal radio spectral index longward of 3-mm wavelength. This is not a unique model, however.

HARPER: Pertaining to your comments on the extent of the infrared sources, we have recently measured the size of NGC 1068 at 60 μm and find that it is approximately 25" in diameter. This corresponds to a linear size of about 2000 pc. Also, preliminary results on several less luminous galaxies are consistent with similar scale sizes. On the basis of these data, I expect that we will find far infrared activity and high rates of star formation extending to rather large radii in the nuclear regions of many galaxies.

RIEKE: I might add that when we talk of an evolutionary sequence of galaxies the spectral information is in some sense the most straightforward to obtain, but we may also find that there is an evolutionary sequence as regards structure and size scales. All these things need to be folded in in order to really get an understanding of how one type of galaxy relates to another.

A. S. WILSON: I think it's important to remember that not all of the properties of Type 2 Seyfert galaxies, to which you alluded, can be explained in terms of a burst of star formation. Firstly, the X-ray emission in at least 3 galaxies with Type 2 optical spectra are variable on a time scale of months. Secondly, a few Type 2's have double radio sources in their nuclei. Both of these results indicate (the first more conclusively than the second) that there is something ultracompact in these objects.

RIEKE: What I meant to suggest is that there is an overlap; and that some Type 2 Seyferts may well be manifestations of star formation, and some of them will have more active sources in their nuclei. Part of getting an evolutionary sequence in these galaxies is to sort out which are which.

BECKLIN: I think it would be a mistake not to point out that studies of the Galactic Center will be very important for understanding the phenomena in galactic nuclei discussed here. This is true because of its relatively close proximity to us. We can get better velocity and mass determinations, star formation rates, gas, dust, and star densities, and potentially a more detailed understanding of the interaction with the active source in the very center.

TOVMASSIAN: I would like to remark that all data presented here by Rieke are in excellent agreement with Ambartsumian's hypotheses on the activity of galactic nuclei and formation of stars by explosion of superdense protostellar matter.

THE INFRARED PROPERTIES OF ACTIVE EXTRAGALACTIC NUCLEI

B. T. Soifer and G. Neugebauer
Palomar Observatory
California Institute of Technology
Pasadena, California 91125

INTRODUCTION

In this paper we review the observed infrared properties of the general classes of active extragalactic nuclei with the purpose of relating the observations to the mechanisms responsible for the emission processes. We will first give a summary of those observations which define the energy distributions and emission line ratios of broad groups of objects. We will intersperse measurements of specific features throughout the discussion that illustrate definite emission mechanisms.

OBSERVATIONAL DATA

The shapes of the observed continuum energy distributions of the nuclei of active galaxies are well established and have been available in the published literature of the past few years. A summary of the typical energy distributions of extra galactic sources has been given by Rieke and Lebofsky (1979a) (Figure 1), who give extensive references to early work.

Although the general shapes of the energy distributions are given in Figure 1, studies of extensive numbers of objects show significant differences between the spectra of different objects within one class. Figure 2, from Rieke (1978), shows the spectral distributions of a number of Seyfert galaxies out to 10 μm from his study of some 50 Seyferts. Although these show the same general increase in flux density into the near infrared, significant differences are seen. In particular, as noted by Neugebauer *et al*. (1976), the energy distributions of type 2 Seyferts rise relatively more steeply into the infrared than those of type 1 Seyferts; significant exceptions are, however, present. McAlary, McLaren and Crabtree (1979), from a smaller sample, have divided the Seyferts into three groups on the basis of a 1-5 μm study as shown in Figure 3; these correspond to (a) dominance by the stellar continuum, (b) dominance by non-thermal power law component

C. G. Wynn-Williams and D. P. Cruikshank (eds.), Infrared Astronomy, 329–349.

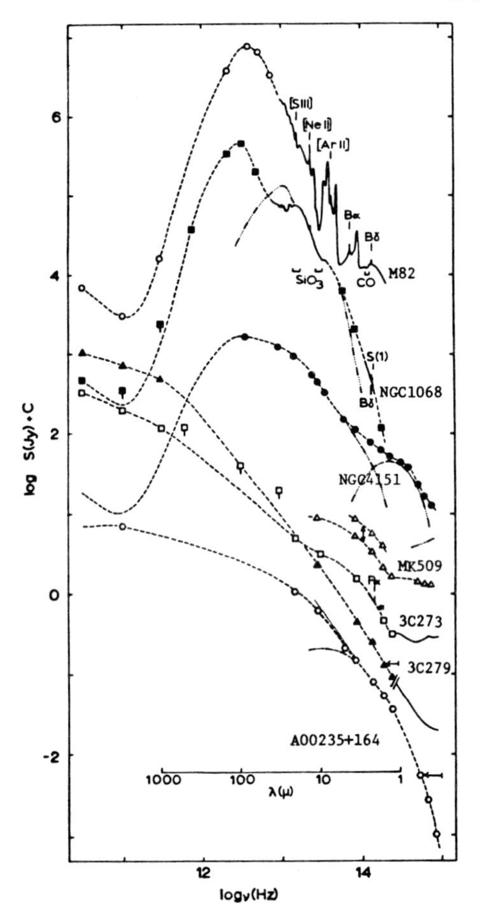

Figure 1. The infrared/optical/radio energy distributions of the major classes of active galactic nuclei (from Rieke and Lebofsky 1979a).

or (c) intermediate cases which contain both components. They find that for $\lambda \sim 2$ μm, type 2 galaxies are largely dominated by stellar continua while the type 1 Seyferts generally show power law spectra in the near infrared. Clearly these results are dependent on the diaphragm used for the observations, since larger diaphragms include more of the stellar radiation, compared to the (nearly) pointlike nucleus. All of the Seyferts which show strong X-ray emission also show evidence of non-stellar emission.

The energy distributions of the quasars are generally typified by that of 3C273 in Figure 1. Detailed near infrared measurements of quasars by Neugebauer et al. (1979) again show significant differences as is clear from Figure 4a and b which show the near infrared energy distributions of low and high redshift quasars. There is no systematic difference between the near infrared energy distributions of radio quiet and radio loud quasars. On the other hand, measurements of quasars at 1 mm correlate extremely closely with the centimeter radio observations (Ennis 1980). The differences between radio quiet and radio loud quasars must therefore show up at about 100 μm.

The colors of a "typical" quasar from 2 μm to 0.5 μm can be generated from the observed [1.6 μm] - [2.2 μm] colors when plotted as a function of red shift. The Caltech data on about 110 quasars are shown in Figure 5, in which the type of survey by which the quasars were found is shown qualitatively. It is seen that within the present sample quasars have similar colors as a function of redshift, independently of how they were discovered, except, perhaps, at the lowest redshifts where those quasars selected from radio surveys have higher H-K values than those found from optical and X-ray surveys. From the histograms it appears that, although there is a spread of ±0.5 mag in quasar H-K colors, the energy distribution of 3C273 is "typical" of quasars at all redshifts. The results of Hyland and Allen presented in this conference agree qualitatively with those of Figure 5.

The continua of the final category of violent extragalactic objects, BL Lac objects, are characterized by a much smoother shape than either the Seyferts or quasars. Observations of their energy distributions are difficult because of their rapid variability which is apparently correlated between the visible and infrared. Figure 6 shows the energy distribution of several BL Lac objects from the visual to radio

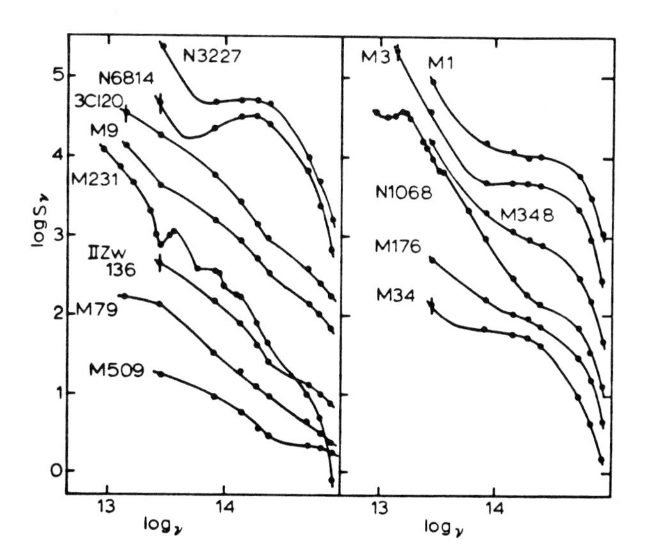

Figure 2. The infrared/optical energy distributions of Seyfert nuclei (from Rieke, 1978).

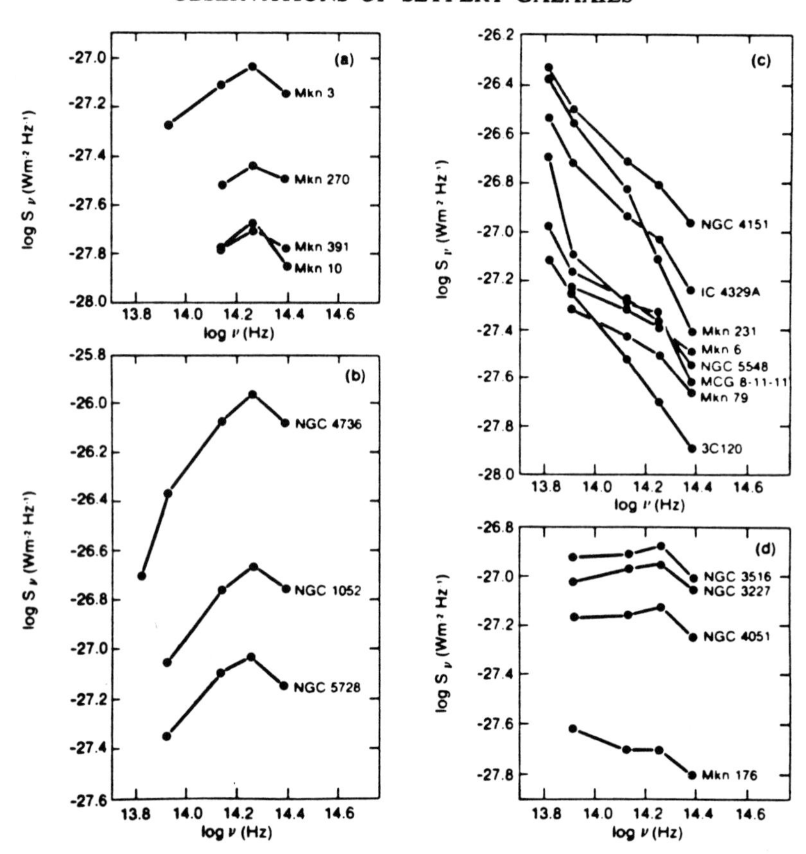

Figure 3. The near infrared energy distributions of Seyfert and emission line galaxies (from McAlary, et al. 1979). Box (a) shows the Seyferts with stellar like energy distributions; (b) shows the energy distributions of emission line galaxies; (c) shows the Seyferts with power law spectra; and (d) shows Seyferts with continua intermediate between (a) and (c).

wavelengths as obtained by O'Dell et al. (1978). The apparently smooth distribution is typical of BL Lac itself and at least several of the brighter BL Lac objects.

Recently Rieke, Lebofsky and Kinman (1979) have observed the near infrared radiation from a number of flat-spectrum radio sources in previously unidentified fields; the fluxes are shown in Figure 7. These objects have far steeper optical-infrared continuum distributions ($F_\nu \alpha \nu^{-3}$) than any other known quasars or BL Lac objects (3C 68.1 comes closest to matching the slope of these sources and has a spectral index $\alpha \sim 2.1$). Rieke et al. noted the large variability of several of these objects and placed them in the class of the optically violently variable quasars. Recently Aaronson and Boroson (1980) have reported visual spectroscopy of one of these objects, 0406 + 121, and Soifer, Neugebauer and Matthews have obtained near infrared spectra of two of these objects (1413 + 135 and 0406 + 12). None of the spectroscopic work indicates any evidence for emission lines; this, plus the large variability, suggests that these objects form the extreme red end of the BL Lac class of objects. Another interesting property of these objects is the significant change in continuum slope as a function of brightness illustrated in Figure 8.

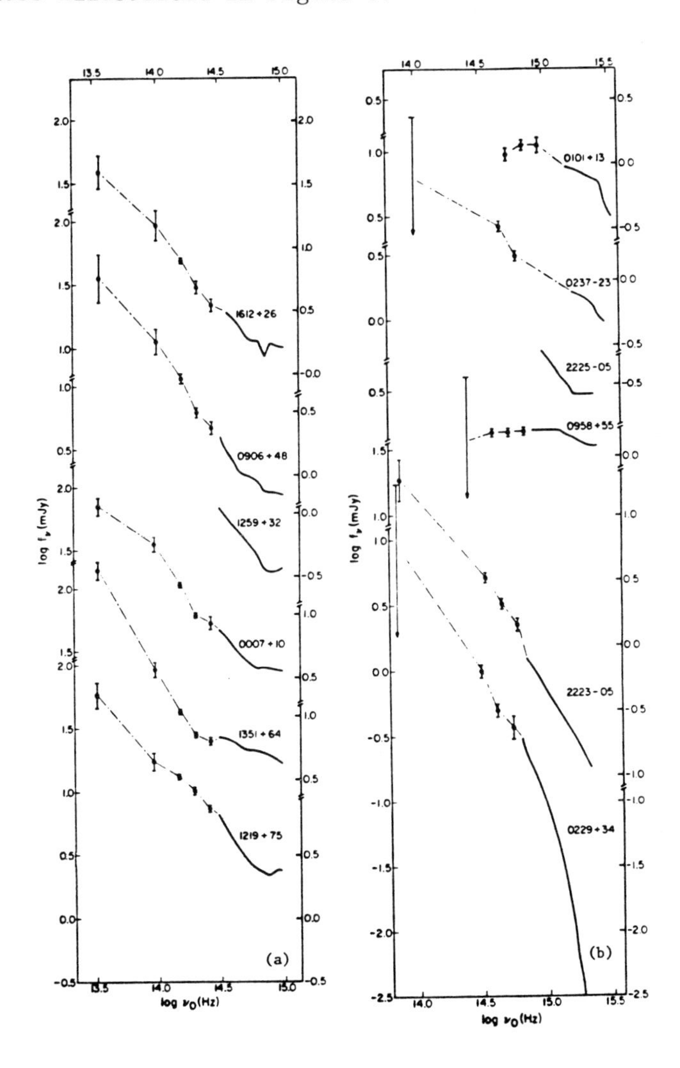

Figure 4. The infrared/optical energy distributions of low (box a) and high (box b) redshift quasars (from Neugebauer, et al.1979). The fluxes are plotted in the rest frame of the quasars.

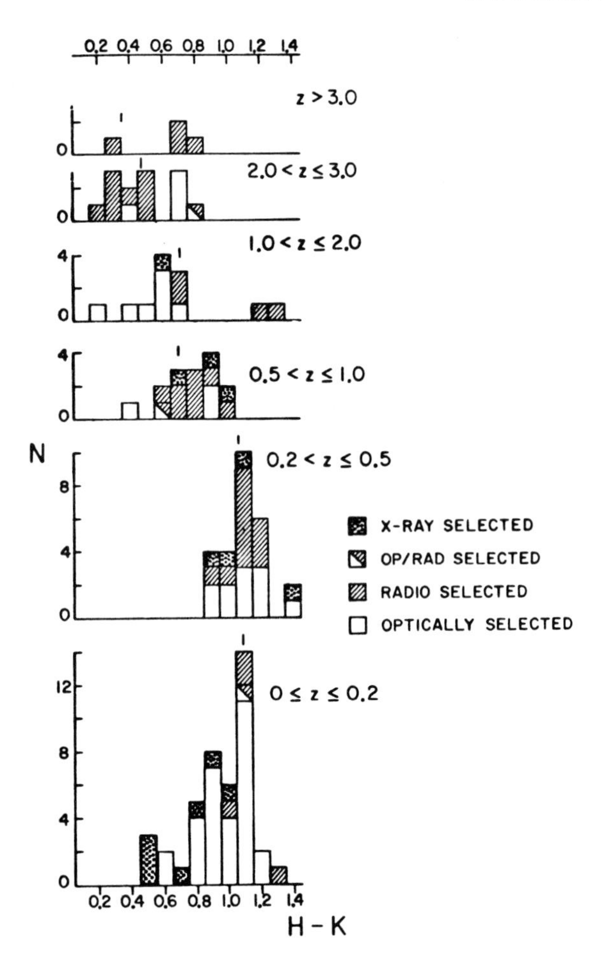

Figure 5. The observed H-K colors of quasars plotted in histograms for the redshift ranges shown. Those quasars discovered via optical, radio, and X-ray techniques are shown in separate boxes. For quasars in the appropriate redshift ranges, color corrections have been applied for Hα, Hβ, and [OIII] emission in the H and/or K bands, taking the mean equivalent widths of these lines from the work of Soifer, et al. (1980a). The vertical tic mark indicates the H-K color 3C273 would appear to have at the mid-range redshift of each histogram. The data on 3C273 are from Neugebauer, et al. (1979).

In recent years, increased emphasis has been placed on observations of the hydrogen emission lines in quasars red-shifted into the near infrared wavelengths. These measurements, which follow the pioneering work of Baldwin (1977), should provide a means of studying the clouds where line formation takes place. The most extensive studies in the infrared have been made by Puetter et al. (1980) and by

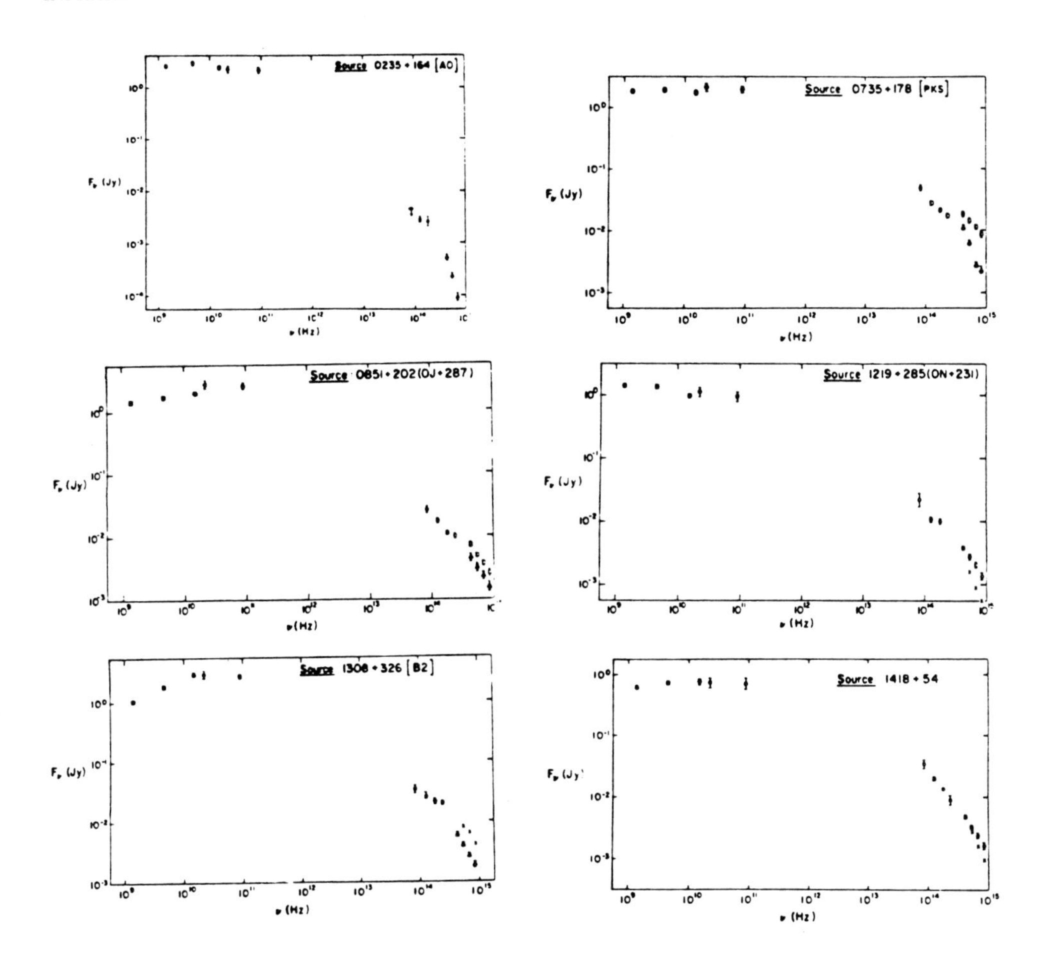

Figure 6. The energy distributions of selected BL Lac objects from the radio through the optical range (from O'Dell, et al. 1978) All the observations of a given object reported here were obtained within a few months of each other. Even over this time span substantial variability can be seen in the observations of 0735+178, 0J287, 0N231, and 1308+32.

Soifer et al. (1980a); Figure 9 shows their combined data on the Pα/Hα/Hβ line ratios in quasars. In Figure 10 the Lα/Hα ratio is shown as obtained for both high and low red-shift quasars. Preliminary data on the Pα/Hα/Hβ ratios for Seyfert galaxies obtained from Mauna Kea (Soifer et al. 1980b) are shown in Figure 11. The differences, which will be discussed below in the context of the dust content of the nuclei, are striking.

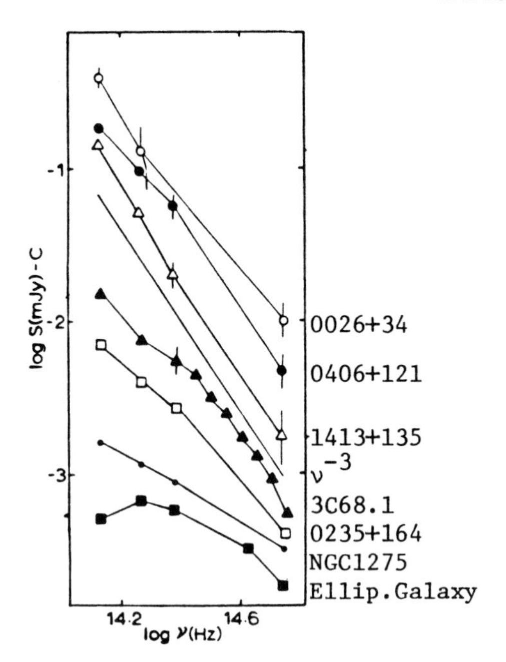

Figure 7. The near infrared/optical energy distributions of a class of radio sources having very steep infrared/optical spectra (from Rieke, Lebofsky, and Kinman, 1979).

DISCUSSION

The main question that we wish to discuss is what mechanisms are responsible for the observed infrared continuum emission in the Seyferts, BL Lac objects, and quasars. The main mechanisms that have been mentioned as potentially producing the infrared emission are thermal emission by dust in the nuclei, and incoherent synchrotron emission. Other emission mechanisms such as synchrotron self-Compton emission and multiple Compton scattering by dense nonrelativistic plasmas are generally thought to be inapplicable as the infrared emission mechanism (e.g., Jones, et al. 1980, O'Dell 1978).

Examples where thermal emission by dust and incoherent synchrotron emission are thought to be the sources of the observed infrared flux can be found in the observations of several classes of active galactic nuclei. The most straightforward example of thermal emission by dust producing the infrared radiation is in the type 2 Seyfert nuclei, archetypical of which is the Seyfert galaxy, NGC 1068.

The energy distribution in NGC 1068, as shown in Figure 1, is quite similar to that seen in M82 and galactic H II regions where there is no doubt that the mechanism producing the infrared radiation is

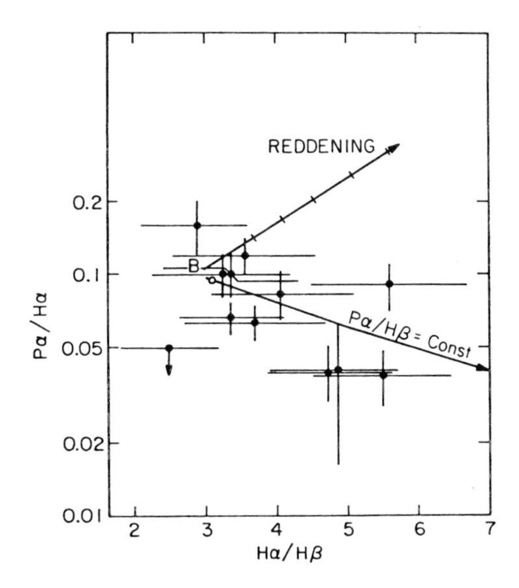

Figure 9. The Pα/Hα/Hβ line ratios in low redshift quasars. The data are from Soifer, et al. (1980a) and Puetter et al. (1980). The B in the figure indicates the position of the Case B recombination valve. The arrow marked reddening indicates the effect normal interstellar reddening would have on the Case B value. The line marked Pα/Hβ = Const is the deviation expected from Case B in the case of collisional excitation when the Balmer lines are optically thin.

NGC 1068 in the infrared, the current consensus is that there is no evidence for variability in the infrared emission in NGC 1068. This is therefore consistent with (but does not prove) thermal dust emission models.

In addition to the photometric and angular size measurements, there is spectroscopic evidence for the presence of dust in the nucleus of NGC 1068. It has been known for over a decade (Wampler, 1968) that the S II lines appeared reddened in the nucleus of NGC 1068. Wampler (1971) and Shield and Oke (1975) have also found that the Balmer decrement is large in NGC 1068, possibly (but again, not necessarily) implying the presence of dust in the nucleus of this galaxy. Recently Neugebauer et al. (1980) have shown that the observed strengths of all of the hydrogen emission lines from Lα to Pα, of He II 1640,4686, and of the S II 4072,10320 lines are consistent with normal line ratios and reddening by ~ 1.5 mag of visual extinction. Kleinmann, et al. (1976) have obtained a spectrum of NCG 1068 from 8 - 13 μm that shows evidence for silicate absorption, indicating that the 10 μm source is viewed through a large column density of obscuring material.

In summary, the observational evidence has pointed to the

thermal emission by dust. The continuum slope from 1 to 20 μm, $\alpha \sim -3$, is much steeper than that seen over the same wavelength range in any source where non-thermal mechanisms are thought to dominate; the continuum slope in the visual is sometimes found to be steeper than this, but even in these cases, e.g. 0235 + 164, 3C 68.1, the continuum slope in the infrared is much less than at visual wavelengths.

Perhaps the strongest argument in favor of a thermal dust model in NGC 1068 is the measured size of the infrared source. Becklin et al. (1973) measured the diameter of the 10 μm source to be 1" or 90 pc. This is quite readily explained if the 10 μm flux is from dust grains heated by a central ionizing source. It is difficult to understand this observation if the infrared source is somehow related to the non-thermal source observed at optical and ultraviolet wavelengths. Hildebrand et al. (1977) have shown that if the far infrared source is a thermal source, it must be at least 5" in extent, and Telesco, et al. (1980) have shown that 20% of the 20 μm flux from NGC 1068 is from a region larger than 3" (260 pc) in radius, consistent with the thermal models of the far infrared and submillimeter emission from this object.

Observed variability is often used as an argument against thermal dust emission models, since observed time scales of variability are often much less than that required in such models. Although in the past there has been controversy over the possible variability of

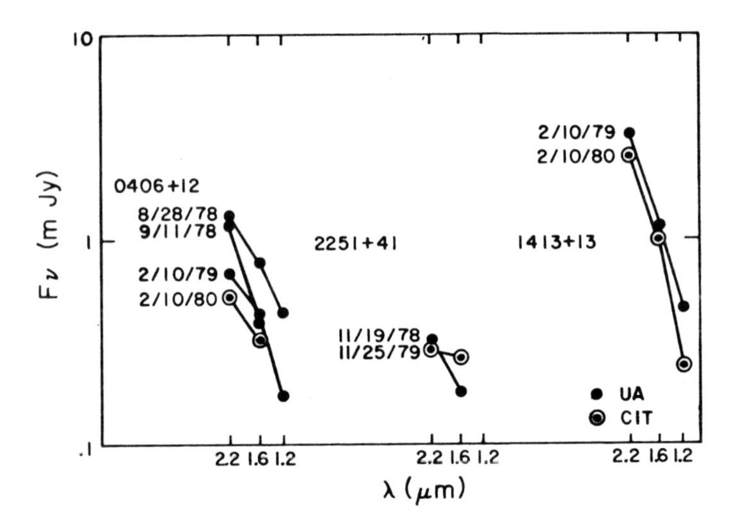

Figure 8. The near infrared energy distributions of three of the objects observed by Rieke, Lebofsky, and Kinman (1979). The energy distributions show significant changes in shape with time. The data noted as UA are from Rieke, et al.; the data noted as CIT are unpublished observations of Soifer, Neugebauer, and Matthews.

conclusion that the mechanism producing the infrared emission in the archetypical type 2 Seyfert galaxy NGC 1068 is thermal emission by dust. In addition, Rieke's (1978) study of Seyfert galaxies shows that the other type 2 Seyfert galaxies are similar to NGC 1068.

In contrast to the case of NGC 1068, the observational evidence over the last few years suggests that in the BL Lac objects the infrared emission is due to incoherent electron synchrotron emission. The evidence for this is basically guilt by association. As best as can be determined, the infrared properties of BL Lac objects are identical to the optical properties of these objects, and since the optical emission is probably due to incoherent electron synchrotron emission, the same must be true for the infrared emission.

Figure 1 shows the energy distribution of the BL Lac object AO 0235 + 164 at maximum light in its outburst of 1976 while Figure 6 illustrates the energy distribution of several BL Lac objects extending from the radio through the optical. All these objects show an energy distribution that is relatively flat through the radio, with a steepening to shorter infrared and optical wavelengths. Such behavior is expected of a synchrotron source which has multiple components at longer wavelengths, becomes optically thin to synchrotron self absorption at ~ 1 mm, and reverts there to the classical power law flux distribution

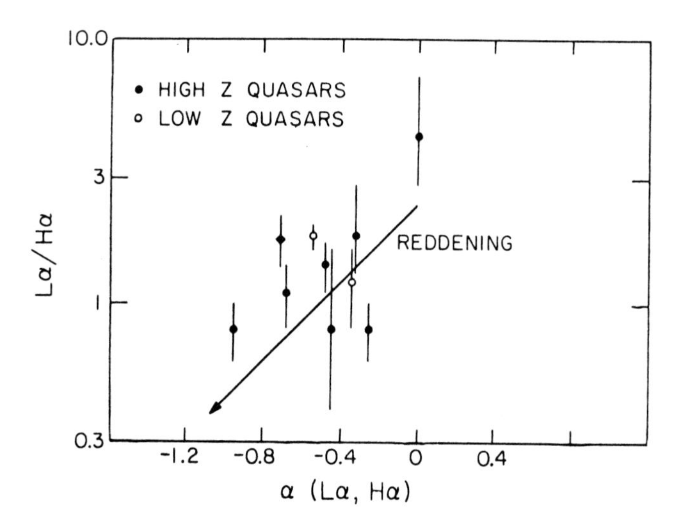

Figure 10. The Lα/Hα line ratio plotted vs. the continuum spectral index α for quasars. The open circles are for the low redshift quasars 3C273 and PG0026+129. The diamond is a high redshift quasar from Puetter, et al. (1980). All other data are from Soifer, et al (1980a). The arrow indicates the effect increasing external reddening would have on both the lines and continuum, starting with a specific intrinsic spectrum.

of an optically thin synchrotron source. The steepening at near infrared and optical wavelengths is taken as indicative of increasingly rapid energy loss in the highest energy electrons.

Related to the energy distributions are the polarization properties of these objects. Although the polarization properties of BL Lac objects are well studied optically (see Angel and Stockman 1980 for a review) there has been comparatively little work on the polarization properties of the infrared emission in BL Lac objects. Where studies are available (e.g. Knacke, Capps, and Johns 1976, 1979, Rieke, et al. 1977, Moore, et al. 1980, and Puschell and Stein, 1980), it is generally found that the strength and position angle of the infrared radiation agrees with that found visually, although there are notable exceptions.

By far the most stringent arguments against thermal emission from BL Lac objects comes from their observed variability. Figure 12, from Rieke and Lebofsky (1979a), shows the variation at 2 μm of the Bl Lac object AO 0235 + 164. If this object is at a distance inferred from observed absorption line red-shifts, the observed variations are many times faster than allowed for emission from a thermal source. Again the similarity in the optical and infrared variability of AO 0235 + 164 leads to the almost inescapable conclusion that the same

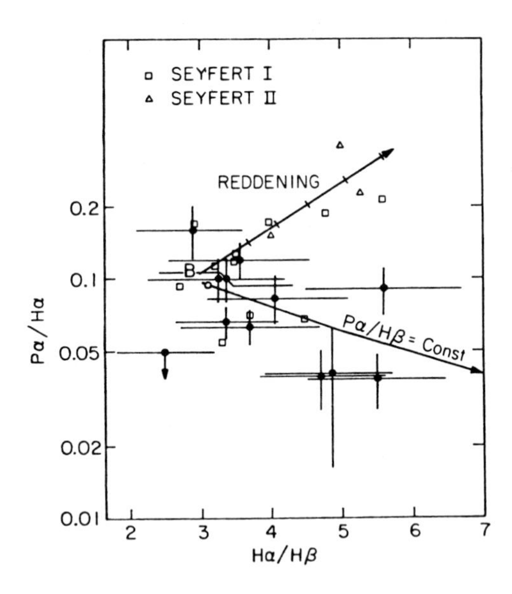

Figure 11. Same as Figure 9 with the addition of observations of Seyfert galaxies from Soifer, et al. (1980b). The tics along the reddening line show successive amounts of ΔE(B-V) = 0.1. Note all the Seyfert II nuclei and many of the Seyfert I nuclei follow this track.

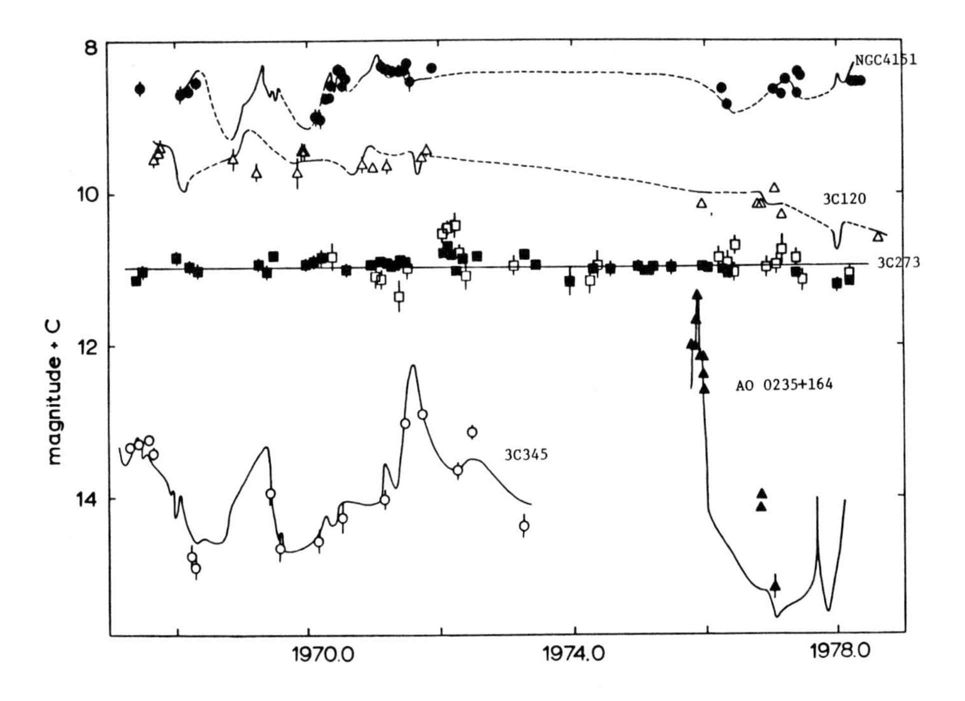

Figure 12. The variability at 2 μm (and 10 μm open boxes for 3C273) of a variety of extragalactic sources (from Rieke and Lebofsky, 1979a).

emission mechanism is at work in both cases. Figure 13 shows Caltech observations of BL Lac itself and OJ287, measured over the same time scales. There is no evidence for variability in the infrared colors, and therefore the spectral energy distributions, as the continua vary by several magnitudes. This is in contrast to the variable energy distributions seen in the objects discovered by Rieke, et al. (Figure 8).

While it is true that the observed variability in objects such as 0235 + 164 (Rieke, et al. 1976) and 1308 + 326 (Moore, et al. 1980) stretches the incoherent synchrotron models to their limits and may in fact require non-isotropic emission processes (e.g. Blandford and Rees, 1978) the evidence seems overwhelming that the same emission process is powering the optical, infrared, and radio emission in these objects.

While type 2 Seyfert galaxies and BL Lac objects present the two extreme types of luminous infrared emitters, and appear to present clear-cut evidence for two distinct infrared emission mechanisms, the situation for type 1 Seyfert galaxies and quasars is by no means as clear-cut.

In the case of the quasars, a distinction can be made in the

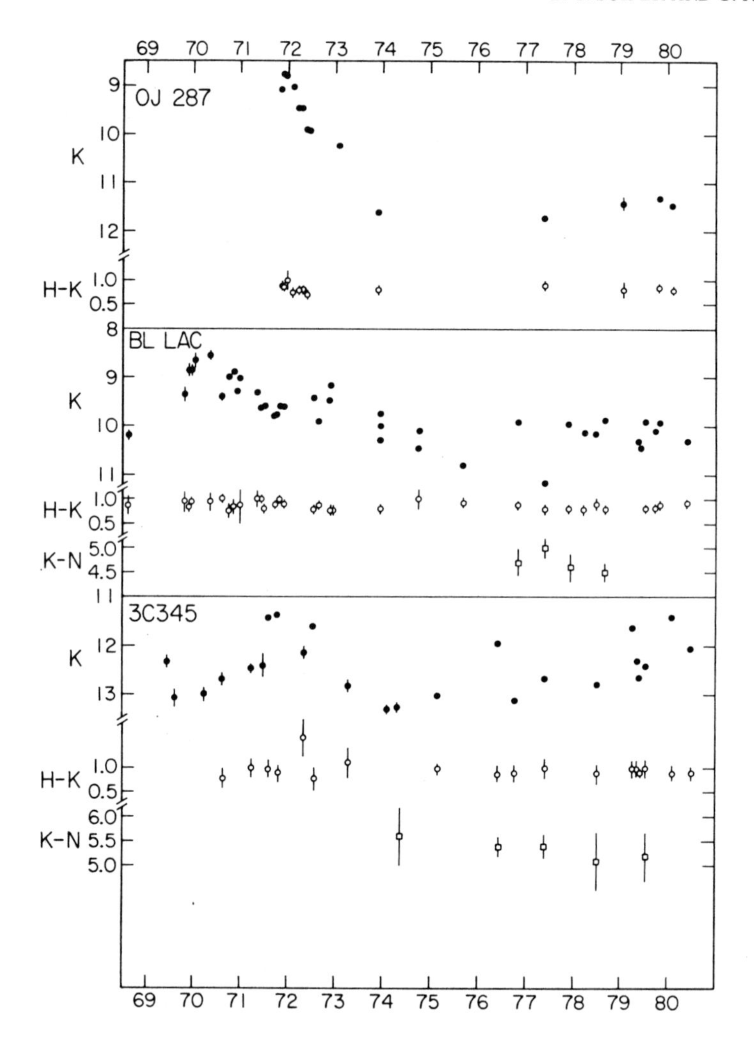

Figure 13. The time history of the K Magnitude and the H-K color for the BL Lac objects BL Lac and OJ287, and the optically violently variable quasar 3C345 (Neugebauer, et al., unpublished). For BL Lac and 3C345 the K-N color is also shown, where it was observed.

class of quasars known as optically violently variable. The members of this class which have been extensively observed at infrared wavelengths, 3C279, 3C345, 3C446, and 3C454.3, all show a high degree of polarization and variability at 2 μm of greater than 1 magnitude over time scales of several years (Neugebauer, et al. 1979); Figure 13 illustrates the 2 μm flux measured by the Caltech group of the quasar 3C345. The range in observed flux is nearly as large as for the BL Lac objects OJ 287 and BL Lac, with similar variability time scales. Angel and Stockman (1980) have argued that these violent variable quasars also have the "simplest" power law energy distributions among the quasars, show significant polarization and thus present an extension of the BL Lac class of objects. Similar arguments apply here for the efficacy of the incoherent synchrotron mechanism to produce the infrared emission.

The non-violently variable quasars by definition do not show large variability, and do not show a large degree of polarization, either at optical or infrared wavelengths (Stockman and Angel 1978). While the infrared energy distributions of these quasars are more complex than those of the highly variable quasars, and show a decrease in slope between 3 and 10 μm, the energy distributions still do not bear any strong similarity to those sources that are believed to show a thermal emission spectrum, and are quite similar to the energy distributions of the "simple spectrum" quasars and BL Lac objects. The energy distribution of 3C273, the best studied quasar at infrared wavelengths, is shown in Figure 1. There is no evidence for any significant excess above an extrapolated power law spectrum $F_\nu \propto \nu^{-1}$ between 20 μm and 1 mm. Again this argues for the incoherent synchrotron mechanism. Figure 12 shows the time history of the 2 μm flux from 3C273. No positive evidence for variability can be seen from these data.

Other observations that have searched for evidence that indicates the association of dust with quasars have been mostly negative. Searches for reddening induced effects on quasar spectra have been negative. Searches of the optical spectra of quasars of moderate to high red-shift have found nearly no evidence for the 2175 Å absorption typical of galactic type dust. Recent studies of the hydrogen lines in quasars (e.g., Soifer, et al. 1980a; Puetter, et al. 1980); (Figure 9), indicate a general weakening of Pα with respect to Hα as the Balmer decrement increases. The arrow in Figure 9 indicated the track that would be followed if the Pα/Hα and Hα/Hβ ratios were defined by extinction due to dust in the line emitting region. The observations clearly seem inconsistent with the presence of dust either mixed with the line emitting gas or in the line of sight to it. Furthermore, the Pα/Hα ratio is generally less than that expected for a typical H II region (Case B), again inconsistent with reddening by dust.

The observations of Figure 10 show that the high and low red-shift quasars share the anomalous Lα/Hα line ratio first suggested by Baldwin (1977). Several authors have suggested that this is in fact due to external reddening. The fact that the low and high red-shift quasars share this property forces the reddening to be local to the quasar. There is a marginal correlation of continuum spectral index with Lα/Hα ratio, seen in Figure 10, that would be consistent with this model, but this is at best a very weak correlation. Furthermore, the lack of a strong infrared excess in 3C273 as seen in Figure 1 appears to be inconsistent with this result.

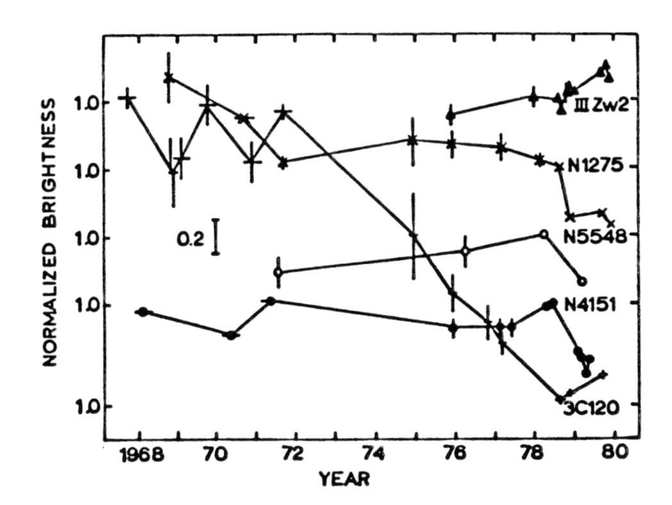

Figure 14. The time history of the infrared flux from Seyfert I galaxies (from Lebofsky and Rieke, 1980).

In contrast to the above arguments, evidence for dust in 3C273 has been given by Allen (1980). The emission is attributed to the 3.3 μm emission feature of unknown origin seen from many galactic dust-containing regions (see e.g. Russell et al. 1977). The detection of this feature in 3C273 is of low enough signal to noise to warrant confirming observations. If confirmed this would provide strong evidence for some thermal dust emission in quasars. Furthermore, Hyland and Allen (this conference) argue that the energy distribution of quasars steepens into the infrared between 1 and 3 μm in a way consistent with a hot dust component.

In summary then, the evidence as it now stands must favor non-thermal emission mechanisms to produce the majority of the infrared flux in quasars, although the evidence is mostly circumstantial, and not as strong as one would like. Furthermore there are data which show that in some quasars dust is in fact present.

The last class of active nuclei that we want to discuss is the type 1 Seyfert galaxies. These galaxies have long been thought to bear significant morphological resemblances to quasars and the energy distribution of the type 1 Seyfert galaxy NCG 4151 is included in Figure 1. This energy distribution is seen in the infrared to resemble more closely that of the quasar 3C273 than that of the Seyfert galaxy NCG 1068. This might be taken as an argument favoring non-thermal emission, and indeed several authors (Stein and Weedman 1976, Neugebauer et al. 1976, McAlary, et al. 1979) have made such arguments. On the other hand, Rieke (1978) has interpreted this distribution as a sum of power law (ultraviolet), stellar, and thermal infrared components and has shown that this interpretation, with a significant thermal component, is usually capable of fitting the observations.

Because both interpretations are possible, the energy distributions cannot be decisive evidence in favor of any emission mechanism, and other evidence must be found. Since the polarization in Seyferts is generally weak (Angel and Stockman 1980), this cannot be used as positive evidence for any emission mechanism.

High variability has often been used as an argument in favor of non-thermal emission mechanisms, but in type 1 Seyfert galaxies the observed time scales for infrared variability are compatible with the thermal dust emission mechanism. Recently Rieke and Lebofsky (1979b) and Lebofsky and Rieke (1980) have reported variability of Seyferts that far exceeds the errors in the observations. This variability is shown in Figure 14; unpublished observations at Caltech appear to confirm this variability. Rieke and Lebofsky and Lebofsky and Rieke have argued that the observed variability time scale in type 1 Seyfert galaxies is consistent with thermal emission models. Indeed, they find an increase in the infrared emission from the nucleus of IIIZw2 in late 1979, about two years after the large optical -UV- radio outburst in this galaxy, to be consistent with thermal dust emission at a distance of 2/3 pc from the central source.

Spectral features in the infrared emission itself support the model of thermal dust emission providing a significant contribution to the infrared flux in type 1 Seyfert nuclei. Rieke (1976) has reported that the 10 μm spectrum of Mk 231 shows absorption characteristic of silicate dust. While this only proves the 10 μm source is obscured by dust, it can be taken as evidence for thermal emission at 10 μm. Recent spectroscopic observations of Mk 231 at 10 μm by Jones, et al. (1980) and Gillett (private communication) show a suggestion of an emission band at 11.3 μm in the rest frame of Mk 231. If this is confirmed, it would provide direct evidence for thermal emission in this type 1 Seyfert galaxy, since the 11.3 μm feature is identified with thermal dust emission in galactic infrared sources.

More potential evidence for thermal emission in NGC 4151 comes from Cutri and Rudy (1980) who report observations of an emission feature at 3.3 μm which they attribute to the feature commonly associated with thermal emission from dust in galactic sources. The most straightforward interpretation of this observation, as thermal emission by material similar to that found in infrared sources in the Galaxy, is made difficult by the absence of a corresponding 11.3 μm feature (Jones et al. 1980 Gillett, private communication). If, however, this interpretation is confirmed, it would again provide proof of significant thermal emission in a type 1 Seyfert galaxy.

Indirect evidence that thermal dust emission may be important in Seyfert nuclei comes from spectroscopic observations of the hydrogen lines in the spectra of Seyfert galaxies. Figure 11 shows the Pα/Hα/Hβ ratio for a sample of Seyfert galaxies (Soifer, et al. (1980b). This plot shows that a significant number of the hydrogen line ratios appear to be affected by reddening in the line of sight to the line-emitting regions. This behavior is in fact distinctly different from that of quasars, which show no such effect (see Figure 9).

Thus, in the case of the type 1 Seyfert nuclei, the evidence seems to be increasing that in many instances at least a significant fraction of the emitted infrared radiation is due to thermal dust emission. This is almost certainly a contrast to quasars, where there is no compelling evidence for thermal dust emission contributing predominately to the observed infrared flux.

ACKNOWLEDGEMENTS

We would like to thank our colleagues who provided preprints of unpublished work. We especially thank A. R. Hyland, Steve Willner, Fred Gillett and Barbara Jones for making available to us and discussing unpublished data. Infrared astronomy at the California Institute of Technology is supported by grants from the NSF and from NASA.

REFERENCES

Aaronson, M. and Boroson, T.: 1980, Nature 283, p. 746.
Allen, D. A.: 1980, Nature 284, p. 323.
Angel, J. R. P. and Stockman, H. S.: 1980, Ann.Rev.Astron.Astrophys. 18, in press.
Baldwin, J. A.: 1977, M.N.R.A.S. 178, p. 67P.
Becklin, E. E., Matthews, K., Neugebauer, G., and Wynn-Williams, C. G.: 1973, Astrophys.J. 186, p. L69.
Blandford, R. D. and Rees, M. J.: 1978 in Pittsburgh Conference on BL Lac Objects, A. M. Wolfe, ed., Univ. of Pittsburgh Press, p. 328.
Cutri, R. and Rudy, R.: 1980, preprint.
Ennis, D. J.: 1980, Ph.D. Thesis, California Institute of Technology, in preparation.
Hildebrand, R. H., Whitcomb, S. E., Winston, R., Stiening, R. F., Harper, D. A., and Moseley, S. H.: 1977, Astrophys.J. 216, p. 698.
Jones, B., Merrill, K. M., Stein, W. A., and Willner, S. P.: 1980, private communication.
Jones, T. W., Rudnick, L., Owen, F. N., Puschell, J. J., Ennis, D. J., & Werner, M. W.: 1980, Astrophys.J., in press.
Kleinmann, D. E., Gillett, F. C., and Wright, E. L.: 1976, Astrophys.J. 208, p. 42.
Knacke, R. F., Capps, R. W., and Johns, M.: 1976, Astrophys.J. 210, p. L69.
Knacke, R. F., Capps, R. W., Johns, M.: 1979, Nature 280, p. 215.
Lebofsky, M. J., and Rieke, G. H.: 1980, Nature 284, p. 410.

McAlary, C. W., McLaren, R. A., and Crabtree, D. R.: 1979, Astrophys.J. 234, p. 471.
Moore, R. L., Angel, J. R. P., Rieke, G. H., Lebofsky, M. J., Wisniewski, W. Z., Mufson, S. L., Vrba, R. J., Miller, H. R., McGimsey, B. Q., Williamon, R. M.: 1980, preprint.
Neugebauer, G., Morton, D., Oke, J. B., Becklin, E. E., Daltabuit, E., Matthews, K., Persson, S. E., Smith, A. M., Soifer, B. T., Torres-Peimbert, S., Wynn-Williams, C. G.: 1980, Astrophys.J. 238, p. 502.
Neugebauer, G., Becklin, E. E., Oke, J. B., and Searle, L.: 1976, Astrophys.J. 205, p. 29.
Neugebauer, G., Oke, J. B., Becklin, E. E., and Matthews, K.: 1979, Astrophys.J. 230, p. 79.
O'Dell, S. L.: 1978, in Active Galactic Nuclei, Hazard, C. and Mitton, S., eds., Cambridge Univ. Press, p. 95.
O'Dell, S. L., Puschell, J. J., Stein, W. A., Owen, F., Porcas, R. W., Mufson, S., Moffett, T. J., and Ulrich, M-H.: 1978, Astrophys.J. 224, p. 22.
Puetter, R. C., Smith, H. E., Willner, S. P., and Pipher, J. L.: 1980, Astrophys.J., in press.
Puschell, J. J. and Stein, W. A.: 1980, preprint.
Rieke, G. H.: 1976, Astrophys.J. 210, p. L5.
Rieke, G. H.: 1978, Astrophys.J. 226, p. 550.
Rieke, G. H., Grasdalen, G. L., Kinman, T. D., Hintzen, P., Wills, B. J., and Wills, D.: 1976, Nature 260, p. 754.
Rieke, G. H. and Lebofsky, M. J.: 1979a, Ann.Rev.Astron.Astrophys. 17, p. 477.
Rieke, G. H. and Lebofsky, M. J.: 1979b, Astrophys.J. 227, p. 710.
Rieke, G. H., Lebofsky, M. J., Kemp, J. C., Coyne, G. V., Tapia, S.: 1977, Astropys.J. 218, p. L37.
Rieke, G. H., Lebofsky, M. J., and Kinman, T. D.: 1979, Astrophys.J. 232, p. L151.
Russell, R. W., Soifer, B. T., and Merrill, K. M.: 1977, Astrophys.J. 213, p. 66.
Shields, G. A. and Oke, J. B.: 1975, Astrophys.J. 197, p. 5.
Soifer, B. T., Neugebauer, G., Becklin, E. E., Matthews, K., Oke, J. B., and Malkan, M.: 1980b, in preparation.
Soifer, B. T., Neugebauer, G., Oke, J. B., and Matthews, K.: 1980a, Astrophys.J., in press.
Stein, W. A. and Weedman, D. W.: 1976, Astrophys.J. 205, p. 44.
Stockman, H. S. and Angel, J. R. P.: 1978, Astrophys.J. 220, p. L67.
Telesco, C. M., Becklin, E. E., and Wynn-Williams, C. G.: 1980, Astrophys.J. (Letters), in press.
Wampler, E. J.: 1968, Astrophys.J. 154, p. L53.
Wampler, E. J.: 1971, Astrophys.J. 164, p. 1.

DISCUSSION FOLLOWING PAPER DELIVERED BY B. T. SOIFER

STEIN: Optical spectropolarization studies of the nucleus of NGC 4151 indicate the presence of dust. However the amount of dust is much less than in the case of NGC 1068, for example, from various points of view.

THOMPSON: To your list of properties that indicate dust in NGC 1068 you should add what Steve Beckwith stated in his review, that the ratio of 2 μm H_2 emission-line luminosity to the far infrared luminosity was the same in Orion as in NGC 1068. NGC 4151, which probably has less dust than NGC 1068, does not show that line.

RIEKE: The tendency for the QSO spectral indices between optical and infrared to lie near -1 is at least partly due to selection. When indices are measured for flat-spectrum radio sources selected for radio brightness (irrespective of optical-infrared brightness), the indices have a broad distribution from 0 to -3.

SOIFER: We may be seeing that effect in the histogram I showed [Figure 5], where there is some suggestion of a tendency for the radio-selected quasars to be slightly redder on average than the optically-selected ones.

BECKLIN: Has anyone measured at $\lambda > 2.2$ μm the steep blank-field QSOs first measured at 2.2 μm by Rieke et al.?

SOIFER: Beichman has attempted to measure several of them, but has not detected any.

LEBOFSKY: One of the very red QSOs has been measured at wavelengths longer than 2 μm; 1413+135 has been measured from 3.5 μm to 10 μm, and the spectrum at 3.5 μm already falls significantly below a power-law extrapolated from the 1-2.2 μm data.

BECKLIN: You should not leave the impression that all of the infrared emission in Seyfert Galaxies is thermal radiation from dust, especially at $\lambda < 10$ μm. A significant infrared nonthermal source is not ruled out.

SOIFER: At least a majority of the infrared flux is probably thermal, although there is clearly a non-thermal source as well, especially in the optical and U-V.

RICKARD: There was once a problem with the steepness of the submillimeter spectrum of NGC 1068, which because of strong upper limits at 1 mm was too steep to be thermal emission.

SOIFER: To the best of my understanding the infrared spectrum is completely consistent with thermal emission.

TOVMASSIAN: You presented convincing evidence of nonthermal emission of BL Lac objects and quasars. Are there any explanations of the origin of synchrotron emission in these objects? Is it not the explosion in the nucleus which causes their radio emission?

SOIFER: I refer this question to Wayne Stein.

STEIN: I would rather not be burdened with the task of explaining the ultimate source of energy at this time.

AN OUTSIDER'S VIEW OF EXTRAGALACTIC INFRARED ASTRONOMY

M. S. Longair*
Mullard Radio Astronomy Observatory
Cavendish Laboratory, Cambridge

1. INTRODUCTION

No one could be more of an outsider to infrared astronomy than myself. Like all outsiders, one forms preconceived notions about disciplines outside one's own speciality. In my own case, I can summarise these under three headings.

(i) Infrared astronomy is "obscured" by dust. Nobody really knows what dust is made of or what the shapes of the dust grains are.

(ii) Infrared astronomy will solve the problems of star formation.

(iii) Ultimately, many of the most important cosmological problems will be solved in the infrared waveband.

It is always pleasant to find that one's preconceived notions are either totally shattered or confirmed beyond one's expectations and this has happened to me several times during this week. The scope of the subject as it unfolded throughout the week is vast. Zuckerman urged infrared astronomers to claim what is rightly theirs in the fields of star formation and the distance scale in our own Galaxy. I would go further and claim that many of the fundamental questions of extragalactic astronomy and cosmology are part of the birthright of infrared astronomers. It was apparent that no field of astronomy will remain unaffected by the great advances and discoveries of infrared astronomy. Infrared astronomers must expect an influx of outsiders such as myself into their area - it will soon acquire the ultimate accolade of being regarded as a conventional diagnostic tool in the equipment of all astronomers.

* Present address: Royal Observatory, Blackford Hill, Edinburgh.

C. G. Wynn-Williams and D. P. Cruikshank (eds.), Infrared Astronomy, 351–360.

What has happened to my preconceived notions as the week has gone on? To be honest, I still don't like dust but I am prepared to come to an accommodation with it. I was particularly impressed by the prospects of understanding more about the chemical and molecular constituents from very high resolution infrared studies and by the idea that simultaneous observation of the solid state and gas phase constituents would lead to real knowledge of the dust and its environment. It is a very ambitious goal but one which is so important that it is worth a major observational and laboratory effort.

So far as star formation is concerned, I found things had somewhat changed direction from questions to which extragalactic astronomers naively expect infrared astronomers to provide answers - what determines the rate of star formation? What is the initial mass function? How does it depend upon density, chemical composition, dust-to-gas ratio and so on? The new view of the Orion Nebula and its various constituents revealed how much we can potentially learn about regions of star formation but raised a whole new range of problems summarised by Zuckerman rather than solving the old. The beautiful observations and analyses of the hot molecular hydrogen in OMC1 adds to the complexity of the region and it was intriguing how the focus of attention shifted from BN to IRc2 to IRc4 as the protostellar objects or extremely young HII regions responsible for the shock waves and expansion of the molecular maser sources. I found myself wondering if we are really certain that we know which objects are protostars.

An outsider can only be deeply impressed by this wealth of new information about regions of star formation and the intriguing problems summarised by Zuckerman. Nonetheless, it is worthwhile remembering that we look to infrared astronomers to give us insight into the fundamental problems which I briefly mentioned above. They are crucial for our understanding of galaxy formation and evolution. I will return to this point repeatedly.

So much for the preliminaries - what about the central questions of importance for extragalactic astronomy?

2. OUR OWN GALAXY AND NORMAL GALAXIES

In all branches of astronomy, an understanding of the properties of our own Galaxy has proved crucial in defining the framework within which we attempt to understand extragalactic systems. On the small scale, I have already referred to the importance of studies of regions of star formation. On the large scale, the new results on the large scale distribution of infrared continuum emission, and its relation to giant HII regions, molecular clouds and radio continuum emission described by Okuda illustrate very clearly how we may hope to relate the infrared morphology of the Galaxy to other tracers of gas, dust, stars and regions of star formation. Drapatz showed the way in which we may eventually hope to relate all these large scale features to the general picture of the evolution of our own Galaxy. However, he also indicated clearly the grave problems of interpretation which have to be solved. It is studies of these types which will generate the prejudices which we will adopt in contrasting our own Galaxy with others.

Gatley gave us a remarkable review of the crucial evidence on the Galactic Centre and told us that there was a black hole there. Rather, I should say, he "guided our intuition," as Zwicky would have said, towards a position in which no other interpretation was reasonable. Why did nobody object? Partly because the wrong sorts of people were present in the audience. More important, I believe, is the fact that black holes are now very much "part of the furniture" of Galactic and extragalactic astronomy. Black holes are very reasonable things to form in astronomical systems and galactic nuclei are particularly natural places for the massive varieties to form. As Gatley emphasised, the centre of our own Galaxy is the closest active galactic nucleus and at infrared wavelengths we can obtain a higher resolution picture than in any other Galaxy. We should recall that the Schwartzschild radius of a 10^7 $M_\odot$ black hole at the Galactic centre subtends an angle of 0.02 milliarcsec which is very small, but not inconceivably small, for study by interferometry at radio and infrared wavelengths.

The studies of the stellar content of nearby normal galaxies by Aaronson, Persson and their colleagues indicate the direction in which studies should proceed. I found the evidence for the intermediate age population rather convincing but it clearly requires us to modify our view of galactic evolution and the birthrate function of stars as a function of age. This necessarily complicates our picture of Galactic evolution which, in any case, is not in a particularly healthy state.

The most ambitious attempt to tie together all the observations of a strong infrared normal galaxy was that of M82 by Rieke. His lecture was a classic example of the wealth of diverse information which can be wholly derived from observations in the infrared waveband, the infrared bolometric luminosity, the integrated light from red giants from the deep CO bands at 2 μm, the extinction from the intensities of the Brackett lines, the ionising flux from the hydrogen recombination lines, and the mass of the central regions from the Ne II rotation curve. Rieke then showed how all of these observations could be reconciled with a simple picture for active recent star formation in M82 involving a mass of $\sim 2 \times 10^8$ $M_\odot$ in new stars. It is certainly encouraging that this can be done but one wonders how unique the procedure is and in particular what the most important constraints on the models are. Specifically, how many free parameters do you need to build an integrated picture of M82? What happens if you add more? One also wonders what will happen when more detailed observations are available at all wavelengths? None of this is criticism of what was done. This type of modelling exercise is essential and will lead to a refinement of the questions we can reasonably hope to answer observationally.

An important goal is to reach a position in which similar types of analyses can be performed for all types of normal galaxy.

3. ACTIVE GALACTIC NUCLEI

We heard a large number of presentations about different classes of galaxies with active galactic nuclei - Seyfert galaxies, X-ray and radio galaxies, quasars and BL Lac objects. We were asked to assimilate a vast amount of data and we should remember what the astrophysical aims of these studies are. To put it crudely, we are trying to distinguish the various contributions to the total spectrum - the stellar component, that due to dust, the continuum and line emission of ionised gas clouds in the nucleus and what I will call the "other components". The contributions of Rieke and Lebofsky and of Scoville and his colleagues were particularly impressive accounts of how high resolution infrared spectroscopy can add to the interpretation of the optical spectrum of the Seyfert galaxies NGC 4151 and 1068 respectively. Other authors showed convincingly how the infrared colours of galaxies with active nuclei fall along a more or less continuous sequence from pure galaxy spectra to pure quasar spectra.

My own view is that the most intriguing aspects of these studies are those which shed some light on the properties of the gas close to the active nucleus and the emission mechanism responsible for the "other components". In turn, I believe these must shed light on the supply of fuel to the galactic nucleus and the structure of its innermost regions.

According to the conventional view, it is the broad line components of the line-emitting regions which originate closest to the nucleus. The infrared observations of the hydrogen recombination lines are crucial in this respect in providing good measures of the extinction in the broad-line regions and, as we heard from Allen and his colleagues, it appears that the Balmer line intensities are considerably enhanced with respect to Lyman α, confirming earlier work on much smaller numbers of objects. The most reasonable interpretation of these data are that collisional excitation of the Balmer lines enhances their line intensities at the expense of Lyman-α and this requires high particle densities, $N \sim 10^8$-10^{10} cm^{-3}. This means that the gas clouds are of relatively small size and mass which is consistent with the variability seen in some of the broad emission line profiles. These are the most compact regions about the nucleus from which we observe line emission and we would like to know how much infrared spectroscopy can add to this picture. Is there dust associated with these regions with normal gas-to-dust ratio?

A second important aspect of the infrared observations is the extraction of the underlying continuum spectrum. Neugebauer and Soifer made an excellent case for the infrared emission in the case of Seyfert II galaxies being dust emission, including a few cases where the variable component could be associated with dust. I would emphasise the importance of delineating as precisely as possible the spectrum of what is normally called the non-thermal component which I have preferred to call the "other" component. The essential point is that any emission mechanism involving ultrarelativistic electrons such as synchrotron radiation or inverse Compton emission results in a broad-band emission spectrum having $\Delta\nu/\nu \approx 1$. This is because the continuum emission is the Fourier transform of the beaming pattern of the relativistic electron and this is more or less independent of how the particle is accelerated in the emission process. Thus, even if the electrons all had the same energy, it would be very difficult to attribute any sharp feature in the observed spectrum to the emission of ultrarelativistic electrons.

I believe we have seen some spectra which are supposed to represent the "other" component which possess features which are too sharp to be explained by the emission of ultrarelativistic electrons. This is not a problem confined to the infrared waveband. A "blue bump" is observed in the ultraviolet spectrum of 3C273 and the spectra of many quasars show wiggles which look real. It is very important to find out whether these features are real or not. This requires a very careful assessment of all the possible contributions to the integrated spectrum from stars, dust and regions of ionised hydrogen as well as the various absorption processes, by interstellar gas and dust, which convert smooth spectra into jagged spectra.

When this is done, we may indeed find a smooth spectrum in which case the emission process may be attributed to ultrarelativistic particles. I would find this result a bit disappointing because the information we can hope to derive from the emission of relativistic particles is very limited. We have to specify how the particles were accelerated to relate them to specific regions in the nucleus. I believe we are still far from being able to understand this.

I find the alternative that the spectrum ends up not being smooth much more intriguing. For example, it is a real possibility that the continuum emission is the thermal emission of an accretion disc about a central massive black hole. The standard model of an accretion disc results in a spectrum which rises to ultraviolet wavelengths as $\nu^{1/3}$. However, there are ways of distorting this by injecting "fuel" in a non-steady manner to the disc. In general, the types of information which can potentially be derived from such studies are of much more direct relevance to the structure of the accretion disc.

We heard of two cases where the above type of analysis seems particularly relevant. Lebofsky showed us the spectrum of a quasar with a steep spectrum in the near infrared which then showed a cut-off at longer wavelengths. Sherwood showed us the millimetre spectrum of a radio quiet quasar which looked dangerously high as compared with the radio and optical continuum spectra. We need to know in more detail the precise spectra of these objects and the nature of the emission mechanism.

Finally, let me emphasise the importance of studying variations in the "other" component over a wide frequency range, extending from the ultraviolet to the far infrared. It is customary to assume that the "other" component varies up and down contemporaneously at all wavelengths in this waveband. If there are "sharp" features in the spectrum these may or may not vary in the same way as the underlying continuum. Rieke's analysis of the spectrum of NGC4151 is precisely the type of analysis needed to throw some light on this question. It might, for example, be that all the variability could be attributed to the "bump" component which would be a very important result.

4. CLASSICAL COSMOLOGY

By classical cosmology, I mean attempts to find the Hubble constant H_o and the deceleration parameter q_o (or the density parameter Ω) from observations of galaxies. The Tully-Fisher method of estimating the distances to spiral galaxies must be Hawaii's greatest contribution to cosmology. Aaronson showed how close the correlation is between the 21-cm velocity width of the neutral hydrogen distribution and the infrared luminosity of spiral galaxies. It is remarkable how close his result agrees with the classical methods of Sandage and Tammann within the Local Supercluster and how his inferred velocity of infall towards the supercluster is in general agreement with that determined from the dipole anisotropy of the microwave background radiation. The major problem which I see with the values of about 95 km s^{-1} Mpc^{-1} is that it results in a rather short cosmological time-scale, $T \leq H_o^{-1} = 1.1 \times 10^{10}$ years, compared with the ages of the oldest globular clusters. I am no expert in the latter field and it requires a very detailed investigation to find out how much room for manoeuvre there is in the globular cluster ages. Equally, how much flexibility is there on Aaronson's value of H_o?

The other urgent requirement is an independent analysis of the Aaronson, Huchra and Mould version of the Tully-Fisher method. Nonetheless, it seems that this infrared-radio technique may provide us with the most accurate estimate of the extragalactic distance scale.

The other aspect of the redshift-distance relation is its extension to large redshifts where differences in the global geometry and dynamics become important, i.e. the redshift-magnitude relation depends upon the world model. Grasdalen's infrared redshift-magnitude relation showed how powerful this technique is in general terms. To put it crudely, the K-correction which dims galaxies in the optical waveband makes them brighter in the 2-3 μm waveband as they move to larger redshifts because of the maximum in the spectrum of a giant elliptical galaxy at about 1 μm. No one is guessing values of q_o yet but the fact that a respectable redshift-magnitude relation can be constructed is encouraging. Lebofsky and Spinrad showed that there is little evidence of colour evolution in the giant ellipticals and this is encouraging if you are interested in using this as a possible route to the geometry of the Universe.

I would make three comments about this work. First of all, there is the question of how you find such distant giant elliptical galaxies. Most of them have been found as a result of searches for radio source identifications. It is legitimate to worry whether or not a giant elliptical galaxy which is a strong radio source really is a typical giant elliptical and whether or not the total light is entirely star-light. My impression from studying the optical spectra of radio galaxies is that their optical spectra can be properly decomposed into a nuclear component consisting of continuum emission and emission line spectrum and a standard giant elliptical galaxy spectrum. Indeed a significant fraction of the most powerful radio galaxies do not contain the nuclear component at all. So, I am quite optimistic about the use of radio galaxies although we need the direct observational evidence that they are not significantly different from normal giant ellipticals.

Second, if you need any convincing that the study of distant galaxies is the birthright of infrared astronomers, you need only look at the identification content of the faintest radio source identifications which we are now making. Jim Gunn and I have been trying to complete the identifications of 3CR radio sources for the last 8 years. At the beginning of this year, we effectively completed this project thanks to the development of the TI/JPL CCD camera. All the new identifications, which could not be made with conventional photography using IIIaJ plates, are with very distant galaxies. All of these are much brighter in i than they are in r. We know that this continues to hold into the infrared waveband for some of the galaxies studied by Grasdalen and by Rieke and Lebofsky which have been easily detected at 2 μm. Indeed, Grasdalen detected 3C65 at 2 μm before we could identify it optically - it is barely detectable in r with the CCD camera. Another example is 3C184 which is about 1.2 magnitudes brighter in i than in r and for which we have measured a redshift of 1 on the basis of a strong emission line spectrum. Thus, as we have always known, as soon as galaxies are observed at large redshifts, they become infrared objects rather than optical objects, the whole of the energy distribution shifting into the 1-5 μm region.

The third point concerns evidence for evolution over cosmological time-scales. We can list a number of separate pieces of evidence all of which suggest that things do change over relatively short time-scales and that there are significant changes between redshifts of 0 and 1. First, the V/V_{max} test for quasars and radio galaxies and the counts of radio sources show that there was very much more high energy astrophysical activity at redshifts $z \sim 2-3$ than there is now. Even over the redshift interval $0 < z < 1$, this activity has increased by a large factor. In exponential models of the evolution, the typical time-scale for decay of these populations is only 10^9 years. Second, significant changes in the colour distribution of the galaxies in distant clusters as compared to similar clusters at the present epoch have been observed by Butcher and Oemler. Third, the counts of faint galaxies have been interpreted as showing an excess of faint galaxies compared with the predictions of uniform models. Fourth, the Westerbork workers have found that the colours of faint radio galaxies at $z \sim 0.5$ may be bluer than those of nearby radio galaxies. Finally, Gunn and Oke find a larger fraction of blue spectra among the brightest cluster members for galaxies with $z > 0.6$ than those with $z < 0.6$.

Obviously some of these pieces of evidence are much stronger than others but the warning is plain. The first thing to be done is to look empirically at the properties of distant galaxies in the infrared and see whether or not they are the same as those of nearby objects. This is a field where theory is not of much help. The models of galactic evolution can explain the above phenomena but they are a posteriori rationalisations rather than firm theoretical predictions.

5. PHYSICAL COSMOLOGY

Ever since the predictions of Peebles and Partridge, it has been clear that if galaxies liberated a large amount of energy when they were first formed, these objects should be infrared sources and contribute to the infrared background emission. These ideas have been revived by various authors from time to time but have floundered for the lack of any firm observational evidence for their existence. I suspect that the main cause for their lack of detection is that they are genuinely infrared objects. As we have seen above, as soon as giant elliptical galaxies are redshifted to $z = 1$, they become invisible in the r waveband. Now, it would be rash to claim that we know what the spectrum of a young galaxy would look like and indeed we can invent a wide range of models for galaxies which would be consistent with all current observations and yet have a vastly different appearance when they were much younger. I believe that this is a field in which we can only find out the answers by direct observation in the infrared waveband.

We could imagine that the programme would proceed in two stages. First of all, the thorough study of current objects as they were when a bit younger than at present i.e. at $z \sim 1$. Then one should proceed to the study of much younger objects having $1 < z \lesssim 10$. This programme will only become feasible once infrared panoramic detectors having reasonably large fields of view become available. However, let us note the importance of these observations for our understanding of how the large scale structure of the Universe came about.

There are two rather extreme views about how galaxies first formed in the Universe. In the strict adiabatic theory propounded by the Moscow group led by Zeldovich, all fine scale structure in the primordial matter at the epoch of recombination is washed out by various damping processes and the first structures to begin forming are large scale regions, on the scale of superclusters. Galaxies form by condensation from this collapsing cloud. The characteristic scales associated with clusters and superclusters can only begin to collapse at redshifts $z \sim 5$ and consequently all galaxies must form at redshifts of 5 or less. An alternative view proposed by Press and Schechter and developed by Rees and White begins with globular cluster size units immediately after the epoch of recombination. Larger systems are built up as a result of hierarchical clustering so that galaxies form very much earlier $z \sim 20$-30.

In this simple example, a direct test of these hypotheses is the search for the young galaxies at redshifts $z \lesssim 5$. One theory predicts that that is when all galaxies must form whereas in the other it takes place at much larger redshifts. Young galaxies at $z \lesssim 5$ should be detectable with the next generation of panoramic infrared detectors if indeed such objects exist.

If we can discover young galaxies at $z \gtrsim 1$, we can imagine all sorts of exciting prospects. Can we compare their clustering with the clustering of galaxies at the present day? Can we detect the remnants of the primordial gas clouds out of which they condensed? How are these phenomena related to the evolution of powerful radio galaxies and quasars? Our ultimate aim is to define direct from observation the sequence of events which led to the present large-scale structure of the Universe. I believe that in many ways these prospects are unique to the infrared waveband.

6. FUTURE PROSPECTS

It is clear that we are on the verge of a huge expansion in infrared astronomy. There are few branches of astronomy which have not been discussed at this conference and infrared astronomy is already making a major impact on all of them. The new facilities here in Hawaii and elsewhere and the new types of detectors and spectrographs under discussion will enable us to capitalise on the breakthroughs summarised at this conference.

A second conclusion is that this must surely be the last conference entitled "Infrared Astronomy". We have all received a marvellous panoramic view of the discipline but we should no longer regard it as an isolated subject. The issues discussed here have spanned essentially the whole of the electromagnetic spectrum. In future, I expect conferences involving infrared astronomers to be much more "mission-oriented" in which infrared astronomy can be seen as one among many different ways of tackling problems such as dust, star-formation, the evolution of galaxies and so on. If this turns out to be the last infrared conference, and I regard that as a token of the maturity of the subject (when was the last "optical astronomy" conference?), it will certainly have gone out with a bang.

Finally, we must look to the future and ask what are the most urgent needs for furthering all these areas of infrared astronomy. On Tuesday evening, space infrared astronomy was discussed and on Monday we talked about astronomy from aircraft. I think we should also ask what the next generation of ground based telescopes ought to be. We hear a great deal about the Next Generation Telescope as large versions of optical telescopes. But might it not be that the next generation ground based telescope should be a very large infrared telescope dedicated to, say, the 1-5 μm window rather than 0.3 to 1 μm. If the choice were to be based entirely upon the importance of the science, it is not clear that the most important questions will be answered in the optical waveband.

I may have started this symposium as an outsider. I have now advanced to the status of "convert". It has been a marvellously stimulating symposium.

DISCUSSION FOLLOWING PAPER DELIVERED BY M. S. LONGAIR

TOVMASSIAN: I would like to make a comment on black holes. You said that black holes exist in astronomy. I would correct this expression. Yes, they exist, but exist in the minds of astronomers, mainly theoreticians. Black holes imply a sense of contraction of a huge mass to a point and the following accretion of the surrounding matter. We have to admit anyhow that everywhere in the universe we see explosions, outflow of matter, but never implosions, infall.

LONGAIR: There are two separate answers to this comment. First of all there are, of course, very good reasons why we expect the presence of a black hole in a galactic nucleus to result in energy generation, outflow of matter and the acceleration of charged particles. The second part of the comment verges on the philosophy of science which means that we are treading on very dangerous territory. The point about black holes is that they provide a very efficient source of energy which is based upon sound physical principles. I believe it is by far the most economical and natural hypothesis for the energy source in active nuclei. I believe this view is shared by a very large number of astronomers. When such a view is held by the majority of astronomers and used as part of the framework for proceeding to solve further problems, I think it is fair to say that black holes have become part of the conventional apparatus of astronomers.

LIST OF PARTICIPANTS

Australia

Aitken, D. K.	Anglo-Australian Observatory
Allen, D. A.	Anglo-Australian Observatory
Ellis, M.	Mt. Stromlo Observatory
Hyland, A. R.	Mt. Stromlo Observatory
Jones, T. J.	Mt. Stromlo Observatory
Mitchell, R. M.	University of Melbourne
Ruelas-Mayorga, R. A.	Mt. Stromlo Observatory
Thomas, J. A.	University of Melbourne

Canada

Avery, L.	Herzberg Institute of Astrophysics
Campbell, B.	Canada-France-Hawaii Telescope
Lowe, R. P.	University of Western Ontario
McAlary, C. W.	University of Toronto
McLaren, R. A.	University of Toronto
Menon, T. K.	University of British Columbia

Chile

Danks, A. C.	European Southern Observatory
Koornneef, J.	European Southern Observatory

France

Baluteau, J.-P.	Observatoire de Meudon
Bensammar, S.	Observatoire de Meudon
Bruston, P.	LPSP/CNRS
Cayrel, G.	Observatoire de Meudon
Cayrel, R.	Canada-France-Hawaii Telescope
Combes, M.	Observatoire de Meudon
de Batz, B.	Observatoire de Meudon
de Bergh, C.	Observatoire de Meudon
de Muizon, M.	Observatoire de Meudon
Encrenaz, T.	Observatoire de Meudon
Falgarone, E.	Observatoire de Meudon
Gispert, R.	LPSP/CNRS
Guibert, J.	Observatoire de Meudon
Heidmann, J.	Observatoire de Meudon
Leger, A	Université Paris
Papoular, R.	CEN Saclay
Puget, J.-L.	Institut d'Astrophysique

Rabbia, Y.	Observatoire du Cerga
Rouan, D.	Observatoire de Meudon
Serra, G.	C.E.S.R., Toulouse
Sibille, F.	Observatoire de Lyon

Germany

Cosmovici, C. B.	MPI Extraterrestrische Physik, Garching
Drapatz, S.	MPI Extraterrestrische Physik, Garching
Elsässer, H.	MPI Astronomie, Heidelberg
Hefele, H.	MPI Astronomie, Heidelberg
Krätschmer, W.	MPI Kernphysik, Heidelberg
Lemke, D.	MPI Astronomie, Heidelberg
Schmid-Burgk, J	MPI Radioastronomie, Bonn
Sherwood, W. A.	MPI Radioastronomie, Bonn
Staude, H. J.	MPI Astronomie, Heidelberg
Wilson, T. L.	MPI Radioastronomie, Bonn

India

Kulkarni, P. V.	Physical Research Laboratory, Ahmedabad
Prabhu, T. P.	Indian Institute of Astrophysics
Rengarajan, T. N.	Tata Institute, Bombay

Italy

di Fazio, A.	Osservatorio Astronomico di Roma
Ferrari Toniolo, M.	Istituto Astrofisica Spaziale, Frascati
Palla, F.	Osservatorio Astrofisica di Arcetri
Persi, P.	Laboratorio Astrofisica Spaziale, Frascati
Saraceno, P.	Laboratorio Plasma Spaziale, Frascati
Tanzi, E. G.	LFCTR, Milan

Japan

Kobayashi, Y.	Kyoto University
Maihara, T.	Kyoto University
Mukai, S.	Kanazawa Institute of Technology
Okuda, H.	Kyoto University
Sato, S.	Kyoto University

Korea

Minn, Y. K.	National Astronomical Observatory, Seoul

Netherlands

Allamandola, L.	University of Leiden
de Graauw, Th.	European Space Agency
Fitton, B.	European Space Agency
Forster, J. R.	Netherlands Foundation for Radio Astronomy

Habing, H. J.	University of Leiden
Tielens, A.G.G.M.	University of Leiden
van Duinen, R.	University of Groningen

South Africa

Glass, I. S.	South African Astronomical Observatory

Sweden

Hjalmarson, Å.	Onsala Space Observatory
Nordh, H. L.	Stockholm Observatory
Olofsson, S. G.	Stockholm Observatory
Sandell, G.	Stockholm Observatory

Switzerland

Moorwood, A. F. M.	European Southern Observatory
Tarenghi, M.	European Southern Observatory

USSR

Tovmassian, G. M.	Byurakan Observatory

United Kingdom

Adamson, A. J.	Leicester University
Arakaki, S. L.	United Kingdom Infrared Telescope
Beattie, D. H.	United Kingdom Infrared Telescope
Beckman, J.	Queen Mary College, London
Bode, M. F.	University of Keele
Evans, A.	University of Keele
Furniss, I.	University College London
Gatley, I.	United Kingdom Infrared Telescope
Giles, A. B.	Leicester University
Harris, S.	Queen Mary College, London
Joseph, R. D.	Imperial College London
Lee, T. J.	United Kingdom Infrared Telescope
Longair, M. S.	University of Cambridge
Longmore, A. J.	United Kingdom Infrared Telescope
Roche, P.	University College London
Rowan-Robinson, M.	Queen Mary College, London
Scarrott, S. M.	University of Durham
Smith, M. G.	Royal Observatory, Edinburgh
Stewart, M.	United Kingdom Infrared Telescope
Ward, M. J.	University of Cambridge
White, G. J.	Queen Mary College, London
Whittet, D. C. B.	University College London
Williams, P. M.	United Kingdom Infrared Telescope
Wolstencroft, R. D.	Royal Obs., Edinburgh, and UKIRT
Zealey, W.	United Kingdom Infrared Telescope

United States

Aaronson, M.	Steward Observatory
Apt, J.	Jet Propulsion Laboratory
Backman, D.	University of Hawaii
Baud, B.	University of California, Berkeley
Beck, S. C.	University of California, Berkeley
Becklin, E.	University of Hawaii
Beckwith, S.	Cornell University
Beer, R.	Jet Propulsion Laboratory
Blitz, L.	University of California, Berkeley
Boggess, N. W.	NASA Headquarters
Cameron, R.	NASA/Ames Research Center
Campbell, M.	Steward Observatory
Capps, R. W.	University of Hawaii
Craine, E. R.	Steward Observatory
Cruikshank, D. P.	University of Hawaii
Cudaback, D.	University of California, Berkeley
Diner, D. J.	Jet Propulsion Laboratory
Dinerstein, H	Lick Observatory
Dyck, H. M.	University of Hawaii
Erickson, E. F.	NASA/Ames Research Center
Evans, N. J., II	University of Texas
Falk, S.	University of Texas
Fazio, G. G.	Center for Astrophysics
Felli, M.	NRAO—VLA
Gautier, T. N.	Steward Observatory
Genzel, R.	Center for Astrophysics
Gillespie, C.	NASA/Ames Research Center
Grasdalen, G. L.	University of Wyoming
Gulkis, S.	Jet Propulsion Laboratory
Haas, M. R.	NASA/Ames Research Center
Hanner, M.	Jet Propulsion Laboratory
Harper, D. A.	University of Chicago
Harvey, P. M.	Steward Observatory
Harwit, M.	Cornell University
Hauser, M.	NASA/Goddard Space Flight Center
Hildebrand, R. H.	University of Chicago
Hoffman, W.	University of Arizona
Hollenbach, D.	NASA/Ames Research Center
Howell, R. R.	Steward Observatory
Jaffe, D.	Center for Astrophysics
Jefferies, J. T.	University of Hawaii
Johnson, P. E.	Jet Propulsion Laboratory
Jones, B.	University of California, San Diego
Keene, J.	University of Chicago
Kemp, J.	University of Oregon
Koch, D.	Smithsonian Astrophysical Observatory
Krisciunas, C.	Kuiper Airborne Observatory
Kunkle, T.	Los Alamos
Lacy, J. H.	Caltech

Lada, C. J.	Steward Observatory
Lester, D. F.	NASA/Ames Research Center
Lebofsky, M.	Steward Observatory
Lindsey, C. A.	University of Hawaii
Lonsdale, C. J.	University of Hawaii
Martin, T. Z.	Jet Propulsion Laboratory
Matthews, K.	Caltech
McCleese, D. J.	Jet Propulsion Laboratory
McCord, T. B.	University of Hawaii
Melnick, G.	Cornell University
Meyer, A.	Kuiper Airborne Observatory
Morrison, D.	University of Hawaii
Natta, A.	Cornell University and Frascati
Neugebauer, G.	Caltech
Nishimura, T.	Steward Observatory
Ogden, P. M.	University of Wisconsin
Orton, G. S.	Jet Propulsion Laboratory
Owensby, P. D.	University of Hawaii
Patel, R. I.	Washington University
Persson, S. E.	Mt. Wilson and Las Campanas Observatories
Potter, A.	NASA Johnson Space Center
Price, S. D.	Air Force Geophysical Laboratory
Puschell, J.	NRAO
Rickard, L. J.	NRAO
Rieke, G. H.	University of Arizona
Russell, R.	Cornell University
Sargent, A. I.	Caltech
Schwartz, P. R.	Naval Research Laboratory
Scoville, N.	University of Massachusetts
Silverberg, R. F.	NASA/Goddard Space Flight Center
Simon, M.	State University of New York at Stony Brook
Sinton, W.	University of Hawaii
Soifer, B. T.	Caltech
Spinrad, H.	University of California, Berkeley
Stein, W. A.	University of Minnesota
Storey, J. W. V.	University of California, Berkeley
Telesco, C. M.	University of Hawaii
Terrile, R. J.	Jet Propulsion Laboratory
Thompson, L. A.	University of Hawaii
Thompson, R.	Steward Observatory
Thronson, H., Jr.	Steward Observatory
Tokunaga, A.	University of Hawaii
Traub, W.	Center for Astrophysics
Tully, B.	University of Hawaii
Werner, M. W.	NASA/Ames Research Center
Wilson, A. S.	University of Maryland
Wing, R. F.	Ohio State University
Wright, E. L.	Massachusetts Institute of Technology
Wynn-Williams, C. G.	University of Hawaii
Young, E.	Steward Observatory
Zeilik, M.	University of New Mexico
Zuckerman, B.	University of Maryland

LIST OF CONTRIBUTED PAPERS

COMPOSITION OF PLANETARY ATMOSPHERES

Beer	The Infrared Spectrum of Venus in the Post-Pioneer Era
de Bergh & Maillard	Near-Infrared Laboratory High-Resolution Spectroscopy for Planetary Applications
Haas, Erickson, McKibbin & Caroff	Far-Infrared Spectrophotometry of Saturn
Rouan, Gautier, D., Baluteau, Marten, Chedin, Scott, N., Husson, Conrath, Hanel, Kunde, & Maguire	Infrared Spectroscopy of the Jovian Atmosphere from Voyager

STRUCTURE OF PLANETARY ATMOSPHERES

Apt	Results from Daily Infrared Imaging of Venus: 1973-1979
*Diner	Morphology and Structure of Venus' North Polar Region by Infrared Imaging from the Pioneer Orbiter
McCleese	Remote Sensing of the Middle Atmosphere of Venus
Mukai, S., Mukai, T., & Sato	Analysis of the Infrared Observations of Venus
Mumma, Buhl, Chin, Deming, Espenak, & Kostiuk	Direct Observation of the Failure of Local Thermodynamic Equilibrium in the CO_2 (001) State in the Lower Atmosphere of Mars
Terrile, Capps, Becklin, & Cruikshank	Correlation of Jovian Cloud Colors with Vertical Structure from Ground-Based Infrared and Voyager Imaging
Tokunaga, Knacke, & Ridgway	High-Resolution 10 Micrometer Spectroscopy of Jupiter and Saturn

THERMAL PROPERTIES AND REMOTE SENSING OF PLANETS AND SATELLITES

Hildebrand, Keene, & Whitcomb	Submillimeter Brightness Temperatures of the Giant Planets

* Poster Presentations

Martin, T.Z.	General Thermal and Albedo Behavior of Mars
Nicholson & Jones, T.J.	Spectrophotometry of Uranus and Its Rings

INTERPLANETARY DUST

Fraundorf, Freeman, Patel, Shirck, & Walker	Laboratory Measurements of Visible and Infrared Optical Absorption in Interplanetary Dust Particles
Hanner	The 10-Micron Emission from Cometary Dust
Potter & Morgan	Mid-Infrared Measurements of Lunar Mineralology

THE SUN

*Lindsey, Hildebrand, Keene, & Whitcomb	Submillimeter Observations of Sunspots

CONTINUUM EMISSION FROM MOLECULAR CLOUDS AND H II REGIONS

Campbell & Hoffman	Large-Scale Far-Infrared Emission from IC 1318 b, c
*Cudlip, Emerson, Furniss, Jennings, King, & Robert	Multiband Far-Infrared Observations of the NGC 6334 Complex
*Erickson, Haas, Caroff, Simpson Goorvitch, & Tokunaga	Far-Infrared Spectra of Compact Galactic Nebulae
*Harris, A., Lemke, Kleiner, & Frey	Near-Infrared Maps of H II Regions M17 and W51
*Jones, T.J. & Hyland	New Infrared Studies of the Chamaeleon Dark Cloud Region
Jaffe, P., Stier, & Fazio	Newly Formed OB-Stars in the Giant Molecular Cloud Complex Southwest of M17
*Keene, Harper, Hildebrand, & Whitcomb	Far-Infrared Observations of Globules
Lada & Wilking	Heat Sources for Bright-Rimmed Molecular Clouds: Infrared Observations of B35
*McBreen, Fazio, & Jaffe, D.	A High-Resolution Far-Infrared Map of M16
Natta, Palla, Preite-Martinez & Panagia	Dust Temperature and IR Emission in High Extinction Molecular Clouds
Nordh & Fridlund	Far-Infrared Emission from Low Mass Star-Forming Regions
Sargent & van Duinen	New Far-Infrared Observations of Molecular Cloud Cores
Schwartz	The MM Wavelength Spectrum of Galactic Far IR Sources

MICROWAVE AND MM-WAVE STUDIES OF MOLECULAR CLOUDS

*Beckman, Watt, White, Phillips, J.P., & Frost	Abundances of $C^{18}O$ and HDO in Molecular Clouds
de Graauw, Lidholm Fitton, Beckman, Israel, Nieuwenhuizen, & van de Standt	CO(J=2-1) Observations of Southern H II Regions
*Falgarone	Star Formation in the Rho Ophiuchi Dark Cloud
*Forster, Goss, de Jong, Norman Habing, Downes, Wilson, T.L., & Dickel	H_2CO Observations of Compact H II Regions
*Harris, S.	A Random View of Cygnus X
Phillips, J.P., White, & Watt	Observations of a Rotating, Expanding Proto-Stellar Molecular Cloud Associated with NGC 2071 (IRS)
*White, Phillips, J.P., Watt, Beckman, & Frost	New Detections of Molecular Transitions at Wavelengths ~1100 Microns
*Zealey & Ninkov	Cometary Globules in the Gum-Vela Region

MASERS, COMPACT IR SOURCES AND YOUNG STARS

*Felli, Johnston, & Churchwell	Compact Radio Component in the Core of M17
Guibert, Epchtein, Nguyen-Quang-Rieu, Turon, & Wamsteker	Infrared Sources Associated with Southern Galactic OH Masers
Howell, McCarthy, & Low	IR Speckle Interferometry of Proto-Stellar Objects
Hyland & McGregor	Infrared Studies of the Young Stellar Population of 30 Doradus
Moorwood & Salinari	An Infrared Survey of H_2O Masers
Persi, Ferrari-Toniolo, & Spada	Photometry of the Obscured IR Object GL490 from 2.3 to 23 Microns
Righini-Cohen, Simon, M. & Felli	VLA Observations of the Becklin-Neugebauer Object and other BN-Object Analogs
Sibille & Lena	Achievement of Diffraction-Limited Spatial Resolution of Large Telescopes with Infrared Speckle Technique
*Tanzi, Tarenghi, & Panagia	Infrared Observations of Stellar Winds from OB Stars
Wright, E.L., Harper, Loewenstein, & Moseley	Far-Infrared Observations of H_2O Masers in NGC 281, NGC 2175 and S255/257

IONIC AND MOLECULAR EMISSION LINES

Aitken & Roche	10-Micron Spatial and Spectral Structure in G333.6-0.2
Beck, Serabyn, Lacy, Geballe, & Smith, H.A.	Observations of the v=0-0 S(2) Line of Molecular Hydrogen in the Orion Molecular Cloud
Baluteau & Moorwood	The Effect of Density Structure on Abundances Derived from Far-Infrared Line Observations of H II Regions
Cosmovici & Strafella	Near-Infrared High-Resolution Spectrophotometry of Interstellar [C I] and C_2
Dinerstein	10-Micron Infrared Line Emission and the Derivation of Chemical Abundances
*Furniss, Jennings, King, Emery, Fitton, & Naylor	Observations of Far-Infrared Fine Structure Lines in M17 and M42
*Hefele & Hoelzle	8-14 μm Spectrophotometry of S 106
*Joseph & Morris, S.A.	He^+/H^+ Recombination Line Ratios and Selective Absorption in the Lyman Continuum by Dust in H II Regions
Lester	Infrared Forbidden Lines in Compact H II Regions
Melnick, Russell, Gull, G.E., & Harwit	Far-Infrared Emission Line and Continuum Observations of NGC 7027
Russell, Melnick, Gull, G.E., & Harwit	Detection of the 157 Micron (1910 GHz) [C II] Emission from the Interstellar Gas Complexes NGC 2024 and M42
Young & Knacke	Observations of the 4.7 μm Molecular Hydrogen Rotational Transition in the Orion Nebula

THE ORION CLOUDS

Baud, Bieging, Plambeck, Welch, & Wright, M.C.H.	Aperture Synthesis Observations of the 3-mm SO Emission from OMC 1
*de Muizon, Rouan, & Baluteau	About a Shock Front in the Orion Nebula; Measurement of the CO J=30-29, 87.2 μm Emission Line
Genzel, Becklin, Wynn-Williams, & Downes	High Velocity Outflow in Orion
*Hjalmarson, Ellder, Friberg, Höglund, Irvine, Johansson, Olofsson, Rydbeck, G., Rydbeck, O.E.H., Guelin, Nguyen-Q-Rieu, & Schloerb	Onsala Molecular Line Studies of the BN/KL Region
Keene, Hildebrand, & Whitcomb	High-Resolution Submillimeter Map of OMC 1

Lee, Gatley, Stewart, Lonsdale, Becklin, & Wynn-Williams	Infrared Reflection Nebulosity in OMC 2
Lonsdale, Becklin, Lee, Gatley, & Stewart	Near Infrared Observations of the Stellar Cluster in OMC 1
Storey, Watson, D.M., Townes, Haller, & Hansen	Molecular Line Observations in the Far Infrared
Traub & Brasunas	Observations of CO Lines in Orion
*Werner, Becklin, Gatley, Neugebauer, & Sellgren	New Far-Infrared Observations of the Orion Nebula/OMC-1 Complex
*Wilson, T.L., Walmsley, Winnewisser, Kislyakov, & Bastien	Quiescent Molecular Clouds Near KL/BN

DUST AND POLARIMETRY

Allamandola, Greenberg, van de Bult, & Hagen	Astrophysical and Laboratory Spectra of Interstellar Grain Mantles
*Greenberg, van de Bult, Allamandola, & Baas	The Nature of Organic Molecules in Interstellar Grain Mantles
*Heckert & Zeilik	Infrared Polarimetry of Compact H II Regions
*Johnson & Kemp	Infrared Polarimetry with Photoelastic Modulators
Jones, B., Merrill, Stein, & Willner	The Dependence of the 8 to 13 Micron Spectrum of NGC 7027 on Position in the Nebula
Krätschmer	Laboratory Study of Amorphous Silicates and Carbonates
Kunkle	A New Relationship Between Polarization and Near-Infrared Extinction
*Leger, Klein, de Cheveigne Guinet, Defourneau, & Belin	Origin of the 3 μm Absorption
*Scarrott, Perkins, Bingham, & Murdin	Optical Polarization Maps of Bipolar Nebulae
*Tielens, Hagen, & Greenberg	Laboratory Study of the 3-micron Ice Band
*Whittet	Infrared Spectroscopy of Dust-Embedded Stars
*Wickramasinghe, D.T., & Allen, D.A.	The 3.4-μm Absorption Band and Organic Material on Interstellar Grains
*Williams, P.M.	Identification of a Strong Emission Line at 3.28 μm

EVOLVED OBJECTS

*Bruston & Gispert	Analysis of Millimetric Photometric Observation Data of the Crab Nebula

*Feast, Catchpole, Carter, & Roberts, G.	A Period-Luminosity Relation for Supergiant Red Variables in the LMC
*Hjalmarson, Irvine, Johansson, Olofsson, & Nguyen-Q-Rieu	Onsala Molecular Line Studies of the IRC+10216 Envelope
*Schmid-Burgk & Scholz	On the Possibility of Dust Formation in M-type Photospheres
*Thronson	Near-Infrared Spectroscopy of Proto-Planetary Nebulae
*Tovmassian	Observations of IR-Emission of Cool Giant Stars
*Tovmassian	On the Variation of IR-Emission of V 915 Aql
*Zealey, Hartl, & Malin	The W28, Lagoon Nebula, and Trifid Nebula Complex

LARGE SCALE EMISSION FROM THE GALAXY

*Danks, Wamsteker, Shaver, & Retallack	A Near-Infrared Study of Region ℓ=305°
*Gispert, Puget, Serra, & Ryter	Far-Infrared Survey of Molecular Clouds along the Galactic Plane
Hauser, Silverberg, Gezari, Kelsall, Mather, Stier, & Cheung	Submillimeter Wave Survey of the Galactic Plane
*Kawara, Kobayashi, Kozasa, Sato, & Okuda	Distribution of Infrared Sources in the Interior of the Galaxy
Nishimura, Low, & Kurtz	Distribution of the Far-Infrared Galactic Emission
Price, S.D.	Large-Scale Infrared Structure of the Galaxy
Serra, Puget, Gispert, & Ryter	Variation of the Emission Mass Function with Galactic Radius?

STELLAR POPULATIONS AND COSMOLOGY

Aaronson	Infrared Magnitudes, H I Velocity Widths, and the Distance Scale
Lebofsky, M.J.	Evolution of Elliptical Galaxies
Persson, Aaronson, Cohen, J.G. Frogel, & Matthews, K.	Infrared Colors of Star Clusters in the Magellanic Clouds
*Puget & Heyvaerts	Population III Stars and the Shape of the Cosmological Black Body Radiation
Spinrad & Bruzual	Evolutionary Predictions for IR Colors of Distant Galaxies

NUCLEI OF GALAXIES

Lebofsky, M.J., Larson, Rieke, Smith, H.A., Tokunaga, & Wollman	High-Resolution Spectroscopy of Galactic Center Sources

Longmore, Sharples, & Hawarden	Infrared and Optical Photometry of Dust-Lane and Radio Galaxies
*Puschell & Heeschen	Preliminary Results of a Search at 10 μm for Nonstellar Emission from Elliptical Galaxies with Compact Radio Sources
Rickard, Harvey, & Thronson	Far-Infrared Studies of Galaxies with Molecular Components
*Russell, Melnick, & Harwit	Narrow-Band Photometry of M 82 at Wavelengths Ranging out to 180 Microns
Telesco & Owensby	A Large 10-μm Source near the Center of M51 (NGC 5194)

SEYFERT GALAXIES AND QSOs

Allen, D.A., Carswell, Ferland, Baldwin, J.A., Barton, & Gillingham	The Hydrogen Emission Lines in Quasars
Bieging, Blitz, Lada, & Stark	Molecular and Infrared Radiation from Seyfert Galaxies
Condon, O'Dell, S.L., Puschell, & Stein	The Spectral-Flux Distribution of Radio "Quiet" and Radio "Loud" QSOs
*Glass	JHK Photometry of Quasars
Hyland & Allen, D.A.	The Infrared Continuum of Quasars
Impey, Brand, Wolstencroft, & Williams, P.M.	JHK Polarimetry and Photometry of BL Lac Objects
McAlary & McLaren	Near-Infrared Spectrophotometry of NGC 4151
Rieke & Lebofsky, M.J.	The Nucleus of NGC 4151
Scoville, Hall, D.N.B., Kleinmann, S.G., & Ridgway	Two-Micron Spectroscopy of the Nucleus of NGC 1068
Sherwood, Schultz, & Kreysa	Mm-Submm Observations of Extragalactic Objects
Ward, Allen, D.A., Smith, M.J., Wilson, A.S., & Wright, A.E.	The Near-Infrared Continua of X-ray and Other Active Galaxies
Wilson, A.S., Ulvestad, & Sramek	Radio Jets in Seyfert Galaxies and the Origin of the Radio-Infrared Relation
*Wolstencroft & Gilmore	Rapid Variations of OJ287 at 1.25 Microns

THEORY

*Bode & Evans, A.	Periodic Infrared Emission by Cosmic Dust Grains
*Bode & Evans, A.	Thermal Infrared Emission from Type II Supernovae
*Di Fazio & Palla	Isothermal Phases in Protostar Collapses

Authors	Title
Falk & Hessman	Infrared Emission from Dust in Gas Lost from Galaxies in Cluster Cores
*Guibert, Epchtein, Nguyen-Quang-Rieu, Turon, & Wamsteker	Infrared Pumping of Circumstellar OH Masers
Harris, S., & Clegg	A Luminosity Function for Molecular Clouds
*Hollenbach & McKee	Dissociation Speeds for Interstellar Shock Waves
*Lebertre & Papoular	The Formation of Spectral Lines in Dust Clouds
*Kulkarni & Ashok	Models for IR Emission from the Rapid Burster
*Rowan-Robinson	Models for Infrared Sources
TECHNIQUES	
*Bensammar & de Batz	The Multiplex Technique for Imagery in the Infrared
*Craine	Potential of Wide-Field Near-IR Imagery
*Gautier, T.N., Low, Poteet, Rieke, Young, Fazio, & Koch	A Small Helium-Cooled Telescope for Spacelab 2
*Giles	Self-Supporting Perfect Masks for 2D Infrared and X-ray Imaging
*Maillard	Cassegrain F.T. Spectrometer for CFH Telescope
*Neugebauer	Latest Estimates of IRAS Sensitivities
*Rabbia	A 15-m Baseline, 11-μm Infrared Heterodyne Interferometer
*Werner & Murphy	A Large Deployable Infrared Space Telescope
*Wing, Rinsland, Hayes, Joyce, & Ridgway	A Coordinated Program of Relative and Absolute Infrared Monochromatic Flux Measurements

INDEX

Entries in this index refer only to the <u>first</u> significant mention of that subject in each chapter.